Clean Energy Production Technologies

Series Editors

Neha Srivastava, Department of Chemical Engineering and Technology
IIT (BHU) Varanasi
Varanasi, India

P. K. Mishra, Department of Chemical Engineering and Technology
IIT (BHU) Varanasi
Varanasi, India

The consumption of fossil fuels has been continuously increasing around the globe and simultaneously becoming the primary cause of global warming as well as environmental pollution. Due to limited life span of fossil fuels and limited alternate energy options, energy crises is important concern faced by the world. Amidst these complex environmental and economic scenarios, renewable energy alternates such as biodiesel, hydrogen, wind, solar and bioenergy sources, which can produce energy with zero carbon residue are emerging as excellent clean energy source. For maximizing the efficiency and productivity of clean fuels via green & renewable methods, it's crucial to understand the configuration, sustainability and techno-economic feasibility of these promising energy alternates. The book series presents a comprehensive coverage combining the domains of exploring clean sources of energy and ensuring its production in an economical as well as ecologically feasible fashion. Series involves renowned experts and academicians as volume-editors and authors, from all the regions of the world. Series brings forth latest research, approaches and perspectives on clean energy production from both developed and developing parts of world under one umbrella. It is curated and developed by authoritative institutions and experts to serves global readership on this theme.

Sushil Kumar Verma • Shailendra Rajput
Jayaramudu Jarugala
Editors

Sustainable and Natural Polymers

Innovations in Clean Energy

Springer

Editors
Sushil Kumar Verma
Centre for Sustainable Polymer
Indian Institute of Technology
Guwahati, Assam, India

Shailendra Rajput
Department of Electrical and Electronic Engineering
Ariel University
Ariel, Israel

Jayaramudu Jarugala
Indian Institute of Chemical Technology
Hyderabad, Telangana, India

ISSN 2662-6861 ISSN 2662-687X (electronic)
Clean Energy Production Technologies
ISBN 978-981-92-0091-7 ISBN 978-981-92-0092-4 (eBook)
https://doi.org/10.1007/978-981-92-0092-4

This Springer imprint is published by the registered company Springer Nature Singapore Pte Ltd.
The registered company address is: 152 Beach Road, #21-01/04 Gateway East, Singapore 189721, Singapore

Preface

The transition toward renewable and sustainable energy systems represents one of the most critical scientific and technological challenges of the twenty-first century. Advanced materials play a pivotal role in this transformation, and sustainable and natural polymers are increasingly becoming central to energy harvesting, storage, and conversion technologies.

This book was conceived with the objective of compiling recent advances in sustainable polymeric materials and their applications in clean energy systems. It explores innovative developments in solar cells, wind energy systems, fuel cells, supercapacitors, batteries, piezoelectric devices, biodegradable composites, and 3D-printed renewable energy components. Special attention has been given to bio-based materials, green synthesis approaches, recyclability, lifecycle assessment, and industrial scalability.

The editors have made deliberate efforts to include contributions from researchers representing diverse institutions and international research communities. This diversity strengthens the interdisciplinary nature of the book and provides a global perspective on sustainable materials research.

We sincerely thank all contributing authors for their scholarly efforts and commitment to excellence. We also acknowledge the reviewers whose constructive feedback helped enhance the quality and clarity of the chapters. Finally, we are grateful to the publishing team at Springer Nature for their guidance and support throughout the development of this project.

It is our hope that this book will stimulate further research, encourage cross-disciplinary collaborations, and contribute meaningfully to the global mission of achieving cleaner and more sustainable energy systems.

Guwahati, Assam, India — Sushil Kumar Verma
Ariel, Israel — Shailendra Rajput
Hyderabad, Telangana, India — Jayaramudu Jarugala

Contents

Editors and Contributors

About the Editors

Sushil Kumar Verma is working as a Research Associate III, Centre for Sustainable Polymer, Technology Complex, IIT Guwahati, Guwahati, Assam, India. He has obtained his M.Sc. degree in Physics from University of Gorakhpur, UP, India. After that, he obtained his M.Tech degree in Polymer Nanotechnology from CIPET Bhubaneswar and Ph.D. degree in Chemical Engineering and Technology in 2018 from Birla Institute of Technology, Ranchi, India. He has industry experience of 10 years. He has been awarded a Research Fellowship (DST-SERB). His current research interests are synthesis and manufacturing and new product development of biodegradable polymer compounds, hydrogel and its nanocomposites for biodegradable plastic, chemical sensors, flexible and rigid packaging applications, energy harvesting, and energy storage

Shailendra Rajput is affiliated with Ariel University, Israel, as a Research Fellow. He was a postdoctoral fellow at Ariel University, Israel (September 2017–April 2021) and Xi'an Jiaotong University, China (May 2015–July 2017). Dr. Rajput received his B.Sc. and M.Sc. degrees from the Dr. Hari Singh Gour University, Sagar, India, in 2006 and 2008, respectively. Dr. Rajput received his Ph.D. degree from the Birla Institute of Technology, Ranchi, India, in 2014. His main research work is associated with energy harvesting, solar energy, energy storage, energy materials, ferroelectricity, piezoelectricity, and biomedical application of electromagnetic waves.

Jayaramudu Jarugala is a Sr. Principal Scientist at CSIR-Indian Institute of Chemical Technology (CSIR-IICT), Hyderabad. He earned his Ph.D. in Polymer Science and Technology from Sri Krishna Devaraya University in 2010, following an MSc and BSc in the same field. His extensive research spans polymers, functional materials, and sustainable technologies, including biodegradable plastics, polymer nanocomposites, and green catalysts. With a notable career, Dr. Jarugala has held positions such as Principal Scientist at the North East Institute of Science and Technology and Assistant Director at the Indian Rubber Manufacturers Research Association. He has collaborated with numerous industry and academic institutions globally and led several significant research projects funded by prestigious agencies like SERB-DST and CSIR. His contributions are recognized through multiple awards and numerous publications in high-impact journals, reflecting his influence and expertise in advancing polymer science and sustainable materials.

Contributors

Sk Mehebub Alom Department of Biotechnology, Brainware University, Barasat, India

Hari. P. Bhasker Department of Physics, Chaudhary Mahadeo Prasad Degree College, University of Allahabad, Allahabad, India

Dipanjan Bhattacharjee Department of Allied Health Sciences, Brainware University, Kolkata, India

Soheli Biswas Department of Molecular Biology and Biotechnology, University of Kalyani, Kalyani, Nadia, India

Sandip Bordoloi Department of Electrical Engineering, Girijananda Chowdhury University, Guwahati, Assam, India

Kwanruthai Butsriruk College of Engineering, Xi'an International University, Shaanxi, China

Dhirendra K. Chaudhary Centre for Renewable Energy, Prof.Rajendra Singh (RajjuBhaiya) Institute of Physical Sciences for Study and Research, V. B. S. Purvanchal University, Jaunpur, India

Interdisciplinary Centre for Water and Energy Research, V. B. S. Purvanchal University, Jaunpur, India

Bimal Das Department of Chemical Engineering, National Institute of Technology Durgapur, Durgapur, India

Dipankar Das Department of Allied Health Sciences, Brainware University, Kolkata, India

Deepshikha Datta Department of Basic Science & Humanities, Institute of Engineering and Management (IEM), Kolkata, University of Engineering and Management (UEM), Newtown, Kolkata, India

Esha Datta Department of Biotechnology, National Institute of Technology Durgapur, Durgapur, India

Sriparna De Department of Allied Health Sciences, Brainware University, Kolkata, India

Nirmala Devi Department of Chemistry, Girijananda Chowdhury University, Guwahati, Assam, India

Punit K. Dhawan Department of Physics, Prof. Rajendra Singh (Rajju Bhaiya) Institute of Physical Sciences for Study and Research, V. B. S. Purvanchal University, Jaunpur, India

Rituparna Duarah Materials Sciences Group, Coal, Energy and Materials Sciences Division, CSIR-North East Institute of Science and Technology, Jorhat, Assam, India

Academy of Scientific and Innovative Research, Ghaziabad, India

Unnayana Gogoi Materials Sciences Group, Coal, Energy and Materials Sciences Division, CSIR-North East Institute of Science and Technology, Jorhat, Assam, India

Ashish Kumar Kaushal Jindal Global Business School, O P Jindal Global University, Sonepat, Haryana, India

Ashutosh Kumar Department of Physics, CSJM University, Kanpur, Christ Church College, Kanpur, Uttar Pradesh, India

Ranjit Kumar Department of Physics, Arignar Anna Government Arts and Science College, Karaikal, Puducherry, India

Department of Physics, Dr. Kalaignar M. Karunanidhi Government Institute for Postgraduate Studies and Research, Karaikal (UT of Puducherry), India

Kaustav Mazumder Department of Chemical Engineering, National Institute of Technology Durgapur, Durgapur, India

Divakar Pamanji Department of Civil Engineering, Mother Theresa Institute of Engineering and Technology, Palamaner, India

Jyotirmoy Panda Department of Pharmaceutical Technology, Brainware University, Kolkata, India

Saista Parvin Department of Biotechnology, Brainware University, Barasat, India

Mingjing Qu School of Business, Xi'an International University, Shaanxi, China

Nilesh Kumar Rai Department of Physics, Chaudhary Mahadeo Prasad Degree College, University of Allahabad, Allahabad, India

Sarojni Rai Department of Applied Science, United University, Prayagraj, Uttar Pradesh, India

Akash Ranjan Faculty of Education, Banaras Hindu University, Kamachha, Varanasi, Uttar Pradesh, India

Gopikishan Sabavath CSJM University, Kanpur, Christ Church College, Kanpur, Prayagraj, Uttar Pradesh, India

Trideep Saikia School of Pharmaceutical Sciences, Girijananda Chowdhury University, Guwahati, Assam, India

Department of Pharmaceutical Sciences, Dibrugarh University, Dibrugarh, Assam, India

Sayantan Sarkar Department of Chemical Engineering, National Institute of Technology Durgapur, Durgapur, India

Jimli Sarma Research Scholar, Centre for Multidisciplinary Research, Tezpur University, Napaam, Sonitpur, Assam, India

Savita Department of Physics, Prof. Rajendra Singh (Rajju Bhaiya) Institute of Physical Sciences for Study and Research, V. B. S. Purvanchal University, Jaunpur, India

Centre for Renewable Energy, Prof.Rajendra Singh (RajjuBhaiya) Institute of Physical Sciences for Study and Research, V. B. S. Purvanchal University, Jaunpur, India

Swasti Saxena Gandhinagar University, Khatraj Kalol Rd, Moti Bhoyan, Kalol, Gujarat, India

Mansi Sharma School of Engineering and Technology, Manav Rachna International Institute of Research and Studies, Faridabad, India

Sarva Shakti Singh Department of Physics, Chaudhary Mahadeo Prasad Degree College, University of Allahabad, Allahabad, India

Sonika Department of Physics, Rajiv Gandhi University, Rono Hills, Doimukh, Papumpare, Arunachal Pradesh, India

Ankit K. Srivastava Indrashil University, Mehsana, Gujarat, India

Neelabh Srivastava Department of Physics, School of Physical Sciences, Mahatma Gandhi Central University, Motihari, East Champaran, Bihar, India

Nitin Srivastava Department of Physics, Shri Chitragupt P. G. College (Affiliated to Dr. B. R. Ambedkar University, Agra), Mainpuri, Uttar Pradesh, India

Chapter 1
Synthesis and Fabrication Techniques of Polymer Nanocomposites

Deepshikha Datta, Jyotirmoy Panda, Kaustav Mazumder, Soheli Biswas, Sk Mehebub Alom, Saista Parvin, Esha Datta, Divakar Pamanji, and Bimal Das

Abstract The recent works emphasizes on the fundamentals, composition, structure and evolution of different polymers used in the production of polymer nanomaterials. It highlights on the polymer matrices, interface interactions and nanoscale effects. The characteristics of the nanomaterials synthesized by various techniques like in-situ polymerization, solution polymerization, melt mixing and sol-gel process are compared to suggest the best technique according to the quality of nanomaterial produced. It is observed that though a high production rate can be achieved by thermoforming process, it lacks maintaining uniform nanofiller dispersion. Electrospinning produces ultrafine nanofibers with high surface area but lacks scalability for industrial applications. Structural, thermal, mechanical and rheological characterization of the polymer nanocomposites is elaborated to substantiate the

---

D. Datta (✉)
Department of Basic Science & Humanities, Institute of Engineering and Management (IEM), Kolkata, University of Engineering and Management (UEM), Newtown, Kolkata, India

J. Panda
Department of Pharmaceutical Technology, Brainware University, Kolkata, India

K. Mazumder · B. Das
Department of Chemical Engineering, National Institute of Technology Durgapur, Durgapur, India

S. Biswas
Department of Molecular Biology and Biotechnology, University of Kalyani, Kalyani, Nadia, India

S. M. Alom · S. Parvin
Department of Biotechnology, Brainware University, Barasat, India

E. Datta
Department of Biotechnology, National Institute of Technology Durgapur, Durgapur, India

D. Pamanji
Department of Civil Engineering, Mother Theresa Institute of Engineering and Technology, Palamaner, India

S. K. Verma et al. (eds.), *Sustainable and Natural Polymers*, Clean Energy Production Technologies, https://doi.org/10.1007/978-981-92-0092-4_1

quality of the product produced by different techniques. The intended application substantiates the unique benefits of each technique. Various factors like mechanical performance, electrical properties, thermal scalability, cost, scalability and application-specific requirement are the factors responsible for selecting the appropriate method for fabrication. The work further highlights on the application of polymer nanocomposites produced by each technique.

1.1 Introduction

1.1.1 Polymer Nanocomposites (PNCs): An Overview

From the beginning of Life, polymers play an important role in developing both nature and our society. From Natural materials such as tortoise shell, horns, tars, and shellac to synthetic offerings such as polyolefins, epoxies, and engineering resins, polymer plays an important part in construction, commerce, transportation, chemical, and entertainment industries across the world. Global polymer production is estimated in the range from 250 billion pounds to more than 400 billion pounds, which is approximately 114 billion Kg to 181 billion Kg. Polymeric materials in their diverse application are not usually chemically or molecularly homogeneous; rather, they are multicomponent systems (Winey and Vaia 2007).

Reinforcement of Plastics using a second organic or inorganic phase filler material for the production of polymer composites is a common practice in the production of modern plastics (Fu et al. 2019; Krishnamoorti and Vaia 2002). Polymer nanocomposites are the radical alternative to polymer composites (Alexandre and Dubois 2000; Fu et al. 2019; Giannelis 1996; Kato and Usuki 2000; Krishnamoorti and Vaia 2001, 2002; Paul and Robeson 2008; Young et al. 2012). In the late 1980s, the term "polymer nanocomposites" evolved to refer to a multicomponent system that majorly consists of polymer and its minor component has a minimum of one dimension in the nano-scale range (<100 nm) (Winey and Vaia 2007). Thus, we can say polymer nanocomposites are polymers reinforced by nanomaterials as nanofillers dispersed in bulk polymer matrices (Tofighy and Mohammadi 2020).

This work highlights the classification of polymer matrices along with its associated nanofillers, synthesis techniques and fabrication processes of polymer nanomaterials. Besides the scalability, cost and industrial feasibility are also elaborated in order to justify its application-specific requirements.

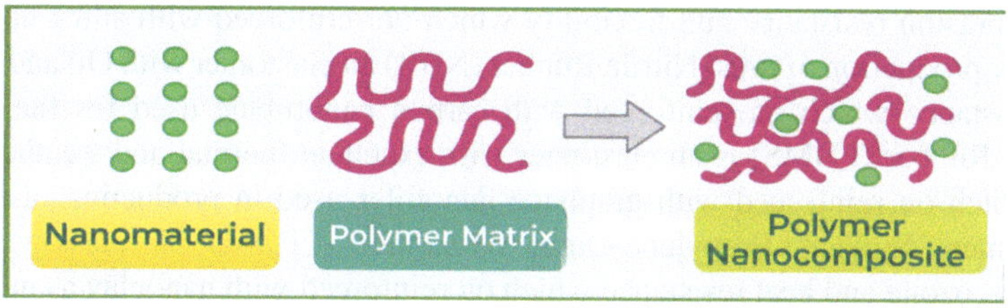

Fig. 1.1 Diagram representing the polymer matrix and nanomaterial as nanofillers constituents in polymer nanocomposites

1.1.2 Fundamental Concepts

1.1.2.1 Polymer Nanocomposites: Compositions

A nanocomposite (Fig. 1.1) represents the amalgamation of two or more phase systems where nanomaterials are incorporated in one phase (Ocaña and del Valle 2018), and polymer nanocomposites are not much of an exception. Polymer nanocomposites are the polymer, reinforced by Nanomaterials as nanofillers, dispersed in bulk polymer matrices (Tofighy and Mohammadi 2020).

1.1.2.2 Polymer Matrix: Types and Roles

Polymer matrices are the bulk polymer part of polymer nanocomposites, they can be classified into thermoplastics, thermosets, and elastomers.

Thermoplastics are often connected with recyclability, flexibility, and ease of processing, whereas Thermosets provide superior thermal and mechanical stability due to their cross-linked network and both matrices, when reinforced with nanofillers such as clay, silica, graphene, and CNTs, exhibit an enhancement in mechanical, thermal, and barrier properties (Paul and Robeson 2008; Wetzel et al. 2006).

Thermosetting polymer matrices such as Epoxy Resin, which have high strength and excellent adhesion, and a brittle nature, when reinforced with clay nanofillers used as aerospace coatings (Wetzel et al. 2006). Polyester resin, which has strong chemical resistance properties when reinforced with nano-silica, is used as a covering for marine structures (Wetzel et al. 2006). Phenolic resin with its strong thermal stability and flame resistance, Polyurethane (PU) with its flexibility and impact resistance and Bismaleimide (BMI) with its high temperature performance on reinforcement with nanofillers, carbon-nanotubes, graphene oxide, and carbon nanofibers, respectively, are used as thermal shields, coating and foams and aerospace composites (Wetzel et al. 2006).

Elastomers are capable of recovering their original shape. Some examples polymer matrix are as follows—Natural Rubber (NR) an elastomer polymer matrix with High Elasticity and Tensile strength which on reinforced with clay nanofillers used for production of flexible seats, Styrene-Butadiene Rubber (SBR) an elastomer with good abrasion resistance and flexibility which on reinforced with silica nanofiller used for production of tires, Nitrile Rubber (NBR) an elastomer with Oil and chemical resistance which on reinforced with carbon nanotubule used for fuel hoses, Silicon Rubber (PDMS) is an elastomer with excellent thermal and weather resistant which on reinforced with graphene nanofiller used in production of flexible electronics, Ethylene-Propylene-Diene Monomer (EPDM) an elastomer with weather, ozone and heat resistance which on reinforced with nanoclay as nanofiller used as roofing membranes (Bokobza 2007). Table 1.1 shows the polymer matrices with its types, associated nanofillers, and property enhancements, and industrial applications:

1.1.2.3 Nanofillers: Types, Size, and Function

Nanofillers are the particles present in nanosizes that in addition to the polymer matrix, introduce unique properties. Based on composition and dimensional structure, Nanofillers can be classified as follows: Carbon-based, Clay-based, Metal oxide, Metal-b, Polymer based, Hybrid fillers. Carbon-based nanofillers like Graphene, Go, and Carbon Nanotubes (CNTs) have 1D/2D tubular or planer in structure (Gawai et al. 2025). Montmorillonite (MMT) and Halloysite (HNTs) are examples of Clay-Based Nanofillers with a Layered or tubular structure (Ray and Okamoto 2003). Metal oxides like TiO_2, ZnO, and Al_2O_3 have a spherical or rod-like structure (Kim et al. 2010). Similarly, metal-based nanofillers like Ag, Au, Cu nanoparticles have a spherical structure (Tran et al. 2013). Where polymer-based nanofillers like Nanocellulose, Chitosan, and Starch nanoparticles are fibrous or crystalline in structure (Habibi et al. 2010). Other than this, different types of nanofillers combine together to form new types of polymers like CNT–TiO_2 and GO-Clay Hybrids, which usually 1D/2D composite structure (Qian et al. 2010). Nanofillers can be classified into four groups based on size and dimensions: The 0D (e.g., TiO_2, Ag), this group particle have 1–100 nm in all direction, it is spherical, The 1D (e.g., CNTs), this group particles have 1–50 nm in diameter and length in micrometres, it is Nanotubes and nanorods in shape, The 2D (e.g., graphene, MMT), this group has thickness in nanometre range but lateral dimensions are greater than 1 μm, it is sheet like in shape, The 3D, this group has aggregated or mesoporous nanostructures with hierarchical property (Hussain et al. 2006). The functions of nanofillers are shown in Table 1.2.

1.1.2.4 Interface Interaction and Nanoscale Effect

The Interface between the Nanofillers and Polymer matrix plays an important role in the development of the unique properties, such as enhanced mechanical, thermal, and barrier properties shown by polymer nanocomposites. As nanofillers present at the nanoscale range, the high surface-to-volume ratio of nanomaterials is maintained, which makes the interface interaction significantly valent pronounced and

Table 1.1 Classification of polymer matrices and common nanofillers

Polymer	Polymer matrix type	Associated nanofillers	Property enhancements	Typical applications	References
Polypropylene (PP)	Thermoplastic	CNTs, TiO_2, Graphene, Nanoclay (MMT)	Tensile strength, thermal resistance, barrier properties	Automotive parts, packaging films	Hossain et al. (2024)
Polyethylene (PE)	Thermoplastic	CNTs, Silica Nanoclay, Nanoparticles	Mechanical strength, flame retardancy, gas barrier	Films, containers, insulation	Hossain et al. (2024)
Polyvinyl Chloride (PVC)	Thermoplastic	CNTs, Graphene, ZnO nanoparticles	Electrical conductivity, impact strength	Pipes, cables, construction materials	Hossain et al. (2024)
Polyamide (Nylon-6,6)	Thermoplastic	CNTs, Graphene Oxide, Nanoclays	Improved mechanical properties, thermal conductivity	Automotive, textiles, electronics	Hossain et al. (2024)
Polycarbonate (PC)	Thermoplastic	Carbon Nanotubes, Graphene	Electrical conductivity, flame retardancy	Optical disks, medical devices, electronics	Hossain et al. (2024)
Epoxy Resin	Thermoset	CNTs, Graphene Oxide, MMT, Metal Oxides (TiO_2)	Mechanical reinforcement, thermal conductivity, corrosion resistance	Aerospace composites, adhesives, coatings	Zaghloul et al. (2023)
Phenolic Resin	Thermoset	CNTs, MMT, Metal Oxides (Al_2O_3)	Thermal stability, flame retardancy	Laminates, brake pads, fire-resistant panels	Zaghloul et al. (2023)
Unsaturated Polyester Resin (UPR)	Thermoset	Nanoclays, CNTs, Graphene	Mechanical strength, flame retardancy	Marine applications, panels, construction	Zaghloul et al. (2023)
Polyurethane (PU) (Rigid)	Thermoset	CNTs, Graphene, Silica Nanoparticles	Mechanical properties, thermal insulation	Insulation foams, coatings	Zaghloul et al. (2023)
Polyurethane (PU) (Elastomer)	Elastomer	CNTs, Nanoclay, Graphene	Abrasion resistance, elasticity, conductivity	Wearable electronics, footwear	Bokobza (2007)
Natural Rubber (NR)	Elastomer	Carbon Black, Graphene, Nanoclays	Tensile strength, abrasion resistance	Tires, seals, gaskets	Bokobza (2007)
Styrene-Butadiene Rubber (SBR)	Elastomer	Graphene, Silica Nanoparticles, Carbon Black	Improved durability, wet grip, elasticity	Tires, shoe soles	Bokobza (2007)

(continued)

Table 1.1 (continued)

Polymer	Polymer matrix type	Associated nanofillers	Property enhancements	Typical applications	References
Silicone Rubber (PDMS)	Elastomer	CNTs, Graphene, ZnO Nanoparticles	Electrical conductivity, stretchability	Medical devices, flexible electronics	Bokobza (2007)
Polylactic Acid (PLA)	Bio-degradable Polymer	Nanocellulose, Graphene, MMT	Mechanical strength, biodegradability	Biodegradable packaging, medical implants	Kayan and Kayan (2023)
Polycaprolactone (PCL)	Bio-degradable Polymer	Nanocellulose, Graphene	Mechanical properties, biodegradation rate control	Tissue engineering, drug delivery	Kayan and Kayan (2023)
Polyaniline (PANI)	Conductive Polymer	CNTs, Graphene, Metal Nanoparticles (Ag, Au)	Electrical conductivity, EMI shielding	Sensors, antistatic coatings	Oladele et al. (2025)
Polypyrrole (PPy)	Conductive Polymer	CNTs, Graphene, Metal Oxides (ZnO)	Conductivity, energy storage	Supercapacitors, biosensors	Oladele et al. (2025)

Table 1.2 Classification of common nanofillers with their properties

Nanofiller	Category	Structure	Key functions	Key references
Carbon Nanotubes (CNTs)	Carbon-based	1D (Tubular)	Mechanical reinforcement, thermal stability, and electrical conductivity	Coleman et al. (2006)
Graphene/GO	Carbon-based	2D (Planar)	High mechanical strength, barrier properties, electrical/thermal conductivity	Kim et al. (2010)
Montmorillonite (MMT)	Clay-based	2D (Layered)	Gas/moisture barrier, flame retardancy, mechanical reinforcement	Alexandre and Dubois (2000), Sinha Ray and Okamoto (2003)
Halloysite Nanotubes (HNTs)	Clay-based	1D (Tubular)	Barrier properties, drug delivery, flame retardancy	Hussain et al. (2006), Ray and Okamoto (2003)
Titanium Dioxide (TiO_2)	Metal oxide	0D (Spherical)	UV protection, thermal stability, photocatalysis, flame retardancy	Kim et al. (2010)
Zinc Oxide (ZnO)	Metal oxide	0D (Spherical)	UV protection, antibacterial, flame retardancy	Kim et al. (2010)
Silver Nanoparticles (AgNPs)	Metal-based	0D (Spherical)	Antimicrobial activity, electrical/optical enhancement	Tran et al. (2013)
Nanocellulose (CNC, CNF)	Bio-based	1D/2D (Fibrous)	Biocompatibility, mechanical reinforcement, barrier improvement	Habibi et al. (2010)
Chitosan Nanoparticles	Bio-based	0D (Spherical)	Biocompatibility, antimicrobial, controlled release	Habibi et al. (2010)
Hybrid (e.g., CNT-TiO_2, GO-MMT)	Hybrid	1D/2D Composite	Multifunctionality—Combined thermal, mechanical, electrical, and barrier enhancements	Qian et al. (2010)

influences the dispersion, alignment, and load transfer capability of the composites. Nanofillers such as Carbon Nanotubes (CNTs), montmorillonite (MMT), graphene, and titanium oxide (TiO_2) interact with the polymer matrix through physical and chemical mechanisms, which includes van-der-waals force, hydrogen bonding, π–π stacking (for aromatic polymers and graphene), electrostatic interaction and covalent bonds in some cases similarly nanofillers enhance interfacial adhesion by introducing reactive groups that bond more efficiently with the polymer matrix, resulting in improved dispersion and reduced agglomeration (Ray and Okamoto 2003). A well-bonded interface ensures effective stress transfer between nanofillers and polymer matrix, leading to enhanced mechanical strength and stiffness. For example, multi-walled carbon nanotubes (MWCNTs) having carboxylic acid functional groups exhibit enhanced compatibility with epoxy resin, showing a substantial improvement in tensile strength and elastic modulus (Kim et al. 2010). Similarly surface-treated TiO_2 nanoparticles can integrate more uniformly within the polymer chain, which results in enhancing thermal stability, UV resistance, and dielectric properties (Chiu et al. 2011).

At the nanoscale, material exhibits size-dependent phenomena which deviate from bulk behaviour, where quantum confinement, surface energy effects, and discrete atomic arrangement become dominant, resulting in the development of unique electrical, mechanical, and optical properties. The nanoconfinement effect in polymer nanocomposites restricts the polymer chain's mobility near the filler surface, resulting in an alteration in glass transition temperature (Tg), crystallinity, and diffusion behaviour (Kashiwagi et al. 2004; Schadler et al. 2007). These changes have a notable impact on the barrier and thermal properties of polymer nanocomposites (Kashiwagi et al. 2004; Schadler et al. 2007). The extent and nature of the interface depend on the nanofiller itself (Thomason and Vlug 1996).

1.1.3 Synthesis Techniques of Polymer Nanomaterials

Polymer nanomaterials are prepared through a range of techniques suited to develop particular structural and functional attributes. Recent breakthroughs have enormously improved efficiency and scalability, which now allows more widespread applications in catalysis, biomedicine, and energy technology. Nanomaterials, in general, are prepared by a wide range of techniques, which are outlined below (Fig. 1.2, Tables 1.3 and 1.5).

1.1.3.1 In-Situ Polymerization

In-situ polymerisation is a chemical encapsulation process quite similar to interfacial polymerisation. The absence of reactants introduced to the core material is what sets in-situ polymerization apart. The procedure is predicated on the idea that the platelets absorb the monomer solution during their reaction. The polymerization of the monomer under the influence of heat, initiators, or UV light comes next. Thus, the creation of intercalated nanocomposites or exfoliated nanocomposites occurs (Nguon et al. 2018). In situ template synthesis is another technique in which the platelet size is regulated as a polymer matrix and the platelets migrate in a high fluid so that the nucleation of the platelets is controlled by the polymer chains entering the do them and spatial distribution considered as homogeneous, thus the nanoparticles are retarded from going back to the standard location like core particles while making the core's condition stable (Gobena and Woldeyonnes 2024). Nevertheless, this process has the disadvantage of amplifying the degradation of polymer chains at high temperature (Patel et al. 2022).

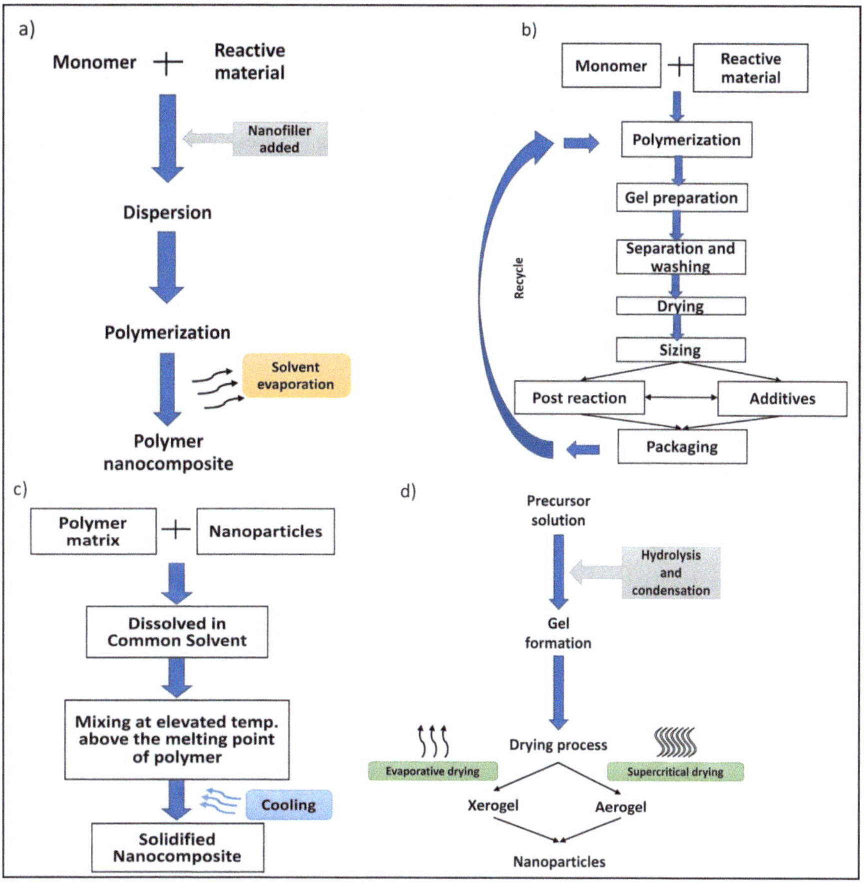

Fig. 1.2 Flowchart of polymer nanocomposite synthesis methods: (**a**) in-situ polymerization, (**b**) solution polymerization, (**c**) melt mixing, and (**d**) sol-gel process

1.1.3.2 Solution Polymerization

Solution polymerization is a commonly employed method of synthesizing polymer nanomaterials, in which monomers are polymerized within a medium of solvent to produce polymers that contain or have well-dispersed nanoparticles. The solvent acts as a medium for dissolving the monomers as well as any nanofillers or additives during this process, which allows for a uniform reaction environment (Gong et al. 2025). This approach enables control over the ultimate polymer's molecular structure and the homogeneous dispersion of nanomaterials with high accuracy, which is vital to obtain optimal physical and chemical properties in the final product. Solution polymerization is especially suitable for the fabrication of polymeric nanoparticles and nanocomposites with functionalized structures for catalytic, biomedical engineering, and energy storage applications (Gobena and Woldeyonnes 2024). The

Table 1.3 Comparison of synthesis techniques

Synthesis method	Process temperature	Dispersion quality	Reaction time	Environmental impact	References
In-Situ Polymerization	Moderate to high (50–90 °C)	Excellent; covalent bonding ensures uniform dispersion and strong interfacial adhesion	Moderate to long (hours)	Moderate; often uses solvents and initiators but can be optimized for greener protocols	Gobena and Woldeyonnes (2024)
Solution Polymerization	Low to moderate (room temp to ~60 °C)	Good; solvent medium facilitates nanoparticle dispersion but solvent removal needed	Moderate (hours)	Moderate to high; solvent use and removal pose environmental concerns, though water-based emulsions reduce impact	Patel et al. (2022)
Melt Mixing	High (above polymer melting point, ~150–250 °C)	Moderate; shear forces disperse fillers but risk of re-agglomeration at high loadings	Short (minutes)	Low; solvent-free and energy-efficient, suitable for large-scale industrial processing	Patel et al. (2022)
Sol-Gel Process	Low to moderate (room temp to ~80 °C)	Excellent; inorganic network forms within polymer matrix, enabling fine dispersion	Long (hours to days)	Moderate; involves hydrolysis and condensation reactions with potential chemical waste, but water-based systems are common	Gobena and Woldeyonnes (2024)

method is critical for the synthesis of advanced polymer nanomaterials with versatility and efficiency in laboratory and industry applications (Patel et al. 2022). This method is widely applied in the synthesis of polymer nanomaterials for numerous advanced applications across various fields like drug delivery system, water treatment and in biosensors also (Gobena and Woldeyonnes 2024).

1.1.3.3 Melt Mixing

This process is followed by most of the industries than solution mixing. In this, two kinds of melt mixers have been employed, which can be classified as batch mixing or continuous mixing. The batch mixing has been inefficient since it could not conduct the large capacities; however, it is suitable for laboratory processes. Additionally, the batch mixer may be referred to as internal mixer and that of Brabender or Haake, while continuous mixers of Banbury or Moriyama as dispersion mixers (Utracki and Kamal 2002). Besides this, this technique is helpful to produce clay-polymer nano composites wherein the polymer matrix may be either elastomeric or thermoplastic (Verma and Goh 2019). Usually, the polymer is melted with some amount of the intercalated clay with the help of extruder or Banbury, this being conducted in the presence of noble gases such as argon, nitrogen or neon. Apart from this, polymer pre-mixed by way of dry mixing with the intercalant upon heating in a mixer and leading to shear, sufficient for the creation of desired polymer nanocomposites (Patel et al. 2022). There are two varieties of melt intercalation: static and dynamic. Static melt intercalation is a process at temperatures 50 °C below the vacuum above Glass Transition Temperature, for instance, melting or glass transition, but not mixing, and is more commonly known as melt annealing. Dynamic is another process of melt intercalation, which is a relatively normal process in extruders and internal mixers (Vasudeo Rane et al. 2017).

1.1.3.4 Sol-Gel Process

"Sol–gel" is a method that involves two reaction stages: sol and gel. Herein, the "sol" represents a colloidal suspension where the solid is dissolved in a liquid. A "gel" represents a cohesive network between these operations. Sol-gel is largely applied to manufacture solids from microscopic particles at reduced temperatures (Haider et al. 2012). Here, the most critical step is the transformation of the precursors to the colloidal by hydrolysis and condensation reaction (Gobena and Woldeyonnes 2024). During this process, particles or fillers are dispersed in a monomer solution (referred to as sol) that comes out from the layers of the particle, as a result, an interacting network of polymers and nanoparticles, known as gel, is created. (Rashid et al. 2022). The polymer, therefore, promotes nanomaterial development and nucleation. The process is comparable to the surface procedure created with chitosan and salicylaldehyde in the presence of TiO_2 nanofillers to build a nanocomposite membrane. (Rashid et al. 2022). The key processes include metal recovery, excess water, solvents, heat, and the use of complicated compounds or catalysts. (Gobena and Woldeyonnes 2024). It is based on the chemical properties of the metal atom and the strong resistance of the alkoxide group. The electrophilic nature of metal atoms and their ability to increase the rate of coordination are the primary considerations (Batra and Pydah 2020).

1.1.4 Fabrication Processes

There are several fabrication techniques for nanomaterials such as.

1.1.4.1 Thermoforming

Thermoforming is a common polymer processing method where a sheet of polymer is heated to above its glass transition temperature and subsequently formed into a desired shape with pressure, vacuum, or mechanical force. Thermoforming is especially appreciated in packaging and automotive industries because it can produce large quantities of components quickly with little post-processing (Kubiak et al. 2022).

Advantages in Production Rate

The high rate of production is an important benefit since thermoforming is capable of continuous or semi-continuous operation, which makes it very efficient in mass production processes. In addition, the technique is adaptable to many different kinds of polymers and makes it possible to produce complex geometries at lower prices compared with other moulding methods (Patil et al. 2019; Sun et al. 2020).

Limitations in Nanofiller Dispersion

In the case of polymer nanocomposite fabrication, thermoforming has certain limitations in the dispersion of nanofillers. The major challenge is that the shear forces developed during heating and shaping operations are usually inadequate to disperse agglomerates of nanofillers like carbon nanotubes, clay, or metal nanoparticles (Patil et al. 2019). Therefore, nanofillers can remain non-uniformly dispersed in the polymer matrix, and localized clusters are formed instead of a uniform dispersion. This non-uniformity can jeopardize the mechanical, thermal, and barrier properties of the end product since the advantages of nanofillers are optimized only when they are properly dispersed at the nano level (Kubiak et al. 2022; Patil et al. 2019). Also, at heating and cooling time periods very rapid in thermoforming, nanofillers may possibly not have had enough time to migrate and distribute completely, thus aggravating inadequacies with dispersions (Sun et al. 2020).

1.1.4.2 Electrospinning

A web of micro- or nanofibers is formed through the electrospinning process. High-voltage electricity and a collector are applied to the liquid solution in this process, enabling the solution to jet from a nozzle and create a jet. Shaped fibers formed through the drying process were then deposited into the collector. This method is an easy and fast means of producing micro- or nanopolymer fibers. Electrospinning from polymers dissolved in the precursor solution can also be utilized to produce inorganic fibers. The presence of polymers can influence the electrospinning solution's viscosity, conductivity, surface tension, and other parameters, all of which are significant for this technique (Babul Reddy et al. 2016; Verreck et al. 2003). In addition, the process is capable of producing fibers with diameters between 2 nm and over 5 nm; however, most fiber diameters lie in the range of 10 to several hundreds of nanometers (Gobena and Woldeyonnes 2024). The production of nano fibers using electrospinning is a relatively simple, easy-to-use, and inexpensive technique. In 1900, Cooley pioneered electrospinning as he was the first one to patent it (Babul Reddy et al. 2016).

1.1.4.3 Additional Fabrication Methods

Other techniques, such as electrodeposition, 3D printing, CVD, and PVD, etc., are explained below (Table 1.4).

Electrodeposition employs electric currents to deposit metal layers on conductive substrates, facilitating applications such as corrosion-resistant coatings and conductive electronics components. Electrodeposition is superior in the production of uniform, thin films and can produce hybrid composites by combining polymers with metals, such as conductive polymer-metal hybrids for sensors, but at the expense of conductive substrates and current control (Jose et al. 2023; Tayyab et al. 2025).

- 3D printing transforms nanocomposite production by allowing layer-by-layer deposition of metal-polymer hybrids with customizable geometries for aerospace structures and biomedical devices. Challenges exist in achieving even

Table 1.4 Comparison of fabrication techniques

Technique	Scalability	Nanofiller dispersion	Cost	Complexity	References
Thermoforming	High	Poor to Moderate	Low	Low	Kubiak et al. (2022)
Electrospinning	Low to Moderate	Excellent	Moderate	Moderate	Babul Reddy et al. (2016)
Electrodeposition	Moderate	Excellent (for coatings)	Moderate	Moderate	Tayyab et al. (2025)
3D Printing	Moderate	Good (depends on feedstock)	Moderate to High	High	Park et al. (2022), Romero-Araya et al. (2021)
CVD	Low to Moderate	Excellent (thin films/coatings)	High	High	Solař et al. (2017)
PVD	Low to Moderate	Excellent (thin films)	High	High	Ma et al. (2021)

nanoparticle distribution in polymer matrices and resolution maintenance due to nozzle limitations (Jose et al. 2023; Park et al. 2022).

- CVD is gas-phase metal–organic precursor decomposition to deposit ultrathin films onto polymers, suited for conductive coatings (such as indium tin oxide) and barrier films but requires high thermal stability and high-energy substrates (Ma et al. 2021; Solař et al. 2017).
- PVD vaporizes metal targets within a vacuum, depositing atoms onto polymers to create wear-resistant or optical coatings, like those for medical devices and solar cells, but tends to need plasma treatment to increase adhesion on untreated polymers (Ma et al. 2021; Solař et al. 2017).

These processes meet various industrial requirements, ranging from scalable coatings (electrodeposition) to precision fabrication (3D printing), though each has trade-offs in cost, scalability, and material compatibility. For example, whereas 3D printing allows for quick prototyping, CVD and PVD provide nanoscale accuracy but at greater operational expense. To reach optimal internal conductivity combined with structural integrity, new efforts of hybrid approaches attempt combinations of methods such as electrodeposition with 3D printing to develop such nanocomposites in the future (Table 1.5).

1.1.5 Factors Influencing Technique Selection

As polymer-based nanocomposites are gaining popularity in current times, new techniques of fabrication are emerging and it is very crucial to understand the various influences that need to be considered before selecting specific fabrication techniques to achieve a desirable polymer nanocomposite. The performance and yield of polymer nanocomposites (PNT) are based on the synergistic relation of polymer matrix and nanofillers, i.e., carbon nanotubes, graphene, nanoclay, etc. (Díez-Pascual 2021). Similarly, factors like dispersion of nanofillers, filler alignments, and interfacial bonding are greatly impacted by the various physical and electrical influences (Khdier et al. 2024). Table 1.6 indicates the different characterization techniques required for the evaluation of the various properties of the nanomaterials.

1.1.5.1 Mechanical Performance

Mechanical Performance Consideration

- **Dispersion of Nanofillers:** Effective functioning of PNTs greatly depends on the proper distribution of the nanofillers in the polymer matrix. During dispersion of nanofillers if phenomenon like agglomerates occur in the distribution matrix, it may lead to stress concentrators at localized areas, which results in reduced mechanical performance or may even early failure of the composite (Jamróz et al. 2019).

Table 1.5 Various fabrication techniques and their images for fabrication of polymer nanocomposites

Sl. no.	Fabrication technique	Image	References
1	Sol-Gel	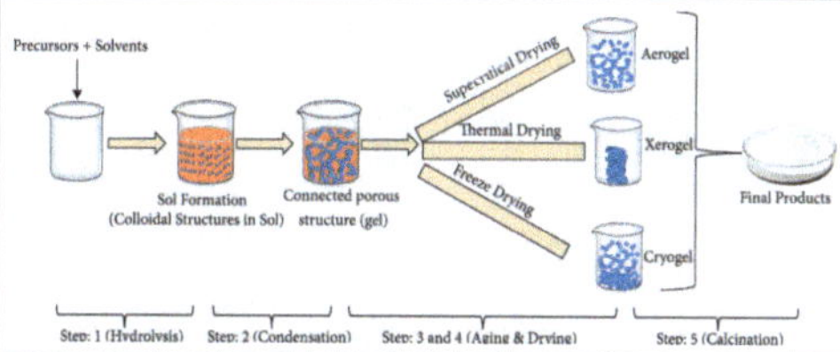	Bokov et al. (2021)
2	Electrospinning	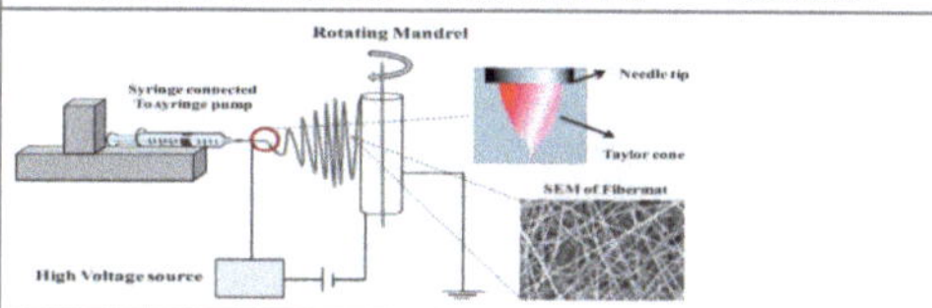	Shah et al. (2022)
3	Melt Mixing	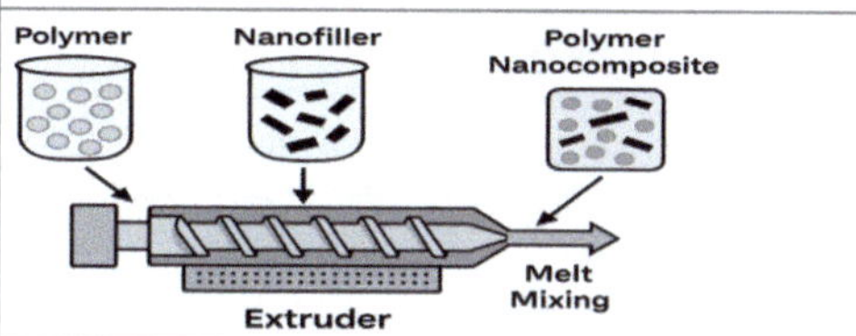	Saleh et al. (2020)
4	Solution Intercalation	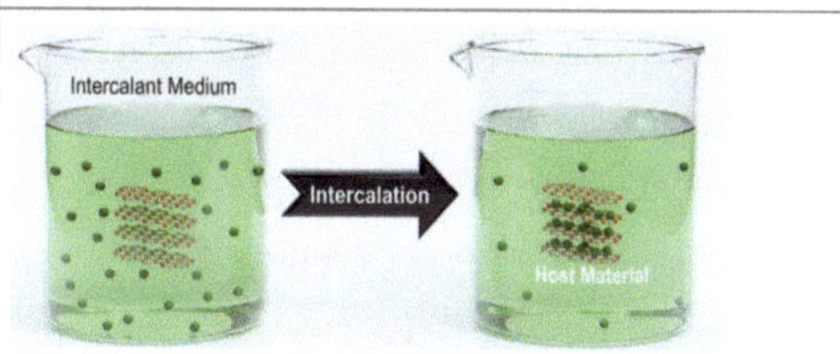	Rajapakse et al. (2021)
5	Magnetic Stirring	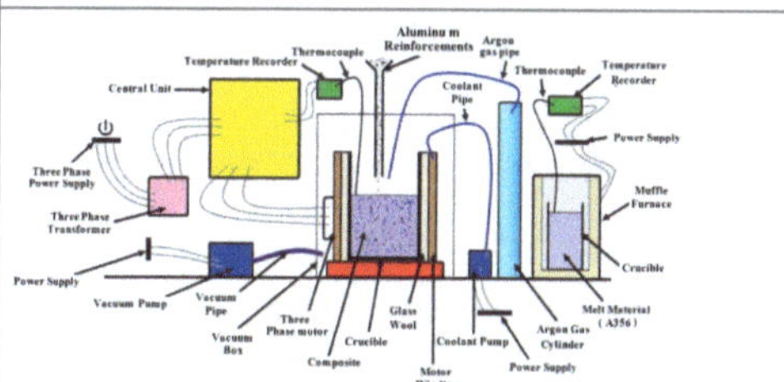	Dwivedi et al. (2014)
6	Compression Moulding	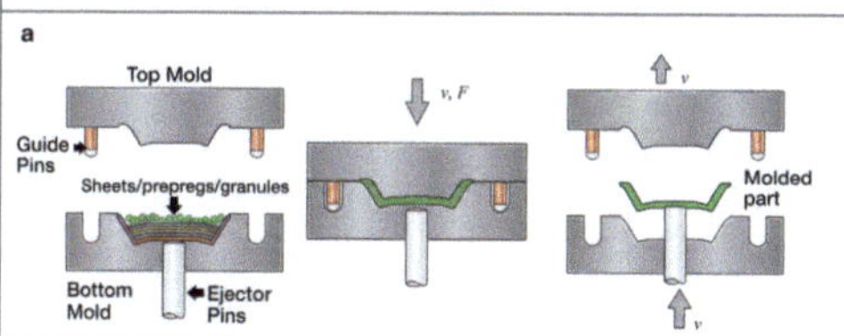	Birat et al. (2015)

(continued)

Table 1.5 (continued)

Sl. no.	Fabrication technique	Image	References
7	Freeze Drying	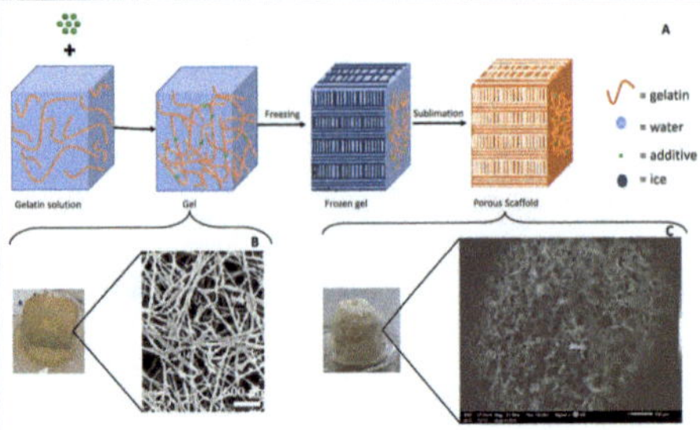	Katrilaka et al. (2023)
8	In-Situ Polymerization	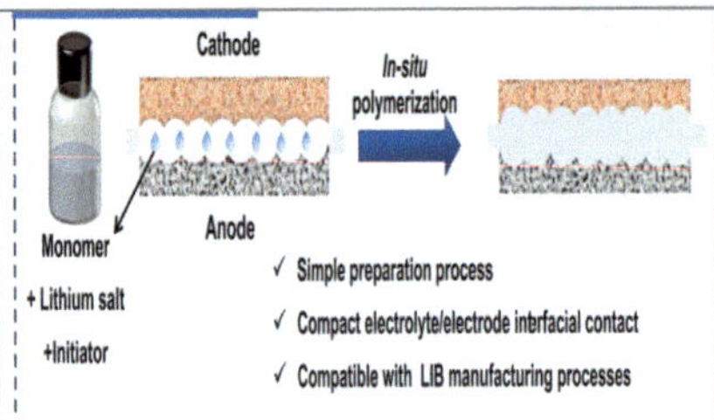	Sun et al. (2022)
9	Ball Milling	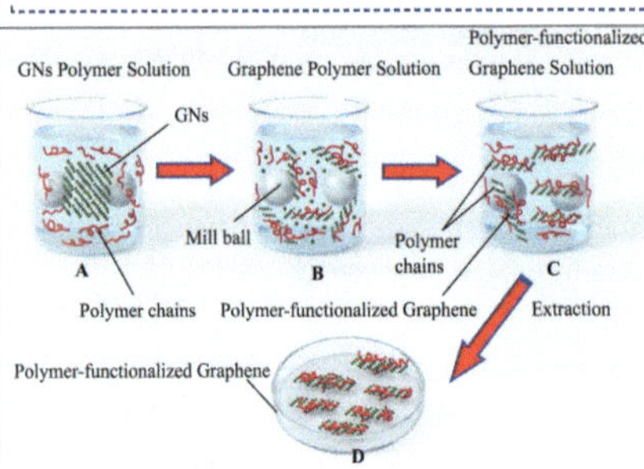	Wu et al. (2011)
10	Intercalation Polymerization	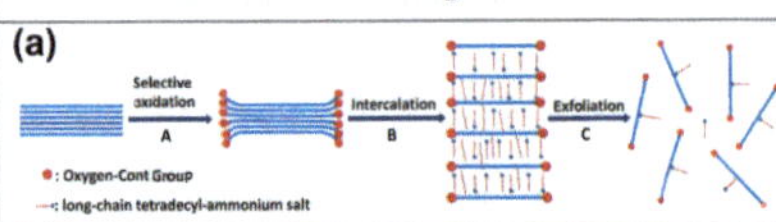	Guo et al. (2018)
11	Thermoforming	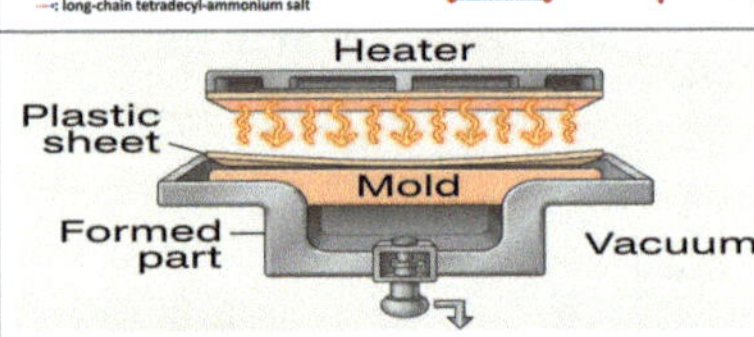	Schüller et al. (2016)
12	Extrusion	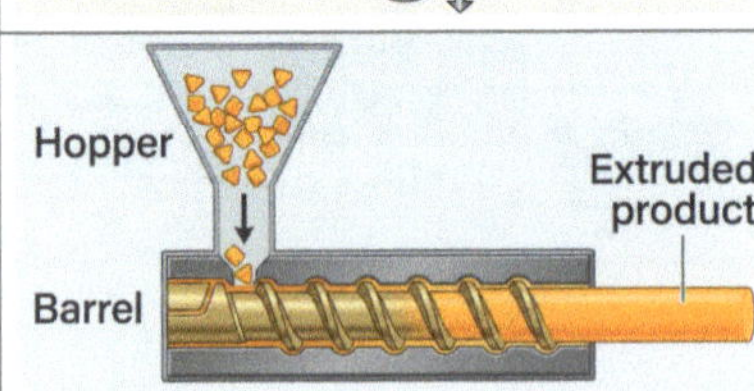	Hoque et al. (2012)

Table 1.6 Summary of characterization techniques

Technique	Property evaluated	Description
Scanning Electron Microscopy (SEM) (Mohammed and Abdullah 2018)	Morphology of the surface, Dispersion	High-resolution images of surface topology and filler distribution
Transmission Electron Microscopy (TEM) (Tang and Yang 2017)	Internal Structure, Nanofiller Dispersion	Visualization of nanofillers within the polymer matrix
X-ray Diffraction (XRD) (Ali et al. 2022)	Crystallinity, Interlayer Spacing	Crystalline phases and assess nanofiller exfoliation or intercalation
Atomic Force Microscopy (AFM) (Landoulsi and Dupres 2013)	Surface Roughness, Topography	Generates 3D surface maps and detects nanoscale features and irregularities
Thermogravimetric Analysis (TGA) (Prime et al. 2008)	Thermal Stability, Behaviour of Decomposition	Evaluates weight loss with temperature to determine degradation temperatures
Differential Scanning Calorimetry (DSC) (Menczel et al. 2008)	Glass Transition, Melting & Crystallization Temperatures	Measures thermal transitions and changes in polymer crystallinity
Dynamic Mechanical Analysis (DMA) (Chartoff et al. 2008)	Viscoelastic Properties, Moduli	Assesses mechanical behaviour under dynamic loading and varying temperatures
Fourier Transform Infrared Spectroscopy (FTIR) (Chen et al. 2015)	Chemical Structure, Functional Groups	Identifies chemical bonds and interactions between nanofillers and polymer matrix
Rheometry (Aho 2011)	Flow Behaviour, Viscosity, Moduli	Measures melt flow and viscoelastic properties to assess processability

- **Filler Alignments:** Mechanical, thermal, and electrical properties of a nanocomposites greatly depend on the alignment of the fillers, as the alignment contributes to the tensile strength, flexural capacity and electrical conductivity of the nanocomposites. In polymer nanocomposite, anisotropic properties can be observed, which may essentially govern the behaviour of the composites in different directions (Afolabi and Ndou 2024).
- **Interfacial Bonding:** In a polymer nanocomposite, interfacial bonding plays the crucial role of adhesion and formation of reinforced composite. This bonding acts as a mediator between two structures in the nanocomposite heterostructure matrix, and their nature determines the efficacy of stress transfer, and it is also needed to note that strong interfacial bonding often leads to improved mechanical performance of the PNT (Kun et al. 2021).

Selection of Fabrication Techniques for Enhanced Mechanical Performance

- **In-Situ Polymerization:** In in-situ polymerization, a homogenous mixture of nanofiller is introduced with the precursor monomer to achieve a uniform dispersion of the nanoparticles in the polymer matrix. This fabrication technique is highly useful for insoluble and thermo-unstable polymers, and it helps to ensure uniform distribution of fillers and strong covalent bonding between fillers and polymer, which results in superior tensile strength and avoidance of agglomeration. For instance, in-situ polymerization of silica nanoparticles into poly(methyl methacrylate) (PMMA) resulted in a 40% improvement in tensile strength over melt-mixed composites because of strong interfacial contacts and less agglomeration (Oehl et al. 2024).

- **Melt Mixing:** In Melt Mixing nanofillers are incorporated with the molten polymer under high heat and shear forces using extruders or internal mixers or any other injection moulder to achieve uniform nanocomposite (Saleh et al. 2020). Though melt mixing fabrication technique has various advantages in using over others like industrial compatibility, cost effectiveness and versatility, but it has some serious disadvantages, thermo-sensitive fillers and polymers can undergo severe degradation under high heat and shear forces resulting in instability of mechanical performance (Eriksson et al. 2020). Despite this melt mixing resultant nanocomposite exhibits increased stiffness and thermal stability, which makes it a suitable candidate for automobile parts (Delli et al. 2024).
- **Electrospinning:** Electrospinning in the fabrication of polymer nanocomposite is emerging as a novel approach to yield anisotropic mechanical properties with better morphological characteristics, resulting in formation of nanocomposites with increased tensile strength (Cho et al. 2025). For instance, the electrospun mats made from polyvinyl alcohol (PVA) and carbon nanotubes (CNT) exhibited a 200% enhancement in Young's modulus in the direction of fiber alignment, rendering them suitable for tendon or ligament repair (Baji et al. 2010).

1.1.5.2 Electrical Performance

Electrical Performance Consideration

- **Percolation Threshold:** Electrical performance of a nanocomposite is crucial in determining its application-specific fabrication. To evaluate this electrical conductivity measurement like percolation threshold used widely, which is a critical concentration of the nanofiller in a matrix to composite become electrically conductive (Al-Saleh and Sundararaj 2008). Deviation or escalation from this percolation threshold changes the electrical performance of the material, which means below percolation threshold creates an insulator nanocomposite and above creates a continuous network of electrical conductivity (Mohan et al. 2019).
- **Filler Aspect Ratio and Conductivity:** Aspect ratio and conductivity of fillers affect the mechanical, electrical and thermal performance of the nanocomposite. Fillers aspect ratio refers to the ratio of a filler particle's longest diameter to its shortest diameter and it affects significantly the conductivity and dielectric behaviour of a nanocomposite, higher aspect ratio fillers enhance the dielectric behaviour and conductivity of the composite in low filler concentration (Mutiso and Winey 2012). On the other hand, type of fillers based on their conductivity

properties contribute significantly on the electrical performance of the polymer nanocomposite, for example the incorporation of graphene, carbon nanotubes (CNTs), and silver nanoparticles (Ag NPs) leads to a notable improvement in electrical properties and the high aspect ratio and inherent conductivity of graphene (~10^6 S/m) facilitate lower percolation thresholds (0.1–1 vol%) in comparison to CNTs (Zare et al. 2024).
- **Dielectric Compatibility:** Dielectric Properties of polymer nanocomposites refers to the capacity of the nanocomposites to store and release energy when placed in an electrical field, and due to their supremacy over traditional polymer, polymer-based nanocomposites are gaining attention in this field (Shao et al. 2022). Though the creation of dielectric compatible material depends on various key factors like permittivity, breakdown strength, and interfacial polarization, still its wide application in the energy and electronics sector pushes forward for scalable fabrication of dielectric compatible polymer-based nanocomposites (You et al. 2023).

Selection of Fabrication Techniques for Enhanced Electrical Performance

- **Solvent Casting:** This process of fabrication is used to ensure uniform filler distribution at low concentration by dispersing nanofillers in polymer solution, followed by solvent evaporation (Wahid et al. 2018). This uniformity in filler distribution is a key to achieve desired conductive nanocomposites. For instance, films made from graphene/polyurethane (PU) through solvent casting demonstrated a percolation threshold of 0.5 vol%, allowing for use in flexible strain sensors (Kanoun et al. 2021). However, this method of fabrication may compromise the mechanical integrity of the nanocomposites, due to the presence of solvent pores not only that, the fabrication process can impart various challenges while optimizing electrical properties of the nanocomposites and polymer selection, due to its limited offering (Sundarram et al. 2015).
- **In-Situ Polymerization:** In-situ polymerization fabrication process creates intimate network contact of conductive nanofillers and polymer matrix, thus increasing the overall electrical performance of the nanocomposites (Sharma et al. 2023). The direct synthesis benefits, i.e., not requiring steps like exfoliation or surface modification of the nanofillers, make it suitable for large-scale fabrication. For instance, the synthesis of Polyaniline (PANI)/graphene composites through this method resulted in a conductivity of 10 S/cm at 2 wt% graphene loading, making them appropriate for anti-static coatings (Beygisangchin et al. 2024).
- **Layer-by-Layer (LbL) Assembly:** A flexible technique for creating polymer nanocomposites with improved electrical performance is layer-by-layer (LbL) assembly. In the LbL fabrication method, alternating layers of nanofillers and polymers are layered to enable precise dielectric properties, as the principle of this assembly relies on the electrostatic interaction and interfacial bonding to add layers (Bhattacharya 2016). Due to tuneable composition and control over the structural architecture of the final nanocomposites, LbL assembly enables the researchers to tailor the nanocomposite, as per the specific electrical performance

requirement (Guzmán et al. 2022). For example, multilayer films of barium titanate ($BaTiO_3$) and polyethyleneimine (PEI) demonstrated a dielectric constant of 45 at 1 kHz, which is perfect for capacitors (Choudhury 2010).

1.1.5.3 Thermal Stability

The selection of fabrication techniques of polymer nanocomposites is greatly influenced by the thermal stability of the polymer and nanofillers and overall scalability of the process of fabrication. These factors are very crucial as they impact the final properties, performance and economical viability of the resulting product.

Thermal Stability Considerations

- **Temperature-Sensitive Polymers:** Selection of fabrication technique must account for the thermal degradation limits of the polymer, specifically while using temperature-sensitive polymers in the nanocomposite fabrication. Processing temperatures for polymers such as polyvinyl chloride (PVC), polyethylene (PE), and polypropylene (PP) must stay below the threshold for polymer degradation (Mansour et al. 2020). Since solution-based techniques function at lower temperatures, they are frequently used to avoid such thermal degradation, also during melting and fabrication, temperature-controlled processes are needed to implement to avoid polymer chain scission (Barra et al. 2023). Such a narrow processing window for fabrication, limits the fabrication techniques while choosing temperature-sensitive polymers.
- **Thermal Behaviour of Nanofillers:** Carbon-based nanofillers like carbon nanotubes, graphene, etc., are capable of withstanding high temperature due to their inherent high thermal conductivity (Shahnaz et al. 2023). But few things need to be considered in this regard like metallic nanoparticles can promote the breakdown of polymers at high temperatures and under thermal stress, surface functionalization of nanofillers can change their processing compatibility and thermal behaviour (Balaji et al. 2024).

Selection of Fabrication Techniques Based on Thermal Stability

The limitations of polymers and nanofillers' heat deterioration property must be taken into consideration during fabrication. For instance:

- In **Melt Mixing** fabrication techniques, nanocomposite materials witnessed temperatures more than 200 °C, which may result in the degradation of the temperature-sensitive polymers like Poly(lactic acid) (PLA), Graphene oxide (GO) by transesterification and polymer chain scission. But the right use of some thermostable nanofillers can improve the thermal stability performance of the nanocomposites in the Melt Mixing fabrication method (Ladavos et al. 2021).

- When compared to melt mixing or solvent casting, **in situ polymerization** usually produces nanocomposites with a more uniform filler dispersion and greater thermal stability, due to lower processing temperatures and better filler integration, thus preserving the integrity of thermally sensitive components (Mapossa et al. 2023). Due to superior thermal stability and uniform filler distribution, in situ polymerization is preferred over melt mixing and solvent casting fabrication techniques (Mapossa et al. 2023).
- In **solvent casting** fabrication methods, thermal stability and integrity of the nanocomposite materials are preserved due to the lower process temperature in comparison to melt mixing and usually, the stability deviation during this fabrication process is very minimal unless high casting temperatures or reactive solvents are used (Jiang et al. 2014).

1.1.6 Scalability

Scalability refers to the ease of converting production from the lab scale to an industrial setup. Techniques like melt blending and extraction are highly adopted in the industrial setup due to their high throughput (up to 1000 kg/h) production and inherent compatibility with the existing production line, which facilitates the ease of integration (Pandey 2020). In scalability issue, Solvent casting faces significant challenges due to their expensive and regulatory precursors like high solvent recovery cost and extensive environmental regulation factors (Pandey 2020). The low production rates of electrospinning and LbL assembly restrict its use to specialized applications, such as medical textiles and they mainly produced in limited batch scale (Tayyab et al. 2024). In situ polymerization is an example of modern processes that can provide outstanding control at small sizes, but when production ramps up, they may have trouble maintaining uniformity and consistency, which makes it challenging for scalability (Tayyab et al. 2024).

1.1.7 Cost and Industrial Feasibility

While choosing specific fabrication technique of polymer nanocomposites, critical factors like cost effectiveness and their industrial feasibility need to consider to achieve desired production. The price of various raw materials for the fabrication may vary from material to material, which can influence the selection of techniques, not only that, but the competitive nature of the cost of materials also opening viable opportunities for the selection and scalability factors of the process. Expensive nanofillers are also key influencers to adopt the technique with minimize filler loading, like in-situ polymerization over melt mixing (Chikwendu et al. 2025). From energy consumption point of view, study reported the energy-saving nature of melt mixing over solvent-based methods, which requires solvent evaporation (Ray and

Okamoto 2003). Cost consideration also affects the selection of techniques which may require high end equipment like vacuum chambers for solvent casting or high-pressure autoclaves for in-situ polymerization, thus favouring techniques like melt mixing for industrial mass production (Darwish et al. 2022). To consider the industrial feasibility, we have already discussed the high throughput and scalability nature of the melt mixing over other techniques due to its ease of industrial integration and mass production and due to its compatibility with thermoplastic like polypropylene and polyethylene, melt mixing technique is widely used in the automotive and packaging industry (Pötschke et al. 2003). Process like in-situ polymerization may offer excellent filler dispersion and interfacial bonding, but due to its complex process and precise control, in-situ polymerization process faces scalability and reproducibility challenge in industrial setup, in spite of this, due to their high performance thermosets, they are preferred in the aerospace and defence equipment industry (Fenta and Mebratie 2024). It is also pertinent to mention that the industrial feasibility favours those techniques, which can be made to align with the current polymer processing technologies like, injection moulding, extrusion, etc. Innovative techniques such as electrohydrodynamic (EHD) processing, nanolithography, and 3D printing offer precise control over nanostructures, thereby gaining popularity in prototyping and customized electronics; however, they encounter challenges related to throughput, environmental sensitivity, and equipment complexity, which hinder their immediate industrial applicability.

1.1.8 Application-Specific Requirements

In spite of all the above-mentioned determinants, the selection of fabrication technique is highly dependent on the end-use application of the polymer nanocomposite. The selection of manufacturing technique directly affects nanofiller distribution, interfacial adhesion, and the overall shape of the composite, all of which are essential for customizing mechanical, thermal, electrical, or barrier characteristics for specific applications (Tayyab et al. 2024).

- Aerospace and defence equipment
  For the applicability of the aerospace and defence equipment application, factors like high strength-to-weight ratio, thermal stability, and EMI shielding are required, to fulfil this, fabrication processes like in-situ polymerization and autoclave curing processes are followed (Ali et al. 2024).
- Electrical, energy storage, or sensor applications
  In electrical, energy storage, or sensor applications, advanced methods such as nanolithography, aerosol jet printing, or laser-assisted deposition may be essential to get the requisite microstructure and performance, although at increased costs and complexity (Tayyab et al. 2024).
- Thin film application
  For thin film-based electronics and other precise applications, such as flexible electronics, solvent-based fabrication techniques play a key role.

- Energy storage components
  Applications like energy storage components require nanocomposites with high dielectric constant and thermal conductivity; for this purpose, LbL assembly fabrication is highly suitable due to their similar characteristics (Díez-Pascual and Rahdar 2022).
- Polymer nanocomposites
  Polymer nanocomposites play a key role due to their biomedical applicability, for such precision applications, electrospinning and solvent casting are ideal for creating porous, conductive scaffolds (Baji et al. 2010).
- Structural or packaging applications
  For general structural or packaging applications and in automobile applications, scalable and cost-effective methods like melt blending or solution mixing are often preferred due to their balance of performance and mass manufacturability.

1.2 Conclusion

In the current advance scenario of material science, polymer nanocomposites can be accessed as a key pillar, due to its wide versatile nature, which makes them preferable in the application of different sectors from aerospace and energy to healthcare and even widely as a sustainable packaging material. But, as highlighted in the chapter, it is very important to consider the various factors like use of nanofillers, choice of synthesis and fabrication methods and even application-specific composition and structure, etc. These key considerations are not merely a technical specification, but also shape the performance, cost and overall potential of the nanocomposite. This chapter also evidentially shows that there is no single superior technique of synthesis and fabrication, when it comes to polymer nanocomposite. Different combination of parameters like processing, scalability, applicability, cost and performance has its own pros and cons for the different sectors. For instance, methods like melt mixing and thermoforming-based fabrication techniques are widely acceptable and popular due its higher potential of industrial level scalability, high throughput and ease of integration but this technique has a huge challenge of optimal nanofiller dispersion. On the other hand, techniques like in-situ polymerization and electrospinning offer great control at the nanoscale, as they offer excellent nanofiller dispersion and robust interfacial bonding, which makes them desirable in high performance sectors like aerospace and electronics. Furthermore, emerging new fabrication techniques are changing the landscape, advance methods like 3D printing and Layer-by-Layer (LBL) techniques are offering novel structural customization and wide flexibility, which infuses the use of polymer nanocomposites in the precision and niche material application sector like healthcare but still need to overcome the challenges of scalability and high-cost constraints. Thus, Polymer nanocomposite fabrication is multidimensional system. As research in this field advances, new avenue of interdisciplinary ideas and innovation will surface, which can then truly unlock the immense potential of this material.

References

Afolabi OA, Ndou N (2024) Synergy of hybrid fillers for emerging composite and nanocomposite materials—a review. Polymers 16(13):1907. https://doi.org/10.3390/polym16131907

Aho J (2011) Rheological characterization of polymer melts in shear and extension: measurement reliability and data for practical processing. 12. https://trepo.tuni.fi/handle/10024/114192

Alexandre M, Dubois P (2000) Polymer-layered silicate nanocomposites: preparation, properties and uses of a new class of materials. Mater Sci Eng R: Rep 28(1):1–63. https://doi.org/10.1016/S0927-796X(00)00012-7

Ali A, Chiang YW, Santos RM (2022) X-ray diffraction techniques for mineral characterization: a review for engineers of the fundamentals, applications, and research directions. Minerals 12(2):205. https://doi.org/10.3390/min12020205

Ali Z, Yaqoob S, Yu J, D'Amore A, Fakhar-E-Alam M (2024) A Comparative review of processing methods for graphene-based hybrid filler polymer composites and enhanced mechanical, thermal, and electrical properties. J King Saud Univ Sci 36(10):103457. https://doi.org/10.1016/j.jksus.2024.103457

Al-Saleh MH, Sundararaj U (2008) A review of vapor grown carbon nanofiber/polymer conductive composites. Carbon 47(1):2–22. https://doi.org/10.1016/j.carbon.2008.09.039

Babul Reddy A, Siva Mohan Reddy G, Sivanjineyulu V, Jayaramudu J, Varaprasad K, Sadiku ER (2016) Hydrophobic/hydrophilic nanostructured polymer blends. In: Design and applications of nanostructured polymer blends and nanocomposite systems, pp 385–411. https://doi.org/10.1016/B978-0-323-39408-6.00016-9

Baji A, Mai Y, Wong S, Abtahi M, Chen P (2010) Electrospinning of polymer nanofibers: effects on oriented morphology, structures and tensile properties. Compos Sci Technol 70(5):703–718. https://doi.org/10.1016/j.compscitech.2010.01.010

Balaji D, Kumar PS, Bhuvaneshwari V, Rajeshkumar L, Singh MK, Sanjay MR, Siengchin S (2024) A review on effect of nanoparticle addition on thermal behavior of natural fiber–reinforced composites. Heliyon 11(1):e41192. https://doi.org/10.1016/j.heliyon.2024.e41192

Barra G, Guadagno L, Raimondo M, Santonicola MG, Toto E, Ciprioti SV (2023) A comprehensive review on the thermal stability assessment of polymers and composites for aeronautics and space applications. Polymers 15(18):3786. https://doi.org/10.3390/polym15183786

Batra RC, Pydah A (2020) Impact analysis of PEEK/ceramic/gelatin composite for finding behind the armor trauma. Compos Struct 237:111863. https://doi.org/10.1016/J.COMPSTRUCT.2020.111863

Beygisangchin M, Kamarudin SK, Rashid SA (2024) Synthesis, properties, and applications of polyaniline–graphene quantum dot nanocomposites: comprehensive review. J Environ Chem Eng 12(5):113460. https://doi.org/10.1016/j.jece.2024.113460

Bhattacharya M (2016) Polymer nanocomposites—a comparison between carbon nanotubes, graphene, and clay as nanofillers. Materials 9(4):262. https://doi.org/10.3390/ma9040262

Birat KC, Pervaiz M, Faruk O, Tjong J, Sain M (2015) Green composite manufacturing via compression molding and thermoforming. In: Manufacturing of natural fibre reinforced polymer composites, pp 45–63

Bokobza L (2007) Multiwall carbon nanotube elastomeric composites: a review. Polymer 48(17):4907–4920. https://doi.org/10.1016/J.POLYMER.2007.06.046

Bokov D, Turki Jalil A, Chupradit S, Suksatan W, Javed Ansari M, Shewael IH, Valiev GH, Kianfar E (2021) Nanomaterial by sol-gel method: synthesis and application. Adv Mater Sci Eng 2021(1):5102014. https://doi.org/10.1155/2021/5102014

Chartoff RP, Menczel JD, Dillman SH (2008) Dynamic mechanical analysis (DMA). In: Thermal analysis of polymers: fundamentals and applications, pp 387–495. https://doi.org/10.1002/9780470423837.CH5

Chen Y, Zou C, Mastalerz M, Hu S, Gasaway C, Tao X (2015) Applications of micro-Fourier transform infrared spectroscopy (FTIR) in the geological sciences—a review. Int J Mol Sci 16(12):30223–30250. https://doi.org/10.3390/IJMS161226227

Chikwendu NOC, Emeka NUC, Onyekachi NE (2025) The optimization of polymer-based nanocomposites for advanced engineering applications. World J Ad Res Rev 25(1):755–763. https://doi.org/10.30574/wjarr.2025.25.1.3820

Chiu CW, Lin CA, Hong PD (2011) Melt-spinning and thermal stability behavior of TiO_2 nanoparticle/polypropylene nanocomposite fibers. J Polym Res 18(3):367–372. https://doi.org/10.1007/S10965-010-9426-0/METRICS

Cho Y, Beak JW, Sagong M, Ahn S, Nam JS, Kim I (2025) Electrospinning and nanofiber technology: fundamentals, innovations, and applications. Adv Mate. https://doi.org/10.1002/adma.202500162

Choudhury A (2010) Dielectric and piezoelectric properties of polyetherimide/$BaTiO_3$ nanocomposites. Mater Chem Phys 121(1–2):280–285. https://doi.org/10.1016/j.matchemphys.2010.01.035

Coleman JN, Khan U, Blau WJ, Gun'ko YK (2006) Small but strong: a review of the mechanical properties of carbon nanotube-polymer composites. Carbon 44(9):1624–1652. https://doi.org/10.1016/j.carbon.2006.02.038

Darwish MSA, Mostafa MH, Al-Harbi LM (2022) Polymeric nanocomposites for environmental and industrial applications. Int J Mol Sci 23(3):1023. https://doi.org/10.3390/ijms23031023

Delli E, Gkiliopoulos D, Vouvoudi E, Bikiaris D, Chrissafis K (2024) Defining the effect of a polymeric compatibilizer on the properties of random polypropylene/glass fibre composites. J Compos Sci 8(2):44. https://doi.org/10.3390/jcs8020044

Díez-Pascual AM (2021) Development of graphene-based polymeric nanocomposites: a brief overview. Polymers 13(17):2978. https://doi.org/10.3390/polym13172978

Díez-Pascual A, Rahdar A (2022) LBL nano-assemblies: a versatile tool for biomedical and healthcare applications. Nanomaterials 12(6):949. https://doi.org/10.3390/nano12060949

Dwivedi SP, Sharma S, Mishra RK (2014) Electromagnetic stir casting and its process parameters for the fabrication and refined the grain structure of metal matrix composites—a review. Int J Adv Res Innov 2(3):125–138. https://doi.org/10.51976/ijari.231421

Eriksson M, Meuwissen M, Peijs T, Goossens H (2020) The influence of melt-mixing conditions and state of dispersion on crystallisation, rheology and mechanical properties of PCL/sepiolite nanocomposites. Int Polym Process 35(3):302–313. https://doi.org/10.3139/217.3890

Fenta EW, Mebratie BA (2024) Advancements in carbon nanotube-polymer composites: enhancing properties and applications through advanced manufacturing techniques. Heliyon 10(16):e36490. https://doi.org/10.1016/j.heliyon.2024.e36490

Fu S, Sun Z, Huang P, Li Y, Hu N (2019) Some basic aspects of polymer nanocomposites: a critical review. Nano Mater Sci 1(1):2–30. https://doi.org/10.1016/J.NANOMS.2019.02.006

Gawai S, Rai S, Bishnoi A (2025) Applications of nano additives in polymeric and rubber composites: a comprehensive review. Orient J Chem 41(2):445–457. https://doi.org/10.13005/ojc/410214

Giannelis EP (1996) Polymer layered silicate nanocomposites. Adv Mater 8(1):29–35. https://doi.org/10.1002/ADMA.19960080104

Gobena ST, Woldeyonnes AD (2024) A review of synthesis methods, and characterization techniques of polymer nanocomposites for diverse applications. Discov Mater 4(1):1–18. https://doi.org/10.1007/S43939-024-00119-0

Gong X, Guo W, Zhao Z, Sun C, Zhang Y, Qu R, Shen Y, Wang Y (2025) Synthesis of aramid nanomaterials by in-situ solution polymerization and application in reinforcement of polyvinyl chloride. Eur Polym J 223:113618. https://doi.org/10.1016/J.EURPOLYMJ.2024.113618

Guzmán E, Ortega F, Rubio RG (2022) Layer-by-layer materials for the fabrication of devices with electrochemical applications. Energies 15(9):3399. https://doi.org/10.3390/en15093399

Guo Y, Xu G, Yang X, Ruan K, Ma T, Zhang Q, Gu J, Wu Y, Liu H, Guo Z (2018) Significantly enhanced and precisely modeled thermal conductivity in polyimide nanocomposites with chemically modified graphene via in situ polymerization and electrospinning-hot press technology. Journal of Materials Chemistry C, 6(12), pp 3004–3015

Habibi Y, Lucia LA, Rojas OJ (2010) Cellulose nanocrystals: chemistry, self-assembly, and applications. Chem Rev 110(6):3479–3500. https://doi.org/10.1021/CR900339W/ASSET/CR900339W.FP.PNG_V03

Haider S, Khan Y, Almasry WA, Haider A, Haider S, Khan Y, Almasry WA, Haider A (2012) Thermoplastic nanocomposites and their processing techniques. Thermoplast Compos Mater. https://doi.org/10.5772/36941

Hoque ME, Chuan YL, Pashby I (2012) Extrusion based rapid prototyping technique: an advanced platform for tissue engineering scaffold fabrication. Biopolymers 97(2):83–93

Hossain MT, Shahid MA, Mahmud N, Habib A, Rana MM, Khan SA, Hossain MD (2024) Research and application of polypropylene: a review. Discov Nano 19(1):1–21. https://doi.org/10.1186/S11671-023-03952-Z

Hussain F, Hojjati M, Okamoto M, Gorga RE (2006) Review article: Polymer-matrix nanocomposites, processing, manufacturing, and application: an overview. J Compos Mater 40(17):1511–1575. https://doi.org/10.1177/0021998306067321

Jamróz E, Kulawik P, Kopel P (2019) The effect of nanofillers on the functional properties of biopolymer-based films: a review. Polymers 11(4):675. https://doi.org/10.3390/polym11040675

Jiang H, Wang LB, Xu DJ, Liu XY (2014) Comparison of solvent-casting and melt-compounding blended biomedical (polylactide)-(polyethylene glycol) mixture as drug carrier. Adv Mater Res 1002:99–104. https://doi.org/10.4028/www.scientific.net/amr.1002.99

Jose S, Sariga, Varghese A (2023) Design and structural characteristics of conducting polymer-metal organic framework composites for energy storage devices. Synth Met 297:117421. https://doi.org/10.1016/J.SYNTHMET.2023.117421

Kanoun O, Bouhamed A, Ramalingame R, Bautista-Quijano JR, Rajendran D, Al-Hamry A (2021) Review on conductive polymer/CNTs nanocomposites based flexible and stretchable strain and pressure sensors. Sensors 21(2):341. https://doi.org/10.3390/s21020341

Kashiwagi T, Harris RH, Zhang X, Briber RM, Cipriano BH, Raghavan SR, Awad WH, Shields JR (2004) Flame retardant mechanism of polyamide 6–clay nanocomposites. Polymer 45(3):881–891. https://doi.org/10.1016/J.POLYMER.2003.11.036

Kato M, Usuki A (2000) Polymerization of organic monomers and biomolecules on hectorite. Polymer-clay nanocomposites. https://2msdumpvxj.us-east-1.awsapprunner.com/en-au/Polymer-Clay+Nanocomposites-p-9780471637004

Katrilaka C, Karipidou N, Petrou N, Manglaris C, Katrilakas G, Tzavellas AN, Pitou M, Tsiridis EE, Choli-Papadopoulou T, Aggeli A (2023) Freeze-drying process for the fabrication of collagen-based sponges as medical devices in biomedical engineering. Materials 16(12). https://doi.org/10.3390/ma16124425

Kayan GÖ, Kayan A (2023) Polycaprolactone composites/blends and their applications especially in water treatment. ChemEngineering 7(6):104. https://doi.org/10.3390/CHEMENGINEERING7060104

Khdier HM, Husham KAF, Shali WM, Al-Atabi HA (2024) Interfacial effects on mechanical, thermal and electrical properties of polymer-based nanocomposites: a review. Ann Chimie Sci Matériaux 48(6):857–869. https://doi.org/10.18280/acsm.480611

Kim H, Abdala AA, MacOsko CW (2010) Graphene/polymer nanocomposites. Macromolecules 43(16):6515–6530. https://doi.org/10.1021/MA100572E/ASSET/IMAGES/MEDIUM/MA-2010-00572E_0012.GIF

Krishnamoorti R, Vaia RA (eds) (2001) Polymer nanocomposites. 804. https://doi.org/10.1021/BK-2002-0804

Krishnamoorti R, Vaia R (2002) Polymer Nanocomposites: Introduction. In: Krishnamoorti R, and Vaia R, Eds., Polymer Nanocomposites-Synthesis, Characterization and Modeling, ACS Symposium Series, Vol. 804, American Chemical Society, Washington, DC.

Kubiak JM, Li B, Suazo M, Macfarlane RJ (2022) Polymer grafted nanoparticle composites with enhanced thermal and mechanical properties. ACS Appl Mater Interfaces 14(18):21535–21543. https://doi.org/10.1021/ACSAMI.2C03797

Kun D, Kárpáti Z, Fekete E, Móczó J (2021) The role of interfacial adhesion in polymer composites engineered from lignocellulosic agricultural waste. Polymers 13(18):3099. https://doi.org/10.3390/polym13183099

Ladavos A, Giannakas AE, Xidas P, Giliopoulos DJ, Baikousi M, Gournis D, Karakassides MA, Triantafyllidis KS (2021) Preparation and characterization of polystyrene hybrid composites reinforced with 2D and 3D inorganic fillers. Micro 1(1):3–14. https://doi.org/10.3390/micro1010002

Landoulsi J, Dupres V (2013) Direct AFM force mapping of surface nanoscale organization and protein adsorption on an aluminum substrate. Phys Chem Chem Phys 15(21):8429. https://doi.org/10.1039/c3cp00137g

Ma L, Yan Z, Zhou X, Pi Y, Du Y, Huang J, Wang K, Wu K, Zhuang C, Han X (2021) A polymer controlled nucleation route towards the generalized growth of organic-inorganic perovskite single crystals. Nature Communications 12(1):1–10. https://doi.org/10.1038/s41467-021-22193-1

Mansour DA, Abdel-Gawad NMK, Dein AZE, Ahmed HM, Darwish MMF, Lehtonen M (2020) Recent advances in polymer nanocomposites based on polyethylene and polyvinylchloride for power cables. Materials 14(1):66. https://doi.org/10.3390/ma14010066

Mapossa AB, Da Silva Júnior AH, De Oliveira CRS, Mhike W (2023) Thermal, morphological and mechanical properties of multifunctional composites based on biodegradable polymers/bentonite clay: a review. Polymers 15(16):3443. https://doi.org/10.3390/polym15163443

Menczel JD, Judovits L, Prime RB, Bair HE, Reading M, Swier S (2008) Differential scanning calorimetry (DSC). In: Thermal analysis of polymers: fundamentals and applications, pp 7–239. https://doi.org/10.1002/9780470423837.CH2;JOURNAL:JOURNAL:BOOKS;WGROUP:STRING:PUBLICATION

Mohammed A, Abdullah A (2018) Scanning electron microscopy (SEM): a review. In: Proceedings of 2018 international conference on hydraulics and pneumatics—HERVEX, pp 7–9

Mohan L, Kumar PN, Karakkad S, Krishnan ST (2019) Determination of electrical percolation threshold of carbon nanotube-based epoxy nanocomposites and its experimental validation. IET Sci Meas Technol 13(9):1299–1304. https://doi.org/10.1049/iet-smt.2019.0011

Mutiso R, Winey K (2012) Electrical conductivity of polymer nanocomposites. In: Elsevier eBooks, pp 327–344. https://doi.org/10.1016/b978-0-444-53349-4.00196-5

Nguon O, Lagugné-Labarthet F, Brandys FA, Li J, Gillies ER (2018) Microencapsulation by in situ polymerization of amino resins. Polym Rev 58(2):326–375. https://doi.org/10.1080/15583724.2017.1364765;WGROUP:STRING:PUBLICATION

Ocaña C, del Valle M (2018) Impedimetric aptasensors using nanomaterials. In: Nanotechnology and biosensors, pp 233–267. https://doi.org/10.1016/B978-0-12-813855-7.00008-8

Oehl G, Naga N, Ziegmann G (2024) Facile synthesis of poly(methyl methacrylate) silica nanocomposite monolith by in situ free radical polymerization of methyl methacrylate in the presence of functionalized silica nanoparticles. ACS Omega 9(14):16279–16287. https://doi.org/10.1021/acsomega.3c10338

Oladele IO, Adelani SO, Taiwo AS, Akinbamiyorin IM, Olanrewaju OF, Orisawayi AO (2025) Polymer-based nanocomposites for supercapacitor applications: a review on principles, production and products. RSC Adv 15(10):7509–7534. https://doi.org/10.1039/D4RA08601E

Pandey PP (2020) Preparation and characterization of polymer nanocomposites. Soft Nanosci Lett 10(01):1–15. https://doi.org/10.4236/snl.2020.101001

Park S, Shou W, Makatura L, Matusik W, Fu K (2022) 3D printing of polymer composites: materials, processes, and applications. Matter 5(1):43–76. https://doi.org/10.1016/J.MATT.2021.10.018

Patel GM, Shah V, Bhaliya J, Pathan P, Nikita KM (2022) Polymer-based nanomaterials: an introduction. In: Smart polymer nanocomposites: design, synthesis, functionalization, properties, and applications, pp 27–59. https://doi.org/10.1016/B978-0-323-91611-0.00018-9

Patil JP, Nandedkar VM, Mishra S (2019) Thermal analysis of the heating stage of the thermoforming process. Int J Recent Technol Eng 8(2):2085–2092. https://doi.org/10.35940/IJRTE.B2301.078219

Paul DR, Robeson LM (2008) Polymer nanotechnology: nanocomposites. Polymer 49(15):3187–3204. https://doi.org/10.1016/J.POLYMER.2008.04.017

Pötschke P, Bhattacharyya AR, Janke A, Goering H (2003) Melt mixing of polycarbonate/multi-wall carbon nanotube composites. Compos Interfaces 10(4–5):389–404. https://doi.org/10.1163/156855403771953650

Prime RB, Bair HE, Vyazovkin S, Gallagher PK, Riga A (2008) Thermogravimetric analysis (TGA). In: Thermal analysis of polymers: fundamentals and applications, pp 241–317. https://doi.org/10.1002/9780470423837.CH3

Qian H, Greenhalgh ES, Shaffer MSP, Bismarck A (2010) Carbon nanotube-based hierarchical composites: a review. J Mater Chem 20(23):4751–4762. https://doi.org/10.1039/C000041H

Rajapakse M, Karki B, Abu UO, Pishgar S, Musa MRK, Riyadh SMS, Yu M, Sumanasekera G, Jasinski JB (2021) Intercalation as a versatile tool for fabrication, property tuning, and phase transitions in 2D materials. Npj 2D Mater Appl 5(1). https://doi.org/10.1038/s41699-021-00211-6

Rashid IA, Tariq A, Shakir HF, Afzal A, Ali F, Abuzar M, Haider T (2022) Electrically conductive epoxy/polyaniline composite fabrication and characterization for electronic applications. J Reinf Plast Compos 41(1–2):34–45. https://doi.org/10.1177/07316844211023991

Ray SS, Okamoto M (2003) Polymer/layered silicate nanocomposites: a review from preparation to processing. Prog Polym Sci 28(11):1539–1641. https://doi.org/10.1016/j.progpolymsci.2003.08.002

Romero-Araya P, Pino V, Nenen A, Cárdenas V, Pavicic F, Ehrenfeld P, Serandour G, Lisoni JG, Moreno-Villoslada I, Flores ME (2021) Combining materials obtained by 3D-printing and electrospinning from commercial polylactide filament to produce biocompatible composites. Polymers 13(21):3806. https://doi.org/10.3390/POLYM13213806/S1

Saleh TA, Shetti NP, Shanbhag MM, Reddy KR, Aminabhavi TM (2020) Recent trends in functionalized nanoparticles loaded polymeric composites: an energy application. Mater Sci Energy Technol 3:515–525. https://doi.org/10.1016/j.mset.2020.05.005

Schadler LS, Brinson LC, Sawyer WG (2007) Polymer nanocomposites: a small part of the story. JOM 59(3):53–60. https://doi.org/10.1007/S11837-007-0040-5/METRICS

Schüller C, Panozzo D, Grundhöfer A, Zimmer H, Sorkine E, Sorkine-Hornung O (2016) Computational thermoforming. ACM Trans Graph (TOG) 35(4):1–9

Shah V, Bhaliya J, Patel GM, Deshmukh K (2022) Advances in polymeric nanocomposites for automotive applications: a review. Polym Adv Technol 33(10):3023–3048. https://doi.org/10.1002/pat.5771

Shahnaz T, Hayder G, Shah MA, Ramli MZ, Ismail N, Hua CK, Zahari NM, Mardi NH, Selamat FE, Kabilmiharbi N, Aziz HA (2023) Graphene-based nanoarchitecture as a potent cushioning/filler in polymer composites and their applications. J Mater Res Technol 28:2671–2698. https://doi.org/10.1016/j.jmrt.2023.12.108

Shao J, Zhou L, Chen Y, Liu X, Ji M (2022) Model-based dielectric constant estimation of polymeric nanocomposite. Polymers 14(6):1121. https://doi.org/10.3390/polym14061121

Sharma S, Verma A, Rangappa SM, Siengchin S, Ogata S (2023) Recent progressive developments in conductive-fillers based polymer nanocomposites (CFPNC's) and conducting polymeric nanocomposites (CPNC's) for multifaceted sensing applications. J Mater Res Technol 26:5921–5974. https://doi.org/10.1016/j.jmrt.2023.08.300

Solař P, Polonskyi O, Olbricht A, Hinz A, Shelemin A, Kylián O, Choukourov A, Faupel F, Biederman H (2017) Single-step generation of metal-plasma polymer multicore@shell nanoparticles from the gas phase. Sci Rep 7(1):1–6. https://doi.org/10.1038/S41598-017-08274-6

Sun Y, Wang Y, Li L, Li M, Fu Y, Zheng J, Chen D, Zhang Y, Zhou H (2020) Direct thermoforming manufacture of cellulose transparent products employing nanospheres. Carbohydr Polym 247:116668. https://doi.org/10.1016/J.CARBPOL.2020.116668

Sundarram S, Kim Y, Li W (2015) Preparation and characterization of poly(ether imide) nanocomposites and nanocomposite foams. In: Elsevier eBooks, pp 61–85. https://doi.org/10.1016/b978-1-78242-308-9.00004-5

Sun M, Zeng Z, Zhong W, Han Z, Peng L, Cheng S and Xie J (2022) In-situ polymerization methods for polymer-based solid-state lithium batteries. Batteries & Supercaps, 5(12), p.e202200338.

Tang CY, Yang Z (2017) Transmission electron microscopy (TEM). In: Membrane characterization, pp 145–159. https://doi.org/10.1016/B978-0-444-63776-5.00008-5
Tayyab QA, Hameed H, Mohiuddin M, Mukhtar A, Shafiq U, Majeed A, Nazli F, Ameen M, Khan MA, Sarwar MT, Ali A (2024) Metal-tolerant Rhizobacteria mediated phytostabilization of heavy metal-contaminated soil: A Review. Plant Environ, 5(01), pp 74–88
Tayyab M, Zizhe L, Rauf S, Xu Z, Sagar RUR, Faiz F, Tayyab Z, Rehman RU, Imran M, Waheed A, Javed R, Surulinathan A, Zafar Z, Fu XZ, Luo JL (2025) Advanced fabrication techniques for polymer–metal nanocomposite films: state-of-the-art innovations in energy and electronic applications. Chem Sci 16(8):3362–3407. https://doi.org/10.1039/D4SC04600E
Thomason JL, Vlug MA (1996) Influence of fibre length and concentration on the properties of glass fibre-reinforced polypropylene: 1. Tensile and flexural modulus. Compos Part A Appl Sci Manuf 27(6):477–484. https://doi.org/10.1016/1359-835X(95)00065-A
Tofighy MA, Mohammadi T (2020) Carbon nanotubes-polymer nanocomposite membranes for pervaporation. In: Polymer nanocomposite membranes for pervaporation, pp 105–133. https://doi.org/10.1016/B978-0-12-816785-4.00005-7
Tran QH, Nguyen VQ, Le AT (2013) Silver nanoparticles: synthesis, properties, toxicology, applications and perspectives. Adv Nat Sci Nanosci Nanotechnol 4(3):033001. https://doi.org/10.1088/2043-6262/4/3/033001
Utracki LA, Kamal MR (2002) Clay-containing polymeric nanocomposites. Arab J Sci Eng Spec Issue 27(1C):43–67
Vasudeo Rane A, Kanny K, Abitha VK, Patil SS, Thomas S (2017) Clay–polymer composites: design of clay polymer nanocomposite by mixing. In: Clay-polymer nanocomposites, pp 113–144. https://doi.org/10.1016/B978-0-323-46153-5.00004-5
Verma D, Goh KL (2019) Functionalized graphene-based nanocomposites for energy applications. In: Functionalized graphene nanocomposites and their derivatives: synthesis, processing and applications, pp 219–243. https://doi.org/10.1016/B978-0-12-814548-7.00011-8
Verreck G, Chun I, Peeters J, Rosenblatt J, Brewster ME (2003) Preparation and characterization of nanofibers containing amorphous drug dispersions generated by electrostatic spinning. Pharm Res 20(5):810–817. https://doi.org/10.1023/A:1023450006281/METRICS
Wahid F, Khan T, Hussain Z, Ullah H (2018) Nanocomposite scaffolds for tissue engineering; properties, preparation and applications. In: Elsevier eBooks, pp 701–735. https://doi.org/10.1016/b978-0-12-813741-3.00031-5
Wetzel B, Rosso P, Haupert F, Friedrich K (2006) Epoxy nanocomposites—fracture and toughening mechanisms. Eng Fract Mech 73(16):2375–2398. https://doi.org/10.1016/J.ENGFRACMECH.2006.05.018
Winey KI, Vaia RA (2007) Polymer nanocomposites. MRS Bull 32(4):314–319. https://doi.org/10.1557/mrs2007.229
Wu H, Zhao W, Hu H, Chen G, (2011) One-step in situ ball milling synthesis of polymer-functionalized graphene nanocomposites. J Mater Chem, 21(24), pp 8626–8632
You L, Liu B, Hua H, Jiang H, Yin C, Wen F (2023) Energy storage performance of polymer-based dielectric composites with two-dimensional fillers. Nanomaterials 13(21):2842. https://doi.org/10.3390/nano13212842
Young RJ, Kinloch IA, Gong L, Novoselov KS (2012) The mechanics of graphene nanocomposites: a review. Compos Sci Technol 72(12):1459–1476. https://doi.org/10.1016/J.COMPSCITECH.2012.05.005
Zaghloul MMY, Zaghloul MMY, Fuseini M (2023) Recent progress in epoxy nanocomposites: corrosion, structural, flame retardancy and applications—a comprehensive review. Polym Adv Technol 34(11):3438–3472. https://doi.org/10.1002/PAT.6144
Zare Y, Munir MT, Rhee KY (2024) Assessment of electrical conductivity of polymer nanocomposites containing a deficient interphase around graphene nanosheet. Sci Rep 14(1). https://doi.org/10.1038/s41598-024-59678-0

Chapter 2
Polymer Electrolytes for Next Generation Energy Storage Applications

Neelabh Srivastava, Nitin Srivastava, and Sonika

Abstract The use of fossil fuel-based energy sources leads to severely impact the environmental pollution and global warming; therefore, it is essential to produce energy from the renewable sources. In energy storage devices, electrolyte plays an important role for deciding the performance of energy storage devices. In battery, liquid electrolyte is commonly used but it has some drawbacks like fast Li dendrite growth, dissolution of electrode ions, etc. Polymer electrolytes have emerged as a pivotal class of materials for energy storage systems, offering a safer and more flexible alternative to conventional liquid electrolytes. This chapter deals with comprehensive overview of polymer electrolytes and their applications in rechargeable batteries. Furthermore, a summary on the challenges and future perspectives associated with the study of polymer electrolytes for next-generation devices is also presented.

Keywords Polymer electrolytes · Environment-friendly · Electrochemical analysis · Li-battery · Ionic conductivity

---

N. Srivastava
Department of Physics, School of Physical Sciences, Mahatma Gandhi Central University, Motihari, East Champaran, Bihar, India

N. Srivastava (✉)
Department of Physics, Shri Chitragupt P. G. College (Affiliated to Dr. B. R. Ambedkar University, Agra), Mainpuri, Uttar Pradesh, India

Sonika
Department of Physics, Rajiv Gandhi University, Rono Hills Doimukh, Papumpare, Arunachal Pradesh, India

S. K. Verma et al. (eds.), *Sustainable and Natural Polymers*, Clean Energy Production Technologies, https://doi.org/10.1007/978-981-92-0092-4_2

2.1 Introduction

The demand for energy consumption has been increasing day by day due to an increase in population. Excessive reliance on fossil fuels is contributing to rising levels of environmental pollution. The energy crisis experienced in recent decades has driven researchers to look for an alternative solution for the development of new, clean, and cost-effective energy storage devices primarily utilizing renewable energy resources (Nitta et al. 2015; Srivastava et al. 2020). Renewable energy sources like solar energy and wind energy are considered unstable towards power generation and uneven energy distribution, hence advanced energy storage technologies are necessary for efficient use of energy (Srivastava et al. 2022a). In the last decades, significant efforts have been put into the development of high--performance energy storage systems, including lithium-ion batteries (LIBs), supercapacitors, and lithium-sulfur batteries. Out of this, Li-ion battery technology has emerged out as one of the most prominent and capable energy storage technologies due to its exotic properties like high energy and high-power density, outstanding cyclic stability. Lithium, with a reduction potential of -3.04 V versus the standard hydrogen electrode, is the most electrochemically active and electronegative metal, offering exceptionally high energy density (Sun et al. 2016). The process of charging and discharging in Lithium-ion batteries involves the de-intercalation and intercalation of Li^+-ions from and into a positive electrode via a redox reaction. Figure 2.1 shows the schematic representation of basic principles of operation of rechargeable Lithium-ion batteries (LIBs) (Badi et al. 2022). Ongoing research focuses on enhancing the synthesis of electrode materials for lithium-ion batteries to optimize charge storage capacity and ensure safety. Many studies have been conducted to improve the high specific capacity, high rate and excellent cycling performance of LIBs. Electrode (mainly cathode) plays a major role for determining the performance of lithium-ion batteries in terms of energy density and cyclability (Mishra et al. 2018). Electrolytes provide the medium for transportation of ions in between the electrode. In commercial battery system liquid electrolytes show high ionic conductivity, better cyclability, and wide electrochemical stability window. Due to highly volatile and flammable nature of such electrolytes, its use can cause electrode corrosion and leakage problem for practical applications. Furthermore, these liquid electrolytes are unsuitable for applications involving higher voltage batteries because of their electrochemical instability at such high voltages (Pigłowska et al. 2021). Therefore, for safety utilization and higher voltage application polymer electrolytes are gaining enormous interest in electrochemical devices such as fuel cells, batteries, supercapacitors, etc. These polymer electrolytes show several advantages over other conventional liquid electrolytes like high energy density, high ionic conductivity, wide electrochemical window, etc. In general, there are two types of polymer electrolytes: solvent-free solid polymer electrolytes (SPEs) and gel polymer electrolytes (GPEs) (Baskoro et al. 2019). Figure 2.2 discusses the schematic representation of comparison of polymer electrolytes in terms of ionic conductivity, mechanical properties, and thermal stability (Ahmed et al. 2024).

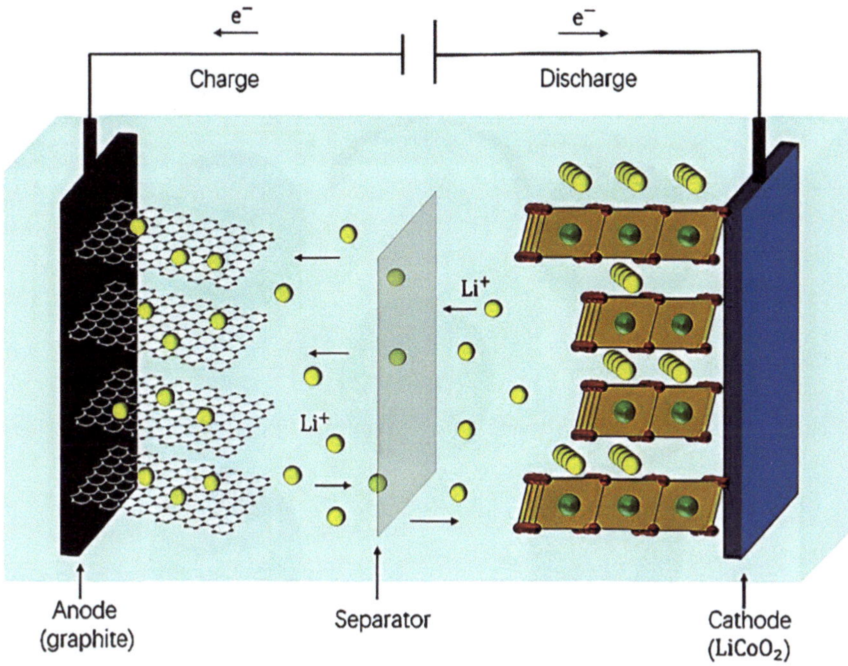

Fig. 2.1 Schematic working principle of rechargeable Li-ion battery (Reproduced from Badi et al. 2022 under CC by 4.0 license)

This chapter presents a comprehensive exploration of polymer electrolytes, their synthesis strategies, ionic transport mechanism, and practical deployment in next generation energy storage devices.

2.2 Solid Polymer Electrolytes (SPEs)

Solid polymer electrolytes are a class of solvent-free electrolytes where ionic conduction occurs through a polymer matrix with a salt. SPEs are synthesized by incorporating organic salt into the polymer matrix. In SPE, ionic conductivity primarily arises from the complexation between lithium salts and electron-donating groups present in the polymer backbones. The electron-rich segments—typically ether oxygens in polymers like poly (ethylene oxide) (PEO)—coordinate with lithium cations to form transient complexes. This coordination facilitates the solvation of lithium ions, weakening the electrostatic interactions between the cation and its counter anion from the dopant salt. As a result, ion dissociation is enhanced, and free lithium ions are more available for conduction.

The movement of these ions follows an ionic hopping mechanism, where Li^+ ions migrate from one coordination site to another, assisted by the segmental motion

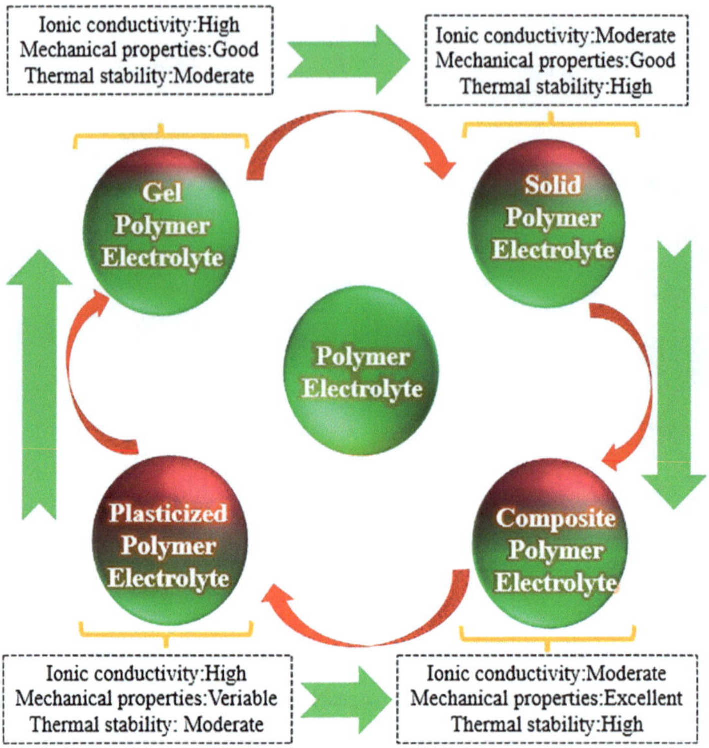

Fig. 2.2 Schematic comparison of different properties of polymer electrolytes (Reproduced from Ahmed et al. 2024 under CC by 4.0 license)

of polymer chains. This dynamic interplay between polymer flexibility and ion mobility underpins the transport behaviour in SPEs, with conductivity typically increasing at elevated temperatures due to enhanced chain mobility. Solid Polymer Electrolytes (SPEs) exhibit numerous intriguing characteristics, including being solvent-free, leak-proof, exhibiting low volatility, and demonstrating high thermal and electrochemical stabilities. They also possess high ionic conductivity, elevated energy density, and ease of processability. Nevertheless, SPEs are not without their limitations, particularly their reduced ionic conductivity at ambient temperatures. Chaurasia et al. (2015) synthesized the PEO-based polymer electrolytes and they found that ionic conductivity of PEO + 10 wt.% $LiPF_6$ polymer electrolytes ~4.4×10^{-7} at ambient temperatures. The low value of ionic conductivity is not favourable for device applications. There are several methods to improve the ionic conductivity of solid polymer electrolytes, including copolymerization, the incorporation of inorganic fillers, the addition of plasticizers, the introduction of ionic liquids, and synthesizing blend polymer electrolytes (Ngai et al. 2016).

2.3 Gel Polymer Electrolytes (GPEs)

It is also referred to as plasticized polymer electrolytes, a term first introduced by Feuillade and Perche in 1975. GPEs are formed by introducing organic molecular solvents known as plasticizers to the polymer alkali metal salt complexes, in which the organic solvents remain trapped within the matrix of polymer. Compared to SPEs, GPEs have great potential application in energy storage devices due to its high ionic conductivity and good mechanical properties. By introducing a liquid plasticizer (EC, PC, DEC, etc.) or solvent into the polymer–salt matrix, the polymer swells, forming a semi-solid gel that combines the mechanical stability of solids with the ionic conductivity of liquids. When a lithium salt (like $LiPF_6$, LiTFSI, LiFSI, etc.) is added to a GPE, it dissociates into lithium ions (Li^+) and its counter-ions. These ions move through the gel via pathways formed by the plasticized polymer network. The solvent or plasticizer (like ethylene carbonate) swells the polymer, increasing free volume and enabling ions to hop or diffuse through. A soft polymer matrix (like poly (vinylidene fluoride), PVDF) allows better ion transport. The molecular motion of polymer chains helps shuttle ions through the material. The gel creates interconnected pathways that enhance ion mobility. GPEs can reach conductivities between *10^{-4} to 10^{-3} S/cm*—good enough for many practical applications like batteries and supercapacitors. There are various approaches through which ionic conductivity of GPEs can be enhanced, such as by adding inorganic filler, copolymerization, etc. In this chapter, we will discuss the above properties (Ngai et al. 2016).

2.4 Synthesis and Structure

2.4.1 Solution Casting Technique

In this technique, the polymer, along with other components such as salts and plasticizers are dissolved in a common solvent and a uniform solution is obtained. As a result, polymer films of various thicknesses (50–300 μm) can be achieved. In the solution casting technique, it is essential that the solvent is devoid of any water content and also acts as a universal solvent capable of dissolving both the salt and the polymer. Before sample preparation, polymers and plasticizers are dried in vacuum (~10^{-3} torr pressure) for 24 h. An appropriate amount of polymer is dissolved in the desired solvent under constant stirring at ambient temperature for some time until we get a clear homogeneous solution. After this, salt is added to this solution in an appropriate amount and after a few hours of constant stirring, different amounts of plasticizers are added to the solution. This solution is stirred for another few hours ~4–5 h until we get a very highly viscous homogeneous solution. The resulting mixture is subsequently transferred into polypropylene petri dishes for the purpose of casting, and the solvent is allowed to evaporate under ambient conditions.

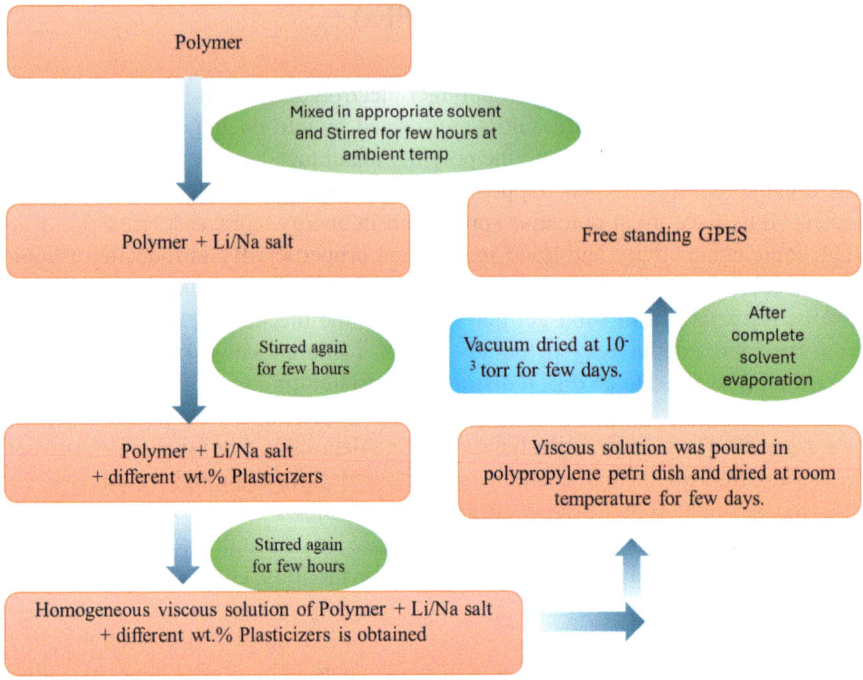

Fig. 2.3 Flow chart for the preparation of gel polymer electrolytes

When the solvent dries, a free-standing film of GPEs is thus obtained. These films were then vacuum-dried for a few days in order to completely evaporate the solvent. A simple flow chart for the preparation of GPEs is shown in Fig. 2.3. The film preparation process is time-consuming, often requiring 10–15 days to achieve full drying, and involves significant chemical consumption, making it relatively costly (Arya and Sharma 2017).

2.4.2 *Hot Press Technique*

Compared to the conventional solution casting method, the hot press technique offers a modern alternative with several benefits—it is faster, more cost-effective, eliminates the need for solvents, and is well-suited for fabricating dense materials. In this method, polymer and liquid electrolyte mixture is placed between two hot plates and then pressure is applied. The interplay between heat and pressure facilitates the diffusion and intermolecular entanglement of polymer chains, resulting in the formation of a gel. This technique is especially advantageous for polymers characterized by a low glass transition temperature (Tg), as they can be shaped into a gel-like state. In this method, the polymer host, salt, and plasticizer are mixed thoroughly in an agate mortar pestle for a few hours. The mixture is then allowed to heat

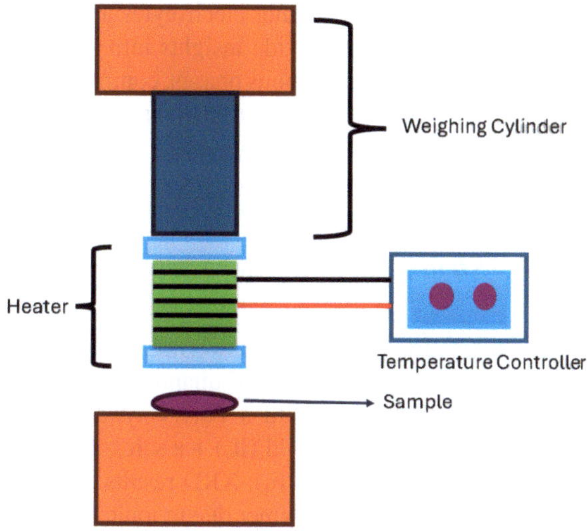

Fig. 2.4 Schematic diagram of hot press technique

slowly below the melting temperature of polymer host in heating chamber. Further, the resulting slurry is put between two stainless steel electrodes under pressure, leading to the formation of a hot slurry. The temperature and the pressure applied need to be chosen based on the polymer that is used (Arya and Sharma 2017). The schematic diagram of hot press technique is shown in Fig. 2.4.

Hot press processing for lithium-ion systems is challenging because lithium salts are highly reactive in ambient conditions, necessitating carefully controlled environments and specialized handling procedures.

2.5 Properties and Characterization of Polymer Electrolytes

Gel polymer electrolytes, a type of polymer electrolyte, are predominantly utilized in energy storage systems, including rechargeable lithium and sodium-ion batteries, supercapacitors, and fuel cells, owing to their exceptional characteristics such as high ionic conductivities, a broad electrochemical stability range, and good thermal stability (Ramesh et al. 2011). Polymer electrolytes are essential in lithium-ion batteries as they enable the advancement of solid-state cells, thereby removing the necessity for flammable liquid electrolytes and promoting the incorporation of high-energy-density electrode materials. They also show promise in emerging battery technologies such as lithium-air, lithium-sulfur, and other solid-state lithium systems. Thanks to their compatibility with various lithium salts, plasticizers, and fillers, polymer electrolytes offer extensive opportunities for customization and performance enhancement. This adaptability underscores their significance in advancing energy storage technologies across a wide range of industrial applications (Saikia et al. 2008). Numerous techniques can be adopted to characterize the polymer electrolytes. Polymer electrolytes are characterized using a range of analytical techniques that provide insights into their surface morphology, microstructure, crystalline and amorphous phases, complexation behaviour, thermal properties, electrical conductivity, ionic transference number, and electrochemical stability window.

2.5.1 *Structural and Microstructural Property Analysis*

X-ray diffraction (XRD) is a powerful and non-invasive technique used to determine the phase composition and quantify the amorphous content within a sample. Shalu et al. (2015) prepared a Poly(vinylidene fluoride-co-hexafluoropropylene) [P(VdF–HFP)] + ionic liquid (IL) + a salt solution (a homogeneous mixture of IL, $BMIMBF_4$, and 0.3 M $LiBF_4$). XRD results suggest the changes in the crystalline nature of P(VdF–HFP) polymer due to ionic liquid-polymer interactions. The addition of salt leads to peak broadening in the XRD spectra, indicating the formation of a complex between the salt and the polymer matrix. With the addition of IL-salt solution in the P(VdF–HFP) polymer, the amorphicity of polymer electrolytes is increased, which further results in an increase in the ionic conductivity. Mohapatra et al. (2008) investigated how incorporating SnO_2 as a filler influences the morphological, electrical, electrochemical, and mechanical stability of composite polymer electrolyte (CPE) films, using PEO as the polymer matrix and NaI as the conducting salt. Gupta et al. (2016) reported the preparation of Solid polymer electrolytes (SPEs) using polymer poly(ethylene oxide) (PEO), salt lithium bis(trifluoromethylsulfonyl)imide (LiTFSI) and ionic liquid (IL)trihexyltetradeylphosphonium bis(trifluoromethylsulfonyl)imide. From XRD results, they have found that upon the addition wt.% of IL polymer electrolytes becomes more amorphous. This will lead to further enhancement in the amorphous phase of polymer electrolytes, which provides a greater number of Li^+-ions that can diffuse between the electrode. Leo et al. (2002) developed a composite polymer electrolyte using PEO and lithium triflate, incorporating LiNasicompound ($Li_{1.4}(Al_{0.4}Ge_{1.6})(PO_4)_3$) as a ceramic filler and EC + PC as plasticizers. The XRD analysis revealed a reduction in peak intensity and noticeable broadening upon filler addition, along with increased background noise—signifying a rise in the amorphous content of the material.

Dzulkipli et al. (2021) discussed the effects of 1-Butyl-3-Methylimidazolium Tetrafluoroborate [BMIM][BF_4] on electrochemical, thermal, and structural characteristics as an ionic liquid gel polymer electrolyte (ILGPE). They have found a similar result showing that upon adding higher amount of ionic liquid, the degree of crystallinity is found to be decreased. This indicates that [BMIM][BF_4] serves not only as a charge carrier but also functions as a plasticizer, which diminishes the crystallinity of the ILGPE system and consequently improves its ionic conductivity (Fig. 2.5).

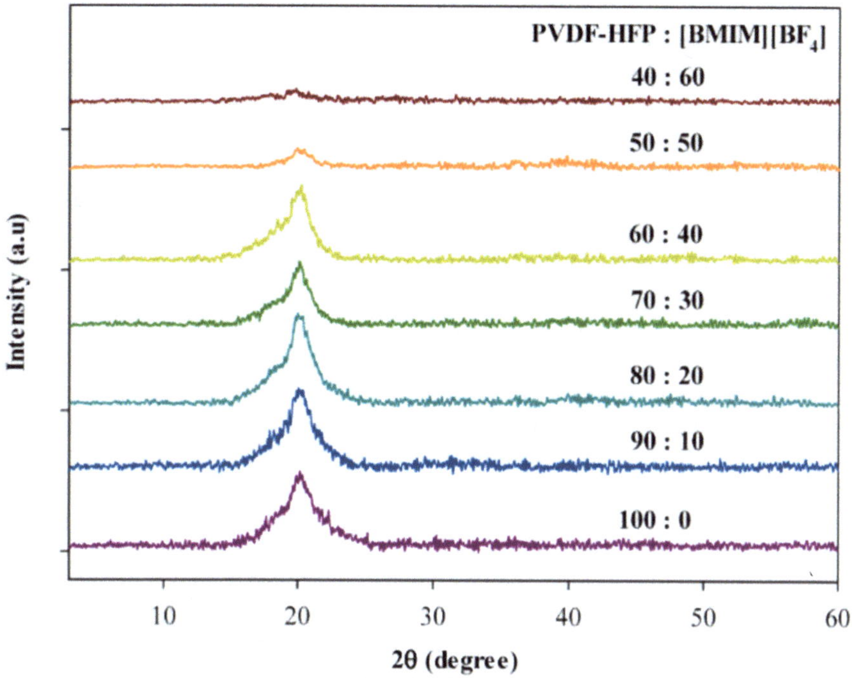

Fig. 2.5 XRD results of pure PVDF-HFP and different percentage ratio of PVDF-HFP to [BMIM] [BF_4] (Reproduced from Dzulkipli et al. 2021under CC by 4.0 license)

2.5.2 *Thermogravimetric Analysis (TGA) and Differential Scanning Calorimetry (DSC) Study*

Thermal study of polymer electrolytes plays an important role in the application of energy storage devices for higher temperature applications. Developing a polymer electrolyte with superior thermal stability is crucial for improving the safety features of energy storage and conversion systems. TGA is the basic technique to study the decomposition temperature, mass change like gas adsorption, phase transitions, etc. The characteristics of the recorded TGA graph depend on various factors such as mass, volume and physical form of the sample. Here, TGA measurement are usually displayed in the form of curve, which is generally plotted between change in mass of the sample (on vertical axis) and temperature (on x-axis). Differential scanning calorimetry (DSC) is one of the important techniques to examine the phase transition changes, thermal stability and glass transition temperature of the sample. Differential Scanning Calorimetry (DSC) quantifies the heat necessary to raise the temperature of a sample as a function of temperature compared to a reference. The sample exhibits physical or chemical changes, which are well indicated by the appearance of an endothermic peak resulting from heat absorption, as well as an exothermic peak that signifies heat release. DSC can provide the information about glass transition temperature of the sample (T_g), crystallization and melting temperature (T_m), and degree of crystallinity (X_c), etc. Pandey and Hashmi (2013) investigated the $LiPF_6$/IL/poly(vinylidine fluoride-co-hexafluoropropylene) (PVdF-HFP) gel electrolyte film for supercapacitor application. It is found that the synthesized GPE exhibits thermal stability up to approximately 325 °C, making it suitable for high-temperature applications as both a separator and electrolyte in electrochemical devices. DSC studies show that an endothermic peak at 135 °C, corresponds to the melting of the host polymer (PVdF-HFP) film. When an ionic liquid (1-ethyl-3-methylimidazolium tris(pentafluoroethyl) trifluorophosphate, EMImFAP) is incorporated into the host polymer, the melting point is reduced to 115 °C. This reduction may be attributed to the higher proportion of liquid components, which enhances the amorphous content. Srivastava et al. (2020) reported the synthesis of ionic liquid-based blend gel polymer electrolyte for Li-ion battery applications. We reported the effect of addition of ionic liquid EMIMTFSI in PVdF-HFP + PMMA +20 wt.% LiTFSI on thermal stability. It is found that blend polymer electrolytes show multistep decomposition at 155 °C, 257 °C and 310 °C and upon increasing the concentration of ionic liquid thermal stability increases. The prepared BGPEs exhibit thermal stability up to approximately 250 °C, making them suitable for high temperature applications. From the DSC studies, it is found that upon the incorporation of IL, the endothermic melting peaks become broader and are displaced towards the lower temperature region. The findings of the DSC study reveal that the addition of IL leads to a reduction in melting temperature, implying a softening effect on the polymer chains and an increase in the motion of the polymer segments. Dzulkipli et al. (2021) investigated the GPE using ionic liquid [BMIM][BF_4]. They have found that the prepared ILGPE shows thermal stability up to ~300 °C. Figure 2.6 illustrates the TGA thermogram and the differential thermogravimetric analysis (DTG) of the ILGPE films for different percentage ratios of PVDF-HFP to [BMIM][BF_4].

Figure 2.7 shows the DSC curve in the temperature range of −100 °C to 180 °C for different percentage ratio of GPE. They have shown similar results as described above, upon increasing the concentration of ionic liquid, the melting temperature (T_m) as well as glass transition temperature (T_g) shifted to lower temperature region. This shifting indicates that an interaction will occur between polymer matrix and ionic liquid. This is due to the plasticization effect of IL, which increases the flexibility of polymer backbone, giving rise to an increase in the ionic conductivity.

2.5.3 Impedance/Electrical Conductivity Studies

In electrochemical energy storage device polymer electrolytes provide the medium for transportation of ions in between the electrodes. The total ionic conductivity of PEs depends upon the flexibility of polymer chain. Achieving high ionic conductivity is a key requirement for PE as it reflects their capacity to facilitate Li^+-ions transport within the polymer matrix. The ionic conductivity (σ) is directly related to

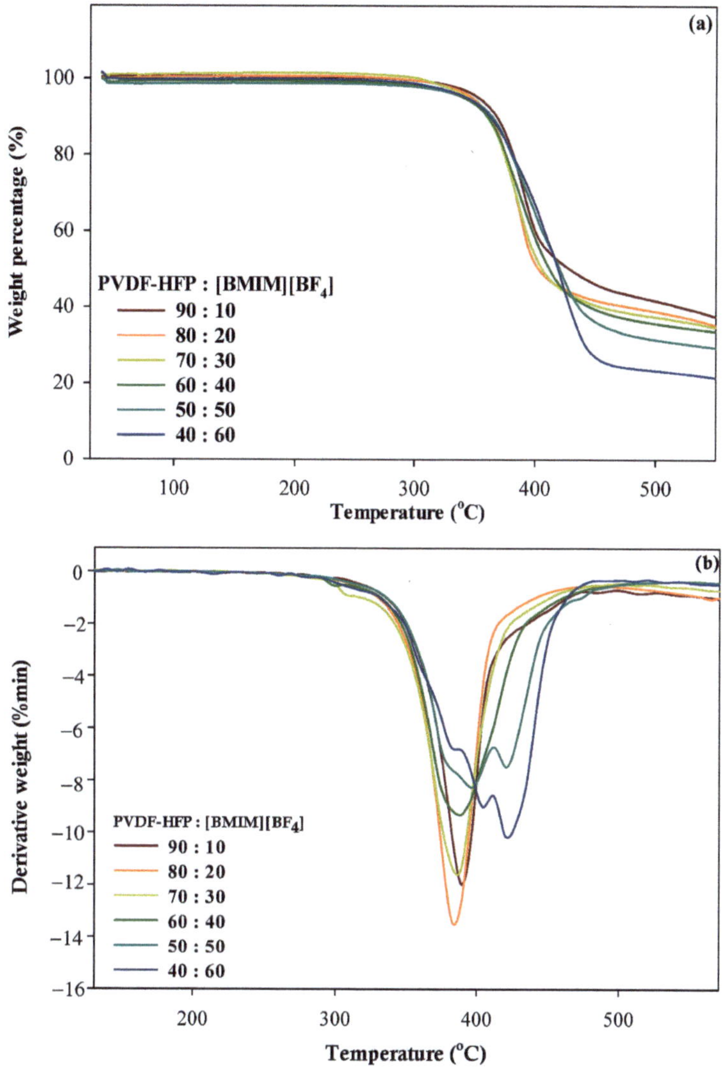

Fig. 2.6 Thermogram of (**a**) TGA and (**b**) DTG of different percentage ratio of PVDF-HFP to [BMIM][BF_4] (Reproduced from Dzulkipli et al. 2021under CC by 4.0 license)

the concentration of the species of charge carriers. The ionic conductivity of polymer electrolytes system has been calculated using complex impedance spectroscopy techniques. In this technique, the prepared disc shaped prepared polymer film is sandwiched between two stainless steel (SS) electrodes. The formula for the calculation of total ionic conductivity is given below (Srivastava et al. 2020):

$$\sigma_{\mathrm{T}} = \frac{1}{R_{\mathrm{b}}} \times \frac{l}{A}$$

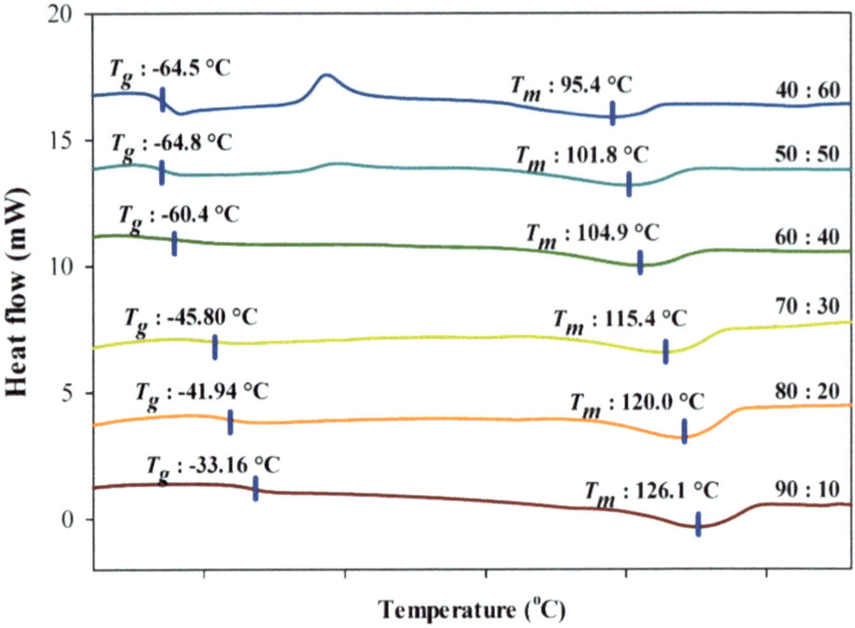

Fig. 2.7 Differential scanning calorimetry (DSC) curves in the temperature range −100 °C to 180 °C for different percentage ratios of PVDF-HFP to [BMIM][BF_4](Reproduced from Dzulkipli et al. 2021 under CC by 4.0 license)

l (cm) and A (cm^2) are thickness and cross-sectional area of the polymer film, respectively and R_b is the bulk resistance of electrolyte obtained from Nyquist plot. Arya et al. (2019) prepared a blended solid polymeric thin film based on PEO-PVP and investigated the structural and electrical properties with different LiBOB variations. The impedance plot comprises two regions based on frequency. Nyquist plot is the semicircle plot which have high frequency region associated with the bulk property of electrolyte, followed by a spike in the low frequency region. The semicircle in the higher frequency is observed due to the parallel combination of bulk resistance and capacitance of electrolyte. An inclined straight line in the low frequency region can be attributed to the double layer capacitance (C_{dl}) formed at electrode electrolyte interface. The intercept on the real axis signifies the bulk resistance of the electrolyte. They have found that the bulk resistance value is minimum for the solid polymer electrolyte with O/Li = 16 salt concentration. They have found that the optimized sample PP16 shows maximum ionic conductivity value ~5.1×10^{-6} S cm^{-1} at 40 °C. The decrease in the bulk resistance value is observed when the temperature is increased. This will provide the flexibility in the polymer chain and promote the faster movement of ions due to thermal activation of charge carriers. With addition of salt more free charge carriers are introduced in the polymer matrix and these carriers are involved in the conduction process, resulting in an observed increase in conductivity. The influence of PEO on the electrochemical characteristics of electrospun PVdF-based polymer electrolytes was examined by

Prasanth et al. (2014), with the maximum ionic conductivity reported as 4.9×10^{-3} S cm^{-1}. Singh et al. (2020) reported the synthesis of IL-GPE based on polymer PVDF-HFP, salt LiTFSI, and ionic liquid $PYR_{13}FSI$. They have prepared IL-GPE; (PVdF-HFP + X wt.% $PYR_{13}FSI$) + 20 wt.% LiTFSI using solution casting techniques. They have found that the maximum ionic conductivity value is ~3.9×10^{-3} S cm^{-1} for 70 wt.% IL containing GPE at 30 °C. The total ionic conductivity value is increased when the ionic liquid wt.% is increased in the PVdF-HFP+ 20 wt.% LiTFSI system. This may be due to the increased amorphous region or free mobile ions of the solid polymer electrolytes. Aziz et al. (2020) investigated the ionic conductivity measurement based on Polyvinyl Alcohol (PVA) Biopolymer Electrolytes. They have prepared the PVA:CS:NH_4I polymer blend electrolyte with different wt.% glycerol plasticizer. The maximum ionic conductivity value is found for higher wt.% content of glycerol, i.e., 42 wt.% and the value is ~1.37×10^{-4} S cm^{-1}. Figure 2.8 shows the Nyquist plot for each electrolyte system at room temperature.

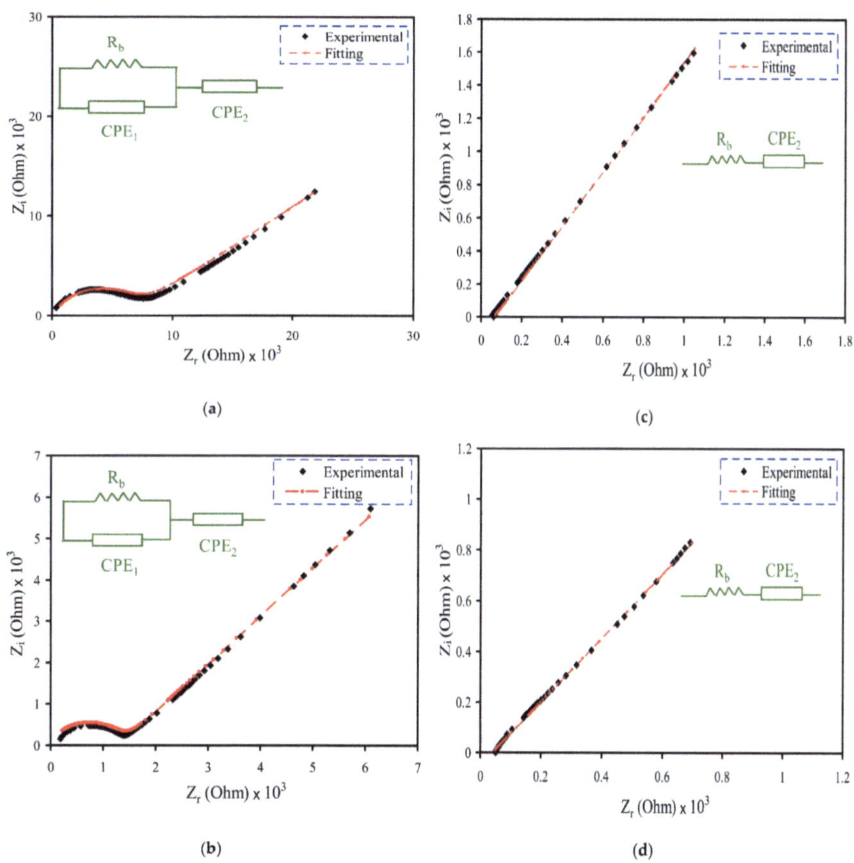

Fig. 2.8 Nyquist plot for prepared blend polymer electrolyte film (Reproduced from Aziz et al. 2020 under CC by 4.0 license)

2.5.4 Transference Number Studies

The transference number determines the type of species that contributes to the conductivity within the polymer electrolyte system. The transference number of ions (t_{ion}) refers to as the ratio of ionic conductivity to total conductivity, which is determined by the ratio of the current carried by the ions to the total current. The ionic transference number of prepared polymer electrolyte is calculated using dc polarization techniques. In this technique, the prepared polymer electrolyte film is sandwiched between two blocking stainless steel (SS) electrode (SS/PEs/SS) and variation of current is monitored as a function of time after applying a small dc voltage. As time progresses, the current diminishes due to the polarization of mobile ions at the working or blocking electrodes. Initially, both ionic and electronic components contribute to the current; however, over time, it either drops to zero in the case of a purely ionic conductor or stabilizes at a residual constant value for mixed ionic–electronic conductors. The ionic transference number can be calculated using the equation given below (Srivastava et al. 2020):

$$t_{\mathrm{ion}} = 1 - \frac{I_{\mathrm{e}}}{I_{\mathrm{t}}}$$

where, I_e is the constant residual current due to electrons contribution only and I_t is the total initial current as a combined result of ion and electrons.

Cationic Transference Number In rechargeable Li/Na-ion batteries, during charging and discharging process mainly Li^+ or Na^+-ions participate in the de-intercalation and intercalation process. So, it is very important to know the contribution of these ions. The cationic transference number (t_{cation}) can be calculated using combined ac/dc techniques. In this techniques polymer electrolytes film is sandwiched between two non-blocking electrodes. A small dc voltage is applied across the cell and the corresponding current as a function of time is measured. The ac impedance response of the cell before and after polarization is determined. The value of cationic transference number is calculated using the equation given below:

$$t_{\mathrm{cation}} = \frac{i_{ss}\left(\Delta V - i_0 R_0\right)}{i_0\left(\Delta V - i_{ss} R_{ss}\right)}$$

where i_0 = initial current values; i_{ss} = steady state current; R_0 and R_{ss} are resistance of the cell before and after polarization, respectively.

Srivastava et al. (2020) reported the synthesis of ionic liquid-based blend gel polymer electrolyte with the composition PVdF-HFP + PMMA +20 wt.% LiTFSI + X wt.% EMIMTFSI. We have reported the ionic transference number value ~0.99 and evidence of good ionic conductor. We have also reported the Li-ion transference number value ~0.43 for the optimized blend gel polymer electrolyte, which

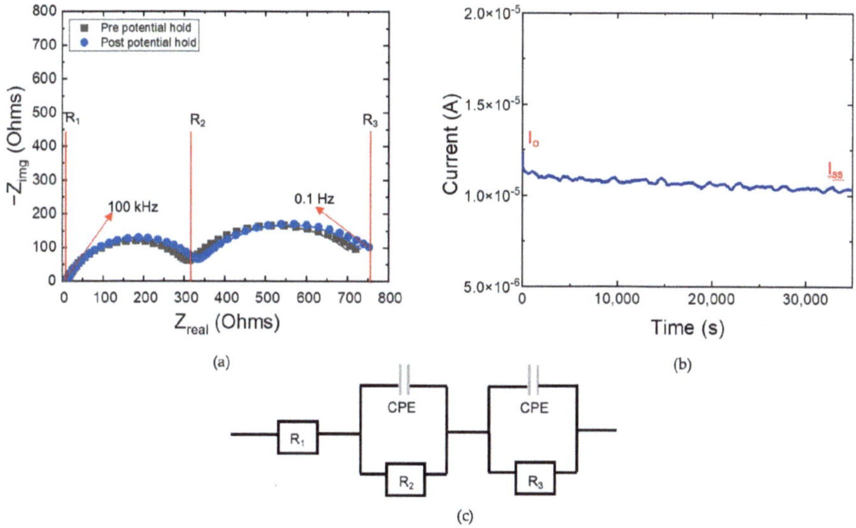

Fig. 2.9 Experimental data and analysis on a Li symmetric cell to determine lithium transference number for O-PEGDMA-VS-0 in 1 M LiPF6 (EC-DEC 1:1 v%): (**a**) Impedance response pre- and post-potentiostatic hold, (**b**) Current response during the potentiostatic hold, (**c**) The equivalent circuit that is used for the fitting of impedance data throughout the study (Reproduced from Boz et al. 2021under CC by 4.0 license)

shows that 43% conductivity is carried out by Li^+-ions. Boz et al. (2021) confined the conventional electrolyte (1 M $LiPF_6$ in EC-DEC) in a microporous polymer network and lithium cation transference number 0.78 was obtained (Fig. 2.9).

2.5.5 Electrochemical Studies

2.5.5.1 Electrochemical Stability Window (ESW)

It is the most important technique used to determine the maximum working voltage limit of polymer electrolyte in electrochemical energy storage devices. This is the region where electrolyte is neither reduced nor oxidized and it is electrochemically stable within this region. The ESW of the prepared polymer electrolyte is examined by linear sweep voltammetry (LSV) techniques. Before charging and discharging process, it is important to check the ESW of polymer electrolyte because battery operates within this voltage limit. In this technique cell is prepared in which polymer electrolyte is sandwiched between stainless steel (SS) and Li electrode (Fig. 2.10). Srivastava et al. (2020) prepared a blend gel polymer electrolyte using PVDF-HFP and PMMA polymer electrolyte, 20 wt.% LiTFSI salt and EMIMTFSI ionic liquid. The optimized BGPE (70 wt.% IL) shows electrochemical stability upto ~4.3 V. When we added in the above system SiO_2 as a nanofiller, the ESW

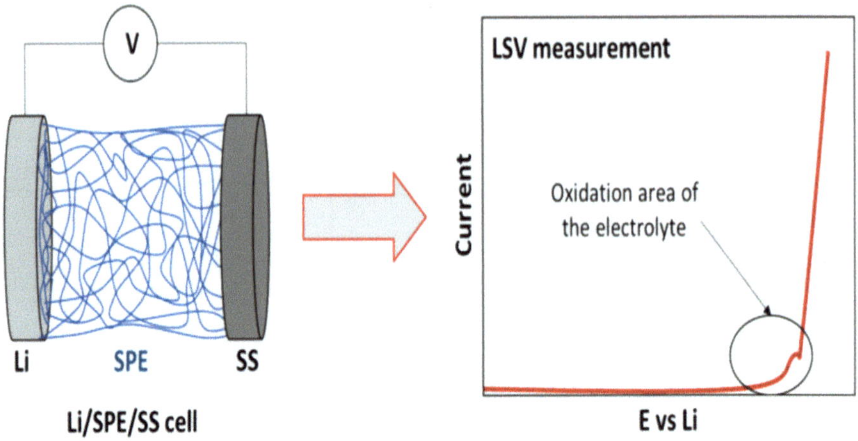

Fig. 2.10 Schematic representation of LSV techniques (Reproduced from Méry et al. 2021under CC by 4.0 license)

value is enhanced and the optimized nanocomposite polymer electrolyte is electrochemically stable up to ~5.4 V, which is good for higher voltage battery application (Srivastava et al. 2022b). The electrochemical characterization, i.e., charge-discharge test is carried out within this voltage limit.

2.5.5.2 Galvanostatic Charge-Discharge Studies

The galvanostatic charge-discharge technique is commonly adopted to assess the specific capacity of rechargeable battery systems by applying a steady current and recording the related voltage response. Srivastava et al. (2020) observed the charge-discharge test for the cell Li/ 70 wt.% EMIMTFSI containing BGPE/$Li_{1.2}Ni_{0.6}Mn_{0.1}Co_{0.1}O_2$ and obtained a maximum discharge capacity ~166 mAh g^{-1} at 0.1C. The cell Li/70 wt.% EMIMTFSI containing BGPE/$Li_{1.2}Ni_{0.6}Mn_{0.1}Co_{0.1}O_2$) retained 67% of maximum discharge capacity up to the 50th cycle and maintained coulombic efficiency (~97%) up to the 140th cycle. The reduction in discharge capacity can be ascribed to the increase in cell resistance followed by cycling.

2.6 Conclusions

Polymer electrolytes have emerged as a pivotal class of materials in the evolution of advanced energy storage technologies, offering a compelling balance between ionic conductivity, mechanical integrity, and electrochemical stability. They offer numerous advantages such as improved safety, higher energy density, and enhanced electrochemical stability. Their versatility in formulation—ranging from solid polymer electrolytes (SPEs) to gel polymer electrolytes (GPEs) and composite systems—has enabled tailored solutions for lithium-ion, lithium-sulfur, and lithium-air batteries, among others. Therefore, in this chapter, we summarize the characteristics properties of polymer electrolytes for the application of energy storage devices.

References

Ahmed MS, Islam M, Raut B, Yun S, Kim HY, Nam KW (2024) A comprehensive review of functional gel polymer electrolytes and applications in lithium-ion battery. Gels 10(9):563

Arya A, Sharma AL (2017) Polymer electrolytes for lithium ion batteries: a critical study. Ionics 23(3):497–540

Arya A, Sadiq M, Sharma AL (2019) Structural, electrical and ion transport properties of free-standing blended solid polymeric thin films. Polym Bull 76(10):5149–5172

Aziz BS, Marf AS, Dannoun EMA, Brza MA, Abdullah RM (2020) The study of the degree of crystallinity, electrical equivalent circuit, and dielectric properties of polyvinyl alcohol (PVA)-based biopolymer electrolytes. Polymers 12(10):2184

Badi N, Theodore AM, Alghamdi SA, Al-Aoh HA, Lakhouit A, Singh PK, Norrrahim MNF, Nath G (2022) The impact of polymer electrolyte properties on lithium-ion batteries. Polymers 14(15):3101

Baskoro F, Wong HQ, Yen HJ (2019) Strategic structural design of a gel polymer electrolyte toward a high efficiency lithium-ion battery. ACS Appl Energy Mater 2(6):3937–3971

Boz B, Ford HO, Salvadori A, Schaefer JL (2021) Porous polymer gel electrolytes influence lithium transference number and cycling in lithium-ion batteries. Electron Mater 2(2):154–173

Chaurasia SK, Saroj AL, Shalu S, Singh VK, Tripathi AK, Gupta AK, Verma YL, Singh RK (2015) Studies on structural, thermal and AC conductivity scaling of PEO-LiPF6 polymer electrolyte with added ionic liquid [BMIMPF6]. AIP Adv 5(7)

Dzulkipli MZ, Karim J, Ahmad A, Dzulkurnain NA, Su'ait MS, Yoshizawa-Fujita M, Tian Khoon L, Hassan NH (2021) The influences of 1-butyl-3-methylimidazolium tetrafluoroborate on electrochemical, thermal and structural studies as ionic liquid gel polymer electrolyte. Polymers 13(8):1277

Gupta H, Balo L, Singh VK, Chaurasia SK, Singh RK (2016) Effect of phosphonium based ionic liquid on structural, electrochemical and thermal behaviour of polymer poly (ethylene oxide) containing salt lithium bis (trifluoromethylsulfonyl) imide. RSC Adv 6(91):87878–87887

Leo CJ, Rao GS, Chowdari BVR (2002) Studies on plasticized PEO–lithium triflate–ceramic filler composite electrolyte system. Solid State Ionics 148(1–2):159–171

Méry A, Rousselot S, Lepage D, Dollé M (2021) A critical review for an accurate electrochemical stability window measurement of solid polymer and composite electrolytes. Materials 14(14):3840

Mishra A, Mehta A, Basu S, Malode SJ, Shetti NP, Shukla SS, Nadagouda MN, Aminabhavi TM (2018) Electrode materials for lithium-ion batteries. Mater Sci Energy Technol 1(2):182–187

Mohapatra SR, Thakur AK, Choudhary RNP (2008) Studies on PEO-based sodium ion conducting composite polymer films. Ionics 14(3):255–262

Ngai KS, Ramesh S, Ramesh K, Juan JC (2016) A review of polymer electrolytes: fundamental, approaches and applications. Ionics 22(8):1259–1279

Nitta N, Feixiang W, Jung L, Yushin Y (2015) Li-ion battery materials: present and future. Mater Today 18:252–264

Pandey GP, Hashmi SA (2013) Solid-state supercapacitors with ionic liquid based gel polymer electrolyte: effect of lithium salt addition. J Power Sources 243:211–218

Pigłowska M, Kurc B, Galiński M, Fuć P, Kamińska M, Szymlet N, Daszkiewicz P (2021) Challenges for safe electrolytes applied in lithium-ion cells—a review. Materials 14(22):6783

Prasanth R, Shubha N, Hng HH, Srinivasan M (2014) Effect of poly (ethylene oxide) on ionic conductivity and electrochemical properties of poly (vinylidenefluoride) based polymer gel electrolytes prepared by electrospinning for lithium ion batteries. J Power Sources 245:283–291

Ramesh S, Liew CW, Ramesh K (2011) Evaluation and investigation on the effect of ionic liquid onto PMMA-PVC gel polymer blend electrolytes. J Non-Cryst Solids 357:2132–2138

Saikia D, Chen-Yang YW, Chen YT, Li YK, Lin SI (2008) Investigation of ionic conductivity of composite gel polymer electrolyte membranes based on P(VDF-HFP), $LiClO_4$ and silica aerogel for lithium ion battery. Desalination 234:24–32

Shalu, Chaurasia SK, Singh RK, Chandra S (2015) Electrical, mechanical, structural, and thermal behaviors of polymeric gel electrolyte membranes of poly (vinylidene fluoride-co-hexafluoropropylene) with the ionic liquid 1-butyl-3-methylimidazolium tetrafluoroborate plus lithium tetrafluoroborate. J Appl Polym Sci 132(7)

Singh SK, Dutta D, Singh RK (2020) Enhanced structural and cycling stability of Li_2CuO_2-coated $LiNi_{0.33}Mn_{0.33}Co_{0.33}O_2$ cathode with flexible ionic liquid-based gel polymer electrolyte for lithium polymer batteries. Electrochimica Acta 343:136122

Srivastava N, Singh SK, Gupta H, Meghnani D, Mishra R, Tiwari RK, Patel A, Tiwari A, Singh RK (2020) Electrochemical performance of Li-rich NMC cathode material using ionic liquid based blend polymer electrolyte for rechargeable Li-ion batteries. J Alloys Compd 843:155615

Srivastava N, Singh SK, Meghnani D, Mishra R, Tiwari RK, Patel A, Tiwari A, Singh RK (2022a) Molybdenum-doped Li/Mn-rich layered transition metal oxide cathode material $Li_{1.2}Mn_{0.6}Ni_{0.1}Co_{0.1}O_2$ with high specific capacity and improved cyclic stability for rechargeable Li-batteries. ACS Appl Energy Mater 5:12183–12195

Srivastava N, Singh SK, Meghnani D, Singh RK (2022b) High-voltage-driven Li/Mn-rich $Li_{1.2}Mn_{0.6}Ni_{0.1}Co_{0.1}O_2$ cathode with nanocomposite blend gel polymer electrolyte for long cyclic stability of rechargeable Li-battery. Energy Storage 4(5):e341

Sun Y, Liu N, Cui Y (2016) Promises and challenges of nanomaterials for lithium-based rechargeable batteries. Nat Energy 1:1–12

Chapter 3
Green Carbon Nanostructure-Based Composites for Environmental Applications and Energy Harvesting

Unnayana Gogoi and Rituparna Duarah

Abstract In recent years, green carbon nanostructure-based composites have attracted significant attention for applications in the field of environmental remediation and energy harvesting. This chapter offers an overview of sustainable carbon nanostructure-based functional composites synthesized through eco-friendly approaches. The discussion begins by introducing the fundamental concepts of carbon nanostructures, highlighting their various types and unique properties. Special emphasis is given to green synthesis strategies, which offer environmentally friendly methods for producing carbon nanostructures while minimizing toxic by-products and energy consumption. Subsequently, the chapter explores the applications of green nanostructure-based composites in areas such as environmental remediation, encompassing air filtration, water purification, and soil detoxification. Beyond environmental remediation, the chapter also discusses their applications in photovoltaic devices, piezoelectric and triboelectric nanogenerators, as well as supercapacitors and rechargeable batteries. Finally, the chapter addresses the key challenges remaining in the green synthesis, large-scale production, and practical applications of these composites. By addressing these gaps, this work aims to provide an overview of current advancements in green carbon nanostructure-based composites and their growing potential in sustainable environmental and energy technologies.

Keywords Carbon nanostructures · Eco-friendly · Sustainability · Environmental remediation · Energy harvesting · Composites

---

U. Gogoi
Materials Sciences Group, Coal, Energy and Materials Sciences Division, CSIR-North East Institute of Science and Technology, Jorhat, Assam, India

R. Duarah (✉)
Materials Sciences Group, Coal, Energy and Materials Sciences Division, CSIR-North East Institute of Science and Technology, Jorhat, Assam, India

Academy of Scientific and Innovative Research, Ghaziabad, India

S. K. Verma et al. (eds.), *Sustainable and Natural Polymers*, Clean Energy Production Technologies, https://doi.org/10.1007/978-981-92-0092-4_3

3.1 Introduction

Rising demand for clean energy and environmental sustainability has led researchers to develop innovative materials that can tackle global environmental challenges. The integration of green nanostructures into functional composites offers promising solutions for both environmental remediation and energy harvesting technologies. Conventional energy harvesting and environmental remediation approaches involve chemical, physical, or mechanical processes that may effectively remove pollutants but often have significant drawbacks related to environmental impact, cost, and secondary pollution risks (United States Environmental Protection Agency 2025). In this context, green carbon nanostructure-based composites have gained immense attention, comprising various nanostructures such as carbon nanotubes, graphene derivatives, carbon quantum dots, and biomass-derived carbon nanomaterials. These composites offer outstanding physical, chemical, and electronic properties that effectively work in environmental remediation and energy harvesting applications (Wang et al. 2019). The high effectiveness of these nanostructure-based composites in environmental remediation and energy harvesting can be attributed to the high surface area, excellent conductivity, chemical stability, and functional tunability (Mahalakshmi et al. 2025). However, their synthesis approaches need to follow green chemistry principles by employing renewable precursors, lowering the use of hazardous reagents, and minimizing energy consumption. The green carbon nanostructures are typically synthesized from precursors such as plant extracts, natural biopolymers, using water as a solvent and low temperatures for processing (Rathod et al. 2024). The synthesized green carbon nanostructures are integrated into biodegradable and synthetic polymers such as chitosan, gelatin, and polylactic acid, ceramics, and metals, to create advanced carbon nanostructure-based composites with multifunctional characteristics. These composites exhibit enhanced properties, like superior mechanical strength, electrical and thermal conductivity, antimicrobial activity, barrier performance, and photocatalytic photolytic degradation, making them highly attractive for applications in various fields of environmental remediation and energy harvesting (Ananda et al. 2025). These green carbon nanostructure-based composites offer a combination of superior performance, biodegradability, and recyclability, contributing to overall environmental sustainability. In this context, this chapter offers a comprehensive overview of green synthesis approaches and the application of green carbon nanostructure-based composites for functional roles in environmental remediation and energy harvesting. It further discusses the challenges and provides a future outlook for the development of scalable and sustainable nanocomposite technologies.

3.2 Sustainable Material Design and the Role of Nanotechnology

In materials science, sustainability focuses on developing resources and processes that are renewable, biodegradable, recyclable, and energy-efficient. In this context, sustainable composites are designed to minimize environmental impact throughout their life cycle, from the extraction of raw material to final disposal. Conventional high-performance composites, often based on synthetic polymers, heavy metals, or fossil-derived compounds, are associated with high carbon footprints and ecological challenges. Nanotechnology offers transformative opportunities by enabling the design of smart, multifunctional composites at the nanoscale (1–100 nm), facilitating reduced raw material consumption, enhanced performance, and improved recyclability. Carbon nanostructures, including carbon nanotubes, graphene, fullerenes, and carbon quantum dots, serve as exceptional reinforcements in sustainable composites due to their high surface-to-volume ratio, superior mechanical, thermal, and electrical properties, and versatile surface chemistry that allows compatibility with a wide range of matrices. Incorporating these nanostructures into biodegradable polymers, ceramics, or metals yields advanced composites with enhanced strength, conductivity, barrier properties, and even antimicrobial and photocatalytic functionalities, thereby extending their applicability to sustainable technologies. A central objective of sustainable composite design is to synthesize and functionalize nanomaterials through eco-friendly green approaches, employing plant extracts, natural biopolymers, and solvent-free or aqueous-based processing. Such strategies ensure that nanocomposites are not only high-performing but also environmentally benign. Leveraging nanotechnology, sustainable composites find applications in solar energy harvesting, water purification, air filtration, antimicrobial and self-cleaning materials, and energy storage devices, offering scalable solutions to global energy and environmental challenges.

3.3 Carbon Nanostructure: Types and Properties

Carbon nanostructures have emerged as one of the most versatile and high--performing classes of advanced materials, offering broad applications in energy conversion and storage, environmental remediation, biosensing, bioimaging, and tissue engineering. Figure 3.1 illustrates the wide range of carbon-based nanostructures that can be obtained from carbon, highlighting the role of bonding states and hybridization (Hu et al. 2006).

Carbon nanostructures exhibit diverse forms due to the different hybridizations of carbon atoms: sp^3 in diamond, sp^2 in graphite, sp^1 in carbyne, and intermediate sp^n states in the fullerene family. Structures such as fullerite and carbon nanotubes display mixed hybridizations ($sp^2 + \pi$, $sp + 2\pi$), as seen in molecular systems like corannulene, ovalene, and bucky-diamond (Hu et al. 2006). When these carbon nanostructures are incorporated into polymeric, ceramic, or metallic matrices, the resulting composites demonstrate significantly enhanced performance compared to their pristine counterparts. Depending on the type of nanostructure and host matrix, carbon nanocomposites can exhibit superior mechanical reinforcement, improved barrier and thermal stability, enhanced electrical and thermal conductivity, as well as functional properties such as photocatalysis, antimicrobial activity, and bio--compatibility. This tunability positions carbon nanostructure-based composites as key materials for next-generation technologies that demand sustainability, multifunctionality, and high efficiency.

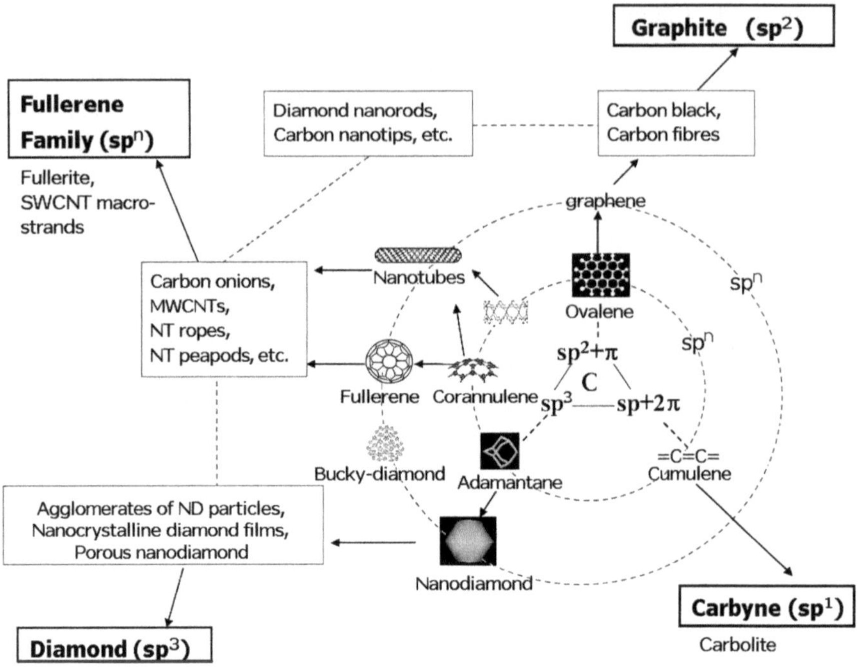

Fig. 3.1 Classification of carbon nanostructures showing typical forms such as sp^3-hybridized diamond, sp^2-hybridized graphite, *sp*-hybridized carbyne, and 'sp^n' indicating intermediate bonding states with non-integer hybridization. (Reproduced with permission from Hu et al. (2006). Copyright 2006 IOPscience)

3.4 Green Synthesis Approaches for Carbon Nanostructures and Their Composites

Green synthesis of carbon nanostructures employs eco-friendly routes, such as plant extract-mediated reduction, biotemplating, and solvent-free or low-energy techniques, to fabricate functional nanomaterials under mild conditions. Biotemplating uses natural templates like biomass to guide morphology in a sustainable, low-waste way. Additionally, solvent-free and low-energy methods such as microwave pyrolysis or catalyst-free CVD using biomass allow energy-efficient production of hollow carbon frameworks or carbon nanotubes without harmful solvents or metals.

3.4.1 Plant Extract-Mediated Synthesis

Synthesis of carbon nanostructures using plant extract offers unique properties such as sustainability, biodegradability, and biocompatibility. Plant extracts serve as reducing and stabilizing agents, enabling the formation of carbon dots or graphene-like structures with antioxidant and fluorescent properties. Numerous studies have demonstrated the synthesis of carbon nanostructures and their composites using plant extracts. Suriani et al. successfully synthesized vertically aligned carbon nanotubes on silicon substrates through thermal catalytic chemical vapor deposition, utilizing palm oil as the carbon precursor. FESEM and Raman analyses confirmed the formation of dense bundles comprising both multi-walled carbon nanotubes (MWCNTs) and single-walled carbon nanotubes (SWCNTs), with the latter exhibiting diameters in the range of 0.6–1.2 nm (Suriani et al. 2009). Similarly, Hakim et al. used coconut shell-derived graphite flakes in a water-assisted, one-step synthesis to produce CNTs and checked their effectiveness in heavy metals (Pb^{2+}) adsorption, demonstrating potential for environmental remediation (Hakim et al. 2018). Additionally, ultra-light graphene/polydopamine-coated multiwalled CNT aerogels were synthesized through a green hydrothermal reduction method involving 2 h of sonication followed by heating at 90 °C for 12 h using GO and MWCNT--PDA. The aerogels exhibited outstanding adsorption capacity towards Pb^{2+} and Cu^{2+} ions in water. The synthesis process did not require any additional reducing agents (Zhan et al. 2019). Paul et al. reported a novel and sustainable strategy for synthesizing multiwalled carbon nanotubes (MWCNTs) using coconut oil as a green, renewable precursor through chemical vapor deposition (CVD). This approach adheres to green chemistry principles by employing an abundant, low-cost natural resource instead of conventional petroleum-based feedstocks. The iron catalyst was prepared via self-propagating high-temperature synthesis, and CNT growth was carried out at 850 °C under a nitrogen atmosphere. Characterization using SEM, TEM, EDX, and Raman spectroscopy analyses confirmed the successful formation of MWCNTs with diameters ranging from 80 to 100 nm, exhibiting relatively high purity and characteristic Raman features, including a D-band positioned at 1327.3 cm^{-1} and a G-band observed at 1576.8 cm^{-1}, with an ID/IG ratio of 1.39, indicating the presence of some structural defects. Interestingly, unlike other plant-derived precursors such as turpentine oil and camphor, coconut oil lacks lauric and oleic acids, making it a distinctive feedstock with strong potential for scalable and eco-friendly CNT production (Paul and Samdarshi 2011).

Guo et al. reported a rapid and contamination-free electrochemical approach for producing high-quality graphene nanosheets (~1.1 nm thick) from exfoliated graphite oxide, demonstrating a scalable and eco-friendly route. The reaction involves electrochemical reduction of the exfoliated GO at a graphite electrode. The process involves no use of chemical reduction using ammonia and hydrazine hydrate (Guo et al. 2009). Alternative green strategies have also been explored, such as utilizing algal extracts (Ahmad et al. 2019) and agricultural residues like sugarcane bagasse (Baweja and Jeet 2019), highlighting the versatility of natural reducing agents. Green-synthesized graphene derivatives and their composites have exhibited promising applications in environmental remediation, particularly for the adsorption of heavy metals such as Pb^{2+}, Cu^{2+}, and Hg^{2+} (Lin et al. 2019), as well as for the photocatalytic degradation of pollutants (Raja et al. 2019). Gurushantha et al. synthesized reduced graphene oxide-ZrO_2 composite by following a green route using reflux method using Cinnamon extract as a reducing agent. The process involves sonicating GO for about 40 min followed by mixing Cinnamon extract with the suspension and refluxing for 45 min. The nanocomposite was synthesized by constant stirring for 24 h (Gurushantha et al. 2017).

Similarly, eco-friendly approaches have been widely employed for the synthesis of carbon quantum dots (CQDs). Zhao et al. synthesized CQDs from pine wood using a green hydrothermal carbonization process, where high-resolution TEM revealed a graphene-like structure with 0.21 nm lattice spacing. Fluorescence quenching studies demonstrated that the pine-derived CQDs showed remarkable selectivity towards Fe^{3+} ions in water (Zhao et al. 2020). Similarly, Zhang et al. demonstrated the successful synthesis of carbon quantum dots (CQDs) from *Rosa roxburghii* pulp through a hydrothermal method at 180 °C for 8 h, followed by centrifugation and filtration. The CQDs demonstrated remarkable sensitivity for the detection of *o*-nitrophenol (Zhang et al. 2020).

Shahba et al. synthesized carbon nanofibers/CQDs from pine fruit using a ball--milling-assisted hydrothermal route. They collected two pine fruits and manually ground them using a mortar. It was then further refined through ball milling to obtain a fine powder. This powder was then dispersed in distilled water, the dispersion was transferred to an autoclave, and subjected to hydrothermal treatment at 180 °C for 4.5 h. The resulting product was repeatedly washed with ethanol and water and then dried at 80 °C for 12 h (Shahba and Sabet 2020). SEM analysis confirmed the presence of nanosized carbon dots, while TEM images revealed one--dimensional carbon nanofibers along with evidence of nanoparticle nucleation and growth during fiber formation. These materials exhibited outstanding photocatalytic activity, efficiently degrading a variety of dyes, including methyl orange, methylene blue, eriochrome black T, eosin Y, and acid blue through reactive oxygen species generation. In addition, they displayed excellent adsorption capacity, completely removing Pb^{2+} and Cd^{2+} ions from aqueous solutions.

In another work, Riaz et al. reported the development of carbon dots-manganese oxide (CDs-MnO_2) and nitrogen-doped carbon dots–manganese oxide (NCDs--MnO2) nanocomposites, which were synthesised via a green ultrasonic approach, employing *Jasminum sambac* leaf extract as both carbon precursor and reducing agent. Their sensing ability towards toxic Cr(VI) ions was investigated using spectrophotometry under varying conditions (pH, ion concentration, temperature, time, and interference). The detection limits achieved were 69 μM for NCDs-MnO_2 and 16 μM for CDs-MnO_2, with the latter requiring milder conditions due to its functionalization and structural features. Both nanocomposites showed excellent recovery (~100%) in real water samples, demonstrating their promise as effective Cr(VI) ion sensors (Riaz et al. 2024).

3.4.2 Biotemplating

Biotemplating is an emerging bottom-up nanofabrication strategy that leverages the structural and functional features of biological systems to direct the synthesis of inorganic materials, including carbon nanostructures. Here, biological entities such as viruses, diatoms, bacteria, and some biomacromolecules serve as templates, which enable replication of their intricate morphologies into carbon nanostructures with controlled geometry, crystallinity, and surface chemistry, characteristics that are often challenging in the case of conventional synthetic techniques. For carbon nanostructure-based composite synthesis, biotemplating typically involves impregnating the biotemplate with a carbon precursor, which is then subjected to pyrolysis, and the obtained carbonaceous replicas preserve the original biological architecture (Sotiropoulou et al. 2008).

Nersisyan et al. used reed bristles (*Andropogon virginicus*) as a bio-template for carbon microtubes (CMTs) synthesis using carbonization in a double-crucible system at 1200 °C. Here, charcoal served as the reducing agent. CMTs thus obtained exhibited uniform tubular morphologies, with inner diameters of 3–7 μm and outer diameters of 4–8 μm. Remarkably, the reed seeds exhibited intrinsic catalytic activity, facilitating the growth of multiwalled carbon nanotubes (MWCNTs) with outer diameters ranging from 10 to 15 nm. The CMTs exhibited a BET surface area of 145 m^2/g with a pore size distribution of 0.5–1.5 nm, highlighting their suitability for adsorption processes and nanostructured material applications (Nersisyan et al. 2013).

Kim et al. used M13 bacteriophage as a template for synthesizing a single-walled carbon nanotube (SWCNT) nanonetwork, and it was integrated with Co_3O_4 nanoflower electrodes (Co_3O_4/SWCNT-M13) for high-power lithium-ion batteries. This composite integrates the superior electrical conductivity of SWCNTs with the large surface area of Co_3O_4 nanoflowers, resulting in exceptional electrochemical performance. The Co_3O_4/SWCNT-M13 anodes achieved a specific capacity of 1283.5 mAh g^{-1} at 2 A g^{-1}, retaining this performance even after 500 cycles (Kim et al. 2020).

Li et al. used diatomite as a template for the synthesis of nepenthes-like N-doped hierarchical graphene (NHG)-based separator for polysulfide scavenging. The material exhibits a 3D porous geometry with large surface area, nitrogen doping, and tortuous inner channels, which together enable physical confinement and chemical immobilization of polysulfides while facilitating efficient Li^+ transport. The NHG-based separator allowed Li-S cells to achieve an initial capacity of 868 mAh g^{-1} with a low decay rate of 0.067% per cycle over 800 cycles at 2 °C, and while retaining a high capacity of 805 mAh g^{-1} at a sulfur loading of approximately 7.2 mg cm^{-2} (Li et al. 2018a). Li et al. synthesized a nitrogen-doped Co_3O_4NP@CNTs@Diatomite//GRusing Diatomite as a biotemplate. Initially, ZIF-8 was deposited onto diatomite using a one-step hydrothermal approach. Owing to the structural similarity between ZIF-67 and ZIF-8, a ZIF-67@ZIF-8 heterostructure was subsequently grown on the diatomite surface through an epitaxial growth method, as reported by

K. Li et al. (Journal of Energy Storage 53, 2022, 105094). The obtained ZIF-67@ ZIF-8@Diatomite composite was calcined under a nitrogen atmosphere at a temperature of 920 °C, during which CNTs were formed via Zn volatilization and the catalytic activity of cobalt nanoparticles. In the final step, exposure to an oxygen atmosphere led to the oxidation of cobalt nanoparticles, yielding cobaltosic oxide (Co_3O_4) nanoparticles (Li et al. 2022).

3.4.3 *Solvent-Free or Low-Energy Processing Techniques*

Most processing techniques for synthesizing carbon nanostructures rely on solvent--mediated routes; however, the extensive use of toxic organic solvents, strong acids, and bases in these methods often results in economic inefficiency, low yield, and environmental hazards, thereby limiting their practical applicability in energy harvesting. To overcome these drawbacks, solvent-free strategies have been explored as sustainable alternatives. For instance, Shihao Feng et al. reported a simple solvent-free block copolymer self-assembly approach for the fabrication of hollow carbon nanostructures (HCNs), such as Janus tadpoles (HCJs), and carbon nanofibers (HCFs), and nanospheres (HCSs). Core-shell polymer structures were initially fabricated and then carbonized to produce hollow carbon structures. The synthesis of the core–shell resin fibers was carried out by combining Pluronic F127 triblock copolymer, hexamethylenetetramine (HMTA), a small amount of water, and *m*-phenylenediamine (mPDA). The mixture was placed in a 50 mL Teflon-lined autoclave and subjected to heating at 200 °C for 24 h. The resin fiber so obtained was subsequently calcined at 1400 °C under argon atmosphere to produce hollow carbon nanofibers (HCF). By reducing the amount of F127 to half or one-quarter while keeping the other reactants constant, the resulting polymers were named resin Janus tadpole or resin sphere, and their corresponding carbonized products were referred to as hollow carbon Janus structures (HCJ) or hollow carbon nanospheres (HCS). By simply tuning the concentration of the triblock copolymer (F127), diverse morphologies were obtained without the need for toxic solvents (Feng et al. 2023). Among these, HCFs exhibited the most promising electrochemical performance as sodium-ion battery anodes, delivering a high reversible capacity of 431 mAh g^{-1} and excellent rate performance, retaining 259 mAh g^{-1} at 800 mA g^{-1} and remarkable long-term cycling stability.

In another study, Bharath et al. developed a green and scalable approach for the synthesis of large-scale, defect-free graphene nanosheets from graphite via mechanochemical ball milling. In the process, $KMnO_4$ and aspartic acid were placed in a stainless-steel vial with a ball-to-powder weight ratio of 10:1. Following this, graphene oxide (GO) was hydrothermally treated with $FeCl_2{\cdot}4H_2O$ to form a Fe_3O_4-reduced graphene oxide (Fe_3O_4-RGO) nanocomposite. The Fe_3O_4-RGO dispersion was then drop-casted onto a glassy carbon electrode for electrochemical studies. Field emission scanning electron microscopy (FESEM) revealed that ball milling reduced the grain size of graphite oxide to 2–5 μm, and

Fe_3O_4 nanoparticles were uniformly anchored across the surface, basal planes, and edges of the graphene sheets. The resulting Fe_3O_4-RGO nanocomposite exhibited excellent electrocatalytic performance and high sensitivity for nitrite detection (Bharath et al. 2015).

3.5 Applications of Green Carbon Nanostructure-Based Composites in Environmental Remediation and Energy Harvesting

Green carbon nanostructure-based composites such as CNTs, GO, and other advanced carbon allotropes have gained significant attention as highly effective materials for environmental remediation owing to their exceptional chemical, physical, and electronic properties (Ogunsola et al. 2024; Sarvalkar et al. 2024; Ashritha and Rufuss 2025). This can be attributed to the large surface area, tuneable surface chemistry, and strong interactions, including, π–π stacking, hydrophobic interactions, and electrostatic attraction, which make them superior adsorbents and catalysts for removing a wide spectrum of pollutants from water, air, and soil. Functionalization of CNTs and GO further enhances their sorption capacity and selectivity, enabling rapid and efficient removal of pollutants such as heavy metals, synthetic dyes, and chlorinated compounds, while minimizing risks of secondary pollution.

3.5.1 *Water Purification*

Several studies have confirmed the strong potential of CNTs as efficient adsorption of diverse contaminants, ranging from heavy metal ions to persistent organic pollutants such as dioxins and polycyclic aromatic hydrocarbons.

Gao et al. reported a magnetic polymer MWCNT nanocomposite for the removal of anionic azo dyes from aqueous solution from ferroferric oxide and poly(1-glycidyl-3-methylimidazolium chloride). The TEM images revealed that both the polymer and Fe_3O_4 nanoparticles were present on the surface of MWCNTs. The nanocomposite was used for various dye removal form aqueous system. The nanocomposite showed maximum adsorption capacities of 47.39, 85.47, and 67.57 mg g^{-1} for amaranth, sunset yellow, and orange II at 25 °C (Zhan et al. 2019). Similarly, Zhan et al. synthesized ultralight graphene/polydopamine-modified MWCNT hybrid aerogels via a green, reducing-agent-free method. These aerogels showed structural stability, abundant active sites, and large surface area, achieving outstanding adsorption capacities of 350.87 mg g^{-1} for Pb(II) and 318.47 mg g^{-1} for Cu(II). This superior activity can be attributed to synergistic mechanisms of surface complexation and chelation, while their ease of separation highlights practical applicability in heavy metal wastewater treatment (Gao et al. 2013).

In another study, GO-$MnFe_2O_4$ composites were fabricated via a facile, green route, where a GO/$MnSO_4$ suspension derived from the Hummers method was directly converted into GO–$MnFe_2O_4$ without toxic reducing agents (Liao et al. 2020). The composites showed excellent adsorption performance for Cu^{2+} and Pb^{2+} and with maximum capacities of 103.41 mg g^{-1} and 263.85 mg g^{-1}, respectively, following the Langmuir isotherm and pseudo-second-order kinetic model and Thermodynamic studies confirmed the adsorption to be spontaneous and endothermic, and the materials exhibited strong reusability over multiple cycles. Furthermore, Lihua et al. reported that spent activated carbon (AC) from a water treatment plant, despite 5 years of prior use, demonstrated exceptional adsorption capacity for Cd(II) and Pb(II) and at concentrations of ~200 μg/L, achieving removal rates of over 86% and 95% and respectively far exceeding those of virgin AC (10–15%) (Li et al. 2018b). Surface analysis showed that spent AC had a lower pH and point of zero charge (PZC), along with an 81% increase in carboxyl groups, which improved its metal-binding capacity. Zhang et al. synthesized an eco-friendly spherical adsorbent (SA@PEI-CDs) via electrostatic self-assembly of sodium alginate (SA) and polyethyleneimine-functionalized carbon dots (PEI-CDs) (Zhang et al. 2022). This composite offered excellent selectivity for Pb^{2+} removal, with an adsorption capacity of 380.39 mg/g at pH 4 and representing an 83.82% increase over bare SA, and maintained 85.67% capacity after five reuse cycles. The adsorption was driven by electrostatic attraction, ion exchange, and coordination of $-NH_2$, –OH, and –COOH groups with Pb^{2+}. In simulated sewage treatment, SA@PEI-CDs efficiently lowered Pb^{2+} concentrations from 1.0 mg/L to levels meeting high-quality water standards.

Carbon nanostructure-based composites have proven to be highly effective adsorbents for pesticide removal from water. In one study, Diel et al. employed multiwalled carbon nanotubes (MWCNTs) as supports for metallic nanoparticles (MNPs) synthesized via a green route using plant extracts from pomegranate, eucalyptus, and pecan leaves (Diel et al. 2021). Characterization confirmed successful impregnation while preserving the structural integrity of the MWCNTs. Notably, MWCNTs impregnated with iron nanoparticles reduced using pecan leaf extract achieved an adsorption capacity of 21.17 mg/g for glyphosate (GLY), corresponding to 84.08% removal efficiency. This work demonstrates the feasibility of eco-friendly, plant-extract-based green synthesis to produce efficient CNT-based adsorbents for pesticide remediation. Similarly, Aris et al. synthesized electrospun graphene oxide/cellulose nanofibre composites and applied them as adsorbents in solid-phase membrane tip adsorption for detection of polar organophosphorus pesticides (OPPs) present in water and food (Aris et al. 2020). The adsorption process was optimized for pH, dosage, and time, yielding linear calibration curves between 0.05 and 10 mg/L, with excellent precision and low detection limits (0.01–0.05 mg L^{-1}). The adsorption of OPPs followed the Freundlich isotherm and pseudo-second-order kinetics with consistently high efficiencies (71–99.9%) across diverse samples. In another study, Suo et al. synthesized activated carbon from sieve-like cellulose/graphene oxide composites (ACCE/G) using corn straw as a source treated with potassium hydroxide. The material demonstrated superior adsorption performance for organophosphorus pesticides, achieving an adsorption capacity of 152.5 mg g^{-1} for chlorpyrifos, outperforming conventional adsorbents such as MWCNTs and activated carbon. Importantly, ACCE/G retained over 80% of its adsorption capacity after eight reuse cycles, underscoring its sustainability for practical water treatment (Suo et al. 2018). Additionally, a reduced graphene oxide–silver (RGO@Ag) composite was reported to catalyze the degradation of the persistent organochlorine pesticide lindane into trichlorobenzene isomers by simultaneously eliminating three HCl molecules, forming aromatic products and silver chloride (Sen Gupta et al. 2015). This transformation, which does not occur with RGO or silver nanoparticles alone, highlights the unique reactivity of the composite. While lindane removal primarily involves physical adsorption, the resulting trichlorobenzenes are stabilized through strong π–π interactions with the composite. The material also demonstrated excellent recyclability and effectiveness in complex natural water matrices, offering a sustainable solution by integrating chemical degradation and adsorption within a single platform. Jamil et al.'s work presents a novel water purification approach by electrospinning polyvinyl alcohol (PVA), TEOS, and green tea solutions to synthesize carbon nanofibers (CNFs), followed by calcination of the CNFs at 280 °C. The nanofibers initially measured ~500 nm in diameter, which decreased by ~50% after carbonization, enhancing their suitability for filtration. CNF-based filters effectively removed contaminants from water, as confirmed by SEM and EDX analyses showing the capture of elements such as Fe, S, Na, Mg, Ti, and Si (Jamil et al. 2021).

3.5.2 Air Filtration

Green carbon nanostructure-based composites have been recognized as strong candidates for air filtration applications owing to their environmentally friendly synthesis, excellent adsorption properties, and robust structural features. CNTs, graphene, and GO-based composites are particularly effective for trapping particulate matter (PM), toxic gases, and other airborne pollutants, attributable to their large surface area, tuneable pore structures, and robust mechanical stability, which together enhance filtration efficiency while maintaining low pressure drops. For instance, Vishwakarma et al. fabricated a lightweight, flexible, and freestanding multiwalled carbon nanotube (MWCNT) membrane by vacuum-assisted filtration of MWCNTs over a Whatman filter paper that achieved over 99% removal efficiency for MP0.3, PM0.3, and BA particles (Vishwakarma et al. 2022). The membrane demonstrated outstanding physicochemical properties, including a contact angle of ~148 ± 7°), pore size of ~16 nm, packing density of ~0.65 g/cc, porosity of ~56%, and a low-pressure drop of ~139.7 Pa. In addition, its reusability, flexibility, and self-cleaning properties enhanced its practicality for real-world applications. Similarly, Jung et al. reported a reduced graphene oxide (rGO) foam-based air filter fabricated on copper mesh that effectively removed PM2.5 particulates with minimal pressure drop (Jung et al. 2018). The filter exhibited exceptional regeneration capacity, maintaining high efficiency through multiple reuse cycles. With a quality factor nearly double that of previously reported filters, the rGO foam-based design offers advantages of simple fabrication, scalability, bidirectional operation, and low energy consumption, making it ideal for both indoor and outdoor air purification. In another study, a nanographene@zinc oxide (NG@ZnO-green) composite was synthesized with hemp stalk bio-extract serving as a sustainable reducing agent (Kutluay 2025). This eco-friendly approach facilitated strong anchoring and homogeneous distribution of ZnO nanoparticles on the nanographene surface, yielding a high specific surface area of 705 m^2g^{-1} and well-structured porosity. The composite exhibited outstanding CO_2 adsorption capacities, ranging from 5.95 to 6.99 mmol/g at 273–298 K and 1 bar pressure, significantly surpassing pure nanographene and conventionally synthesized NG@ZnO. Moreover, it maintained 95% efficiency even after ten cycles, confirming its potential as a sustainable and reusable adsorbent for carbon capture. Kuo et al. further developed a QD/TiO_2@PI/PES electrospun nanofiber membrane with combined air filtration, photocatalytic, and antibacterial functionalities (Kuo et al. 2024). Optimized using the Taguchi method, the membrane achieved a filtration efficiency of 99.6% with a high-quality factor (0.151), while simultaneously degrading methylene blue (45.5%) and acetone (87%), and achieving complete inactivation of *Staphylococcus aureus* under UV irradiation within 4 h. Collectively, these studies highlight the versatility and efficiency of green carbon nanostructure composites for advanced air purification and environmental protection.

3.5.3 Soil Detoxification

Soil detoxification using green carbon nanostructure composites involves the application of eco-friendly CNTs, graphene, and biochar synthesized through sustainable methods and incorporated into composite structures for effective soil remediation (Dhanapal et al. 2024). These green composites utilize the large surface area, excellent adsorption capacity, and catalytic activity of carbon nanostructures to degrade, immobilize, or adsorb a wide range of soil contaminants, such as heavy metals and persistent organic pollutants, through mechanisms of adsorption, reduction, and photocatalysis. Liu et al. synthesized a novel composite by coating magnetic multiwalled carbon nanotubes (Fe_3O_4-MWCNTs) with a zeolitic imidazolate framework (ZIF-8) through a coordination-polymerization technique, using zinc sulfate and 2-methylimidazole. This process produced a material where ZIF-8 crystals were uniformly deposited on the magnetic nanotube surfaces, resulting in a structure with extensive surface area, high porosity, and robust superparamagnetic properties ideal for environmental remediation applications. When applied to contaminated soil, the composite efficiently extracted eight types of organophosphorus pesticides, leveraging electron exchange between the pesticide molecules and the abundant, reactive sites found in the porous M-M-ZIF-8 framework. The interplay of high surface area, tailored pore network, and responsive magnetism enabled rapid and selective adsorption, making the composite highly effective as a soil detoxification agent for persistent organic pollutants (Liu et al. 2018).

Lu et al. developed ferruginated carbon quantum dots (Fe-CQDs) nanozymes capable of mitigating cadmium-induced stress in wheat by mimicking natural antioxidant enzymes such as superoxide dismutase (SOD), peroxidase (POD), and catalase (CAT) to neutralize reactive oxygen species (ROS) such as $^{\bullet}OH$, H_2O_2, and O_2^-. Root application of 10 mg L^{-1} Fe-CQDs effectively reduced oxidative damage, evidenced by lower levels of O_2^-, H_2O_2, and malondialdehyde, while simultaneously increasing glutathione content (Lu et al. 2025). Mandal et al. developed a composite consisting of cornstalk biochar-supported graphene oxide integrated with nanoscale zero-valent iron (CTBC-GO/nZVI) to immobilize copper in soil. The biomass was pyrolyzed under nitrogen at temperatures between 150 and 650 °C, then immersed in a graphene-containing aqueous solution, and finally impregnated with nZVI. The SEM analysis showed that after incorporating graphene sheets, the biochar surface became uniformly structured, forming macro- and micropores. Additionally, a continuous graphene membrane was observed covering the biochar, indicating a cohesive composite structure. The composite demonstrated higher copper immobilization efficiency than pristine biochar and graphene oxide-biochar (GO@BC), achieving over 65% copper removal after 14 days. Sequential extraction analysis showed that CTBC-GO/nZVI transformed more bioavailable and accessible copper forms into less accessible and less toxic ones, thereby reducing copper's toxicity in soil (Mandal et al. 2020).

3.6 Applications in Energy Harvesting

Green carbon nanostructures-based composites have gained attention as versatile materials for energy harvesting applications owing to their outstanding electrical, mechanical, and surface properties, coupled with environmentally friendly synthesis methods. CNTs, graphene, and carbon quantum dots and their composites are extensively employed in flexible piezoelectric nanogenerators, triboelectric nanogenerators, and supercapacitors for harvesting mechanical, solar, and environmental energy. Their large surface area, superior conductivity, and tuneable structural features facilitate efficient energy conversion and storage (Dos Reis et al. 2024; Tuichai et al. 2020).

3.6.1 Photovoltaic Devices (Solar Cells)

Various nanostructures, including quantum dots, nanoparticles, nanotubes, nanowires, and quantum wells, have been extensively applied to solar cells to enhance energy conversion efficiency. These nanostructures can be synthesized in diverse compositions, including semiconductors and metals, and engineered into core-shell architectures with dielectric or semiconductor coatings. Their fabrication can be accomplished through solution-based chemical approaches, chemical vapor deposition (CVD), or physical vapor deposition, offering versatile fabrication routes.

Owing to their remarkable electrical conductivity, mechanical strength, and thermal stability, CNTs have emerged as highly promising materials for photovoltaic applications. Individual CNTs exhibit electrical conductivity reaching 10^6 S cm^{-1} and intrinsic carrier mobility over 10^5 $cm^2 \cdot V^{-1} \cdot s^{-1}$ at room temperature, making them ideal electrode materials (Mann et al. 2003). Mechanically, CNTs possess a tensile strength of 10–100 GPa, Young's modulus of 1–2 TPa, and elongation at break of 6–12% (Zhou et al. 2015). Their hollow tubular geometry provides a large specific surface area of ~1600 m^2/g (Cinke et al. 2002), low density, and thermal conductivity up to ~3500 $W \cdot m^{-1} \cdot K^{-1}$ (Kim et al. 2001). These properties can serve as transparent front electrodes in perovskite solar cells (PSCs), and organic solar cells (OSCs), function as counter electrodes in dye-sensitized solar cells (DSSCs), and act as electron and hole transport materials in DSCs and PSCs, respectively. Semiconducting single-walled CNTs (SWCNTs) can also function as electron acceptors in OSC active layers, providing an alternative to fullerene derivatives. Hosseinnezhad et al. developed DSSCs employing eco-friendly natural dyes extracted from black pomegranate, cornelian cherry, tangerine peel, and ruby grape as photosensitizers. MoS_2/GO hybrid and composite materials were developed as substitutes for costly platinum counter electrodes. Comparative testing showed that DSSCs incorporating the MoS_2/GO composite delivered higher efficiency than those with the MoS_2/GO hybrid and even outperformed platinum-based devices under identical conditions, highlighting a sustainable, low-cost pathway for advancing DSSC technology (Hosseinnezhad et al. 2023). Graphene, with its exceptional electrical, thermal, optical, and mechanical properties, is highly promising for flexible solar cells. Pristine graphene demonstrates electron mobility as high as 2×10^5 $cm^2 \cdot V^{-1} \cdot s^{-1}$ at room temperature (Chen et al. 2008), electrical conductivity of approximately 10^4 $S \cdot cm^{-1}$, and thermal conductivity of around 5000 $W \cdot m^{-1} \cdot K^{-1}$ (Balandin et al. 2008), surpassing single-walled CNTs (~3500 $W \cdot m^{-1} \cdot K^{-1}$) and exceeding copper (~400 $W \cdot m^{-1} \cdot K^{-1}$) by an order of magnitude. With an optical transmittance of up to 97.7% that is independent of wavelength, it emerges as a promising alternative to indium tin oxide (ITO) for transparent electrodes (Selopal et al. 2015). Mechanically, graphene exhibits high strength (~125 GPa), modulus (~1100 GPa), ultralight areal density (0.77 $mg \cdot m^{-2}$), and large specific surface area (~2630 $m^2 \cdot g^{-1}$) (Roy-Mayhew and Aksay 2014). These attributes, combined with high conductivity and transparency, allow graphene to function as a photoanode in DSCs, an electron acceptor in OSCs, and a cost-effective counter electrode in DSCs. Amollo et al. incorporated graphene oxide (GO) into thin-film OSCs (TFOSCs) to enhance photon harvesting and charge transport. GO doped into the PEDOT:PSS hole transport layer improved hole injection, electron blocking, and interfacial charge transport, leading to enhanced short-circuit current density (Jsc) and improved overall power conversion efficiency (PCE). Additionally, embedding GO within the P3HT:PCBM photoactive layer enhanced optical absorption, exciton dissociation, and carrier mobility, resulting in photocurrents up to 18 mA cm^{-2}. Overall,

GO inclusion boosted device efficiency by ~40% in the HTL and ~120% in the active layer, demonstrating its dual function as an interfacial modifier and photoactive enhancer (Amollo et al. 2018). Flexible carbon nanofiber (CNF) mats fabricated from aqueous alkaline lignin/PVA solutions via electrospinning, stabilization, and carbonization exhibit properties strongly dependent on carbonization temperature. At 1500 °C, these CNFs achieved a large specific surface area of 941 m^2g^{-1} and enhanced graphitization, resulting in excellent electrochemical performance as binder-free counter electrodes in DSSCs. Devices employing these CNFs attained a maximum efficiency of 7.60%, closely matching conventional platinum electrodes (7.67%), highlighting lignin-based CNFs as a low-cost, flexible, and sustainable alternative for solar cell applications (Zhao et al. 2018).

3.6.2 Piezoelectric and Triboelectric Nanogenerators

Green carbon nanostructure-based composites have emerged as highly promising materials for piezoelectric and triboelectric nanogenerators (PENGs and TENGs) attributed to their outstanding electrical conductivity, mechanical flexibility, and environmentally friendly synthesis routes. By incorporating carbon nanomaterials such as CNTs, GO, and CQDs into polymeric or biogenic matrices, these composites can efficiently harvest mechanical energy from human motion, environmental vibrations, and other mechanical sources. Hou et al. fabricated a CNT/cellulose nanocomposite film, in which CNTs were embedded within the cellulose type I crystalline fiber network and deposited on the fiber surface. This hybrid material exhibited a maximum decomposition temperature 142 °C higher than that of pure cellulose, indicating enhanced thermal stability. A flexible PENG developed from this film produced an output current as high as 6.19 mA along with a current density of 1.55 mA/cm^2, capable of powering electronic devices, including LED lights, watches, and calculators, using only gravitational energy, without external force. This study highlights a sustainable approach for creating high-performance PENGs from cellulose and conductive CNTs (Hou et al. 2023). Bhunia et al. developed a lead-free, sol–gel-processed rGO/P(VDF-TrFE) nanocomposite, exhibiting strong interfacial interactions that significantly enhanced its piezoelectric, ferroelectric, and triboelectric properties relative to pristine P(VDF-TrFE). PENGs based on this composite achieved over 22-fold higher output voltage, reaching 89.7 V at 0.5 wt% rGO. Hybrid piezo–triboelectric nanogenerators (HPTENGs) displayed an output power density of 0.28 W/cm^3 and an average open-circuit voltage of 227 V, sufficient to directly light 45 blue LEDs with simple finger tapping (Bhunia et al. 2019). Wang et al. reported a soft, degradable, multifunctional biogenic nanogenerator composed of CQD/cellulose nanocrystal (CNC)/polyhydroxybutyrate (PHB) nanofibers fabricated via electrospinning. The synergistic effects of CQDs and CNCs enhanced PHB's degradability and piezoelectric/photoluminescent responses, generating output voltages and currents 6.7 and 4.4 times greater than those of pristine PHB devices. This nanogenerator demonstrated self-adaptive responses to mechanical stimuli, functioning as a tactile sensor that provided real-time motion detection, and was successfully applied as a dual-mode (piezoelectric/photonic) encrypted alarm system, highlighting its multifunctional capabilities (Wang et al. 2023).

3.6.3 Supercapacitors and Batteries

Recent advances in nanotechnology have highlighted carbon nanostructures such as CNTs, GO, and carbon blacks as high-performance electrode materials, owing to their large surface area, excellent electrical conductivity, and outstanding mechanical robustness. Extensive research has explored the potential of composites incorporating green carbon nanostructures for energy storage applications, particularly in supercapacitors and batteries.

Carbon nanospheres have been successfully synthesized from carbon soot using natural, renewable precursors such as virgin green leaf (*Anisomeles malabarica*) and castor oil. These nanospheres exhibit properties comparable to reduced graphene oxide and have been applied in dye-sensitized solar cells (DSSCs) and integrated photo-supercapacitors. The nanospheres were dispersed in *N*-methyl-2-pyrrolidone and polyvinylidene fluoride to form a paste, which was then coated onto an FTO substrate, dried, and employed as a counter electrode. In DSSCs, they achieved a photoconversion efficiency of 8.47%, while in supercapacitor devices, they delivered a power density of 0.3 mW/cm^2 and an energy density of 11.83 mWh/cm^2, sufficient to power an LED for up to 60 s, while retaining 85.2% of their capacitance after 3000 cycles. The integrated photo-supercapacitor demonstrated a cumulative efficiency of 3.3% with a storage efficiency of 38%, underscoring the potential of green-synthesized carbon nanospheres for sustainable energy storage and conversion. The environmentally friendly synthesis method relies on renewable resources like castor oil and plant leaves, offering a cost-effective route (Arjunkumar et al. 2024). Perera et al. reported a facile, green, one-pot alkaline hydrothermal method to synthesize deoxygenated graphene (hGO), achieving deoxygenation levels comparable to hydrazine-reduced graphene oxide without toxic agents. The resulting hGO formed freestanding, binder-free paper electrodes, which were assembled into CR2032 symmetric supercapacitors using LiTFSI and EMITFSI electrolytes. The electrodes exhibited impressive performance, with energy densities of 20 Wh kg^{-1} (LiTFSI) and 50 Wh kg^{-1} (EMITFSI), and maximum power densities of 11 kW kg^{-1} and 14.7 kW kg^{-1}, respectively (Perera et al. 2012). Sykam et al. reported a rapid, eco-friendly method for synthesizing reduced graphene oxide (rGO) by reducing graphene oxide with tulasi (Holy Basil) green tea extract, subsequently exposed to microwave irradiation for 1 min. The prepared rGO was combined with acetylene black as the conductive additive and polytetrafluoroethylene as a binder, followed by adding N-methyl pyrrolidone to obtain the electrode. The resulting rGO served as a highly efficient supercapacitor electrode, delivering a specific capacitance of 164.6 F/g at a current density of 1 A/g, maintaining 59.6% capacitance at 10 A/g, and showing less than 5% capacity loss after 3000 cycles (Sykam et al. 2018). Ugwuoke et al. synthesized carbon dots (CDs) from *Nigella sativa* seeds using an eco-friendly hydrothermal method, providing a sustainable electrode material. The CD-based supercapacitor was fabricated by laser engraving laser-induced graphene (LIG) patterns onto Kapton paper using a CO_2 laser, followed by drop-casting a CD solution to deposit the active material onto the LIG electrodes. TEM analysis confirmed quasi-spherical CDs with sizes of 6.08–9.91 nm, while fluorescence spectra exhibited green emission with excitation-dependent redshift behaviour. Electrochemical characterization revealed electrical double-layer capacitance behaviour, achieving an areal capacitance of 2.82 mF/cm^2 at a current density of 5.21 $\mu A/cm^2$, demonstrating their suitability for supercapacitor applications (Ugwuoke et al. 2024). Tong et al. developed a sustainable method to synthesize hierarchical porous carbon nanofibers (HPCNFs) from coal. In their approach, deionized water served as the solvent, coal as the carbon source, polyvinyl alcohol as a spinning assistant, and Pluronic F127 as a soft template. The electrospun nanofibers were then subjected to carbonization at 900 °C, resulted in HPCNFs. The nanofibers exhibited outstanding electrochemical performance as supercapacitor electrodes, delivering a specific capacitance of 265.2 F/g at a current density of 1 A/g, maintaining 83.1% of this capacitance (220.3 F/g) at 20 A/g, and demonstrated remarkable cycling stability with no loss after 20,000 cycles. The improved performance is due to the mesoporous structure formed via Pluronic F127 decomposition and the increased surface area for charge storage, highlighting a sustainable route for coal valorization and high-performance electrode fabrication (Tong et al. 2019).

3.7 Challenges and Future Outlook

Green carbon nanostructure-based composites are increasingly recognized as sustainable materials with significant potential for environmental remediation and energy harvesting applications. However, their large-scale deployment faces several challenges. Current green synthesis approaches often exhibit low yields, poor reproducibility, and limited control over structural features, while issues of scalability and cost-efficiency remain major obstacles. Additional challenges include achieving uniform dispersion within composite matrices, ensuring long-term stability under operational conditions, and mitigating potential environmental or biological toxicity. Moreover, achieving a balance between enhanced performance and sustainability is challenging, and the lack of standardized testing protocols further complicates the translation of these materials from laboratory research to real-world applications.

Future research should focus on developing eco-friendly, cost-effective, biomass-derived synthesis methods and green surface functionalization techniques for nanostructure-based composites, enabling tailored multifunctionality, while adhering to principles of biodegradability, recyclability, and waste valorization for long-term sustainability. Additionally, standardized life-cycle assessments and thorough toxicity evaluations will be essential. Ultimately, the transition from laboratory-scale innovations to pilot-scale systems will be key to realizing the broader impact of these materials in addressing global sustainability challenges.

3.8 Conclusion

Amid growing global concern, green carbon nanostructure-based composites are emerging as a sustainable solution for addressing global challenges in environmental remediation and renewable energy. Their high surface area, tuneable surface chemistry, superior electrical conductivity, and mechanical robustness make them ideal for applications ranging from water purification and pollutant adsorption to energy harvesting and storage. However, significant challenges remain, including scalable and cost-effective green synthesis, uniform dispersion within matrices, long-term operational stability, and comprehensive understanding of environmental and biological safety. Future progress will depend on integrating biomass-derived and waste-to-carbon nanostructures with advanced green functionalization strategies to create high-performance, environmentally benign composites. The development of hybrid multifunctional systems, coupled with innovations in device integration, circular economy practices such as recyclability and biodegradability, and the use of artificial intelligence-driven design and in-situ characterization, will accelerate optimization of structure–property relationships. Standardized testing, performance benchmarking, and regulatory protocols will further support safe commercialization. Therefore green carbon nanostructure-based composites have the potential to play a transformative role in clean energy generation, sustainable environmental management, and the advancement of global green technologies.

References

Ahmad S, Ahmad A, Khan S, Ahmad S, Khan I, Zada S, Fu P (2019) Algal extracts based biogenic synthesis of reduced graphene oxides (rGO) with enhanced heavy metals adsorption capability. J Ind Eng Chem 72:117–124

Amollo TA, Mola GT, Nyamori VO (2018) High-performance organic solar cells utilizing graphene oxide in the active and hole transport layers. Sol Energy 171:83–91

Ananda B, Krushna BR, Gagana M, Sharma SC, Ray S, Subha VJ, Kumari BN, Manjunatha K, Wu SY, Nagabhushana H (2025) Biodegradable chitosan-based carbon dot-infused intelligent films with UV-blocking and shape memory properties for shrimp preservation and milk freshness monitoring. J Ind Eng Chem

Aris NIF, Rahman NA, Wahid MH, Yahaya N, Keyon ASA, Kamaruzaman S (2020) Superhydrophilic graphene oxide/electrospun cellulose nanofibre for efficient adsorption of organophosphorus pesticides from environmental samples. R Soc Open Sci 7:500

Arjunkumar B, Mangalaraja RV, Ragupathi C, Ramalingam G, Alam MM, Al-Sehemi AG, Anbukumaran SK (2024) Green carbon nanospheres for integrated photo-supercapacitors device with improved efficiency. Diam Relat Mater 150:111699

Ashritha KS, Rufuss DD (2025) Carbon quantum dots for sustainable water treatment: a critical review on synthesis, properties, challenges and applications in forward osmosis desalination technologies. Chem Eng J 514:163059

Balandin AA, Ghosh S, Bao W, Calizo I, Teweldebrhan D, Miao F et al (2008) Superior thermal conductivity of single-layer graphene. Nano Lett 8:902–907

Baweja H, Jeet K (2019) Economical and green synthesis of graphene and carbon quantum dots from agricultural waste. Mater Res Express 6(8):0850g8

Bharath G, Madhu R, Chen SM, Veeramani V, Mangalaraj D, Ponpandian NJ (2015) Solvent-free mechanochemical synthesis of graphene oxide and Fe_3O_4–reduced graphene oxide nanocomposites for sensitive detection of nitrite. J Mater Chem A 3(30):15529–15539

Bhunia R, Gupta S, Fatma B, Prateek GRK, Garg A (2019) Milli-watt power harvesting from dual triboelectric and piezoelectric effects of multifunctional green and robust reduced graphene oxide/P(VDF-TrFE) composite flexible films. ACS Appl Mater Interfaces 11:38177–38189

Chen JH, Jang C, Xiao S, Ishigami M, Fuhrer MS (2008) Intrinsic and extrinsic performance limits of graphene devices on SiO_2. Nat Nanotechnol 3:206–209

Cinke M, Li J, Chen B, Cassell A, Delzeit L, Han J et al (2002) Pore structure of raw and purified HiPco single-walled carbon nanotubes. Chem Phys Lett 365:69–74

Dhanapal AR, Thiruvengadam M, Vairavanathan J, Venkidasamy B, Easwaran M, Ghorbanpour M (2024) Nanotechnology approaches for the remediation of agricultural polluted soils. ACS Omega 9:13522–13533

Diel JC, Franco DS, Igansi AV, Cadaval TR Jr, Pereira HA, Nunes ID, Basso CW, Alves MD, Morais J, Pinto D, Dotto GL (2021) Green synthesis of carbon nanotubes impregnated with metallic nanoparticles: characterization and application in glyphosate adsorption. Chemosphere 283:131193

Dos Reis GS, de Oliveira HP, Candido IC, Freire AL, Molaiyan P, Dotto GL, Grimm A, Mikkola JP (2024) Supercapacitors and triboelectric nanogenerators based on electrodes of greener iron nanoparticles/carbon nanotubes composites. Sci Rep 14:11555

Feng S, Li K, Hu P, Cai C, Liu J, Li X, Zhou L, Mai L, Su B-L, Liu Y (2023) Solvent-free synthesis of hollow carbon nanostructures for efficient sodium storage. ACS Nano 17(22):23152–23159

Gao H, Zhao S, Cheng X, Wang X, Zheng L (2013) Removal of anionic azo dyes from aqueous solution using magnetic polymer multi-wall carbon nanotube nanocomposite as adsorbent. Chem Eng J 223:84–90

Guo HL, Wang XF, Qian QY, Wang FB, Xia XH (2009) A green approach to the synthesis of graphene nanosheets. ACS Nano 3(9):2653–2659

Gurushantha K, Anantharaju KS, Renuka L, Sharma SC, Nagaswarupa HP, Prashantha SC, Vidya YS, Nagabhushana H (2017) New green synthesized reduced graphene oxide–ZrO2 composite as high performance photocatalyst under sunlight. RSC Adv 7(21):12690–12703

Hakim YZ, Yulizar Y, Nurcahyo A, Surya M (2018) Green synthesis of carbon nanotubes from coconut shell waste for the adsorption of Pb(II) ions. Acta Chim Asiana 1(1):6–10

Hosseinnezhad M, Ghahari M, Mobarhan G, Rouhani S, Fathi M (2023) Towards low cost and green photovoltaic devices: using natural photosensitizers and MoS_2/graphene oxide composite counter electrodes. Opt Mater 139:113775

Hou R, Jin Z, Sun D, Shi B, Wang L, Shan X (2023) Carbon nanotubes doped cellulose nanocomposite film for high current flexible piezoelectric nanogenerators. J Alloys Compd 965:171422

Hu Y, Shenderova OA, Hu Z, Padgett CW, Brenner DW (2006) Carbon nanostructures for advanced composites. Rep Prog Phys 69(6):R05

Jamil T, Munir S, Wali Q, Shah GJ, Khan ME, Jose R (2021) Water purification through a novel electrospun carbon nanofiber membrane. ACS Omega 6(50):34744–34751

Jung W, Lee JS, Han S, Ko SH, Kim T, Kim YH (2018) An efficient reduced graphene-oxide filter for PM2.5 removal. J Mater Chem A 6:16975–16982

Kim P, Shi L, Majumdar A, McEuen PL (2001) Thermal transport measurements of individual multiwalled nanotubes. Phys Rev Lett 87:215502
Kim S, Lim Y, Kang TH, Moon J, Choi IS, Lee YJ, Yi H (2020) Biotemplated nanocomposites of transition-metal oxides/carbon nanotubes with highly stable and efficient electrochemical interfaces for high-power lithium-ion batteries. ACS Appl Energy Mater 3(8):7804–7812
Kuo YL, Chen WT, Chen CH, Wu HT, Chiou CS, Chen YC, Chen HW (2024) Antibacterial nanofibers of carbon quantum dot/titanium dioxide by electrospinning for the removal of particulate matter and volatile organic compounds. Mater Chem Phys 316:129079
Kutluay S (2025) Boosting the CO_2 capture and regeneration performance by nanographene@zinc oxide as novel green-synthesized nanocomposites. J Ind Eng Chem
Li Q, Song Y, Xu R, Zhang L, Gao J, Xia Z, Tian Z, Wei N, Rümmeli MH, Zou X, Sun J (2018a) Biotemplating growth of nepenthes-like N-doped graphene as a bifunctional polysulfide scavenger for Li–S batteries. ACS Nano 12(10):10240–10250
Li D, Li'an H, Zhansheng W, Ping G, Guanyi C, Renfu J (2018b) A new function of spent activated carbon in BAC process: removing heavy metals by ion exchange mechanism. J Hazard Mater 359:76–86
Li K, Teng H, Sun Q, Li Y, Wu X, Dai X, Wang Y, Wang S, Zhang Y, Yao K, Bao Z (2022) Engineering active sites on nitrogen-doped carbon nanotubes/cobaltosic oxide heterostructure embedded in biotemplate for high-performance supercapacitors. J Energy Storage 53:105094
Liao F, Diao G, Li H (2020) Green synthesis of graphene oxide-$MnFe_2O_4$ composites and their application in removing heavy metal ions. Micro Nano Lett 15(1):7–12
Lin Z, Weng X, Ma L, Sarkar B, Chen Z (2019) Mechanistic insights into Pb(II) removal from aqueous solution by green reduced graphene oxide. J Colloid Interface Sci 550:1–9
Liu G, Li L, Huang X, Zheng S, Xu X, Liu Z, Zhang Y, Wang J, Lin H, Xu D (2018) Adsorption and removal of organophosphorus pesticides from environmental water and soil samples by using magnetic multi-walled carbon nanotubes@ organic framework ZIF-8. J Mater Sci 53(15):10772–10783
Lu Z, Li Y, Chen K, Chai S, Su G, Wu C, Sun M, Wang Y, Feng S, Hao M, Rao H (2025) Multi-activity ferruginated carbon quantum dots nanozyme improves wheat seedling growth and Cd tolerance. Crop J 13:510–523
Mahalakshmi D, Nandhini J, Karthikeyan E, Karthik KK, Sujaritha J, Vandhana V, Lokeshwar R (2025) Carbon nanomaterials for emerging contaminant remediation: addressing pharmaceutical pollution in the water cycle with precision. Water Cycle 6:449–472
Mandal S, Pu S, He L, Ma H, Hou D (2020) Biochar induced modification of graphene oxide & nZVI and its impact on immobilization of toxic copper in soil. Environ Pollut 259:113851
Mann D, Javey A, Kong J, Wang Q, Dai H (2003) Ballistic transport in metallic nanotubes with reliable Pd ohmic contacts. Nano Lett 3:1541–1544
Nersisyan HH, Lee TH, Lee KH, Maeng DY, Lee JH (2013) Fabrication of tunable carbon micro- and nanotubes using reed as bio-template. Mater Lett 107:79–82
Ogunsola SS, Oladipo ME, Oladoye PO, Kadhom M (2024) Carbon nanotubes for sustainable environmental remediation: a critical and comprehensive review. Nano-Struct Nano-Objects 37:101099
Paul S, Samdarshi SK (2011) A green precursor for carbon nanotube synthesis. New Carbon Mater 26(2):85–88
Perera SD, Mariano RG, Nijem N, Chabal Y, Ferraris JP, Balkus KJ Jr (2012) Alkaline deoxygenated graphene oxide for supercapacitor applications: an effective green alternative for chemically reduced graphene. J Power Sources 215:1–10
Raja A, Rajasekaran P, Selvakumar K, Arivanandhan M, Swaminathan SM (2019) Green approach to the preparation of reduced graphene oxide for photocatalytic and supercapacitor application. Optik 190:21–27
Rathod S, Preetam S, Pandey C, Bera SP (2024) Exploring synthesis and applications of green nanoparticles and the role of nanotechnology in wastewater treatment. Biotechnol Rep 41:e00830

Riaz T, Azam R, Shahzadi T, Shahid S, Mansoor S, Javed M, Bahadur A, Iqbal S, Mahmood S, Alotaibi KM, Alshalwi M (2024) Carbon dots and nitrogen-doped carbon dots-metal oxide nanocomposites: robust agents for effective sensing of ions. J Mater Sci: Mater Electron 35(13):940

Roy-Mayhew JD, Aksay IA (2014) Graphene materials and their use in dye-sensitized solar cells. Chem Rev 114:6323–6348

Sarvalkar PD, Jagtap AS, Kalake SV, Sawant SA, Sutar RB, Deshmukh DV, Kamble SS, Tibe AP, Yadav JB, Prasad NR, Kulkarni SN (2024) Synthesis of graphene oxide anchored copper oxide nanocomposites for biological activity and environmental remediation. ES Energy Environ 26:1313

Selopal GS, Milan R, Ortolani L, Morandi V, Rizzoli R, Sberveglieri G et al (2015) Graphene as transparent front contact for dye-sensitized solar cells. Sol Energy Mater Sol Cells 135:99–105

Sen Gupta S, Chakraborty I, Maliyekkal SM, Adit Mark T, Pandey DK, Das SK, Pradeep T (2015) Simultaneous dehalogenation and removal of persistent halocarbon pesticides from water using graphene nanocomposites: a case study of lindane. ACS Sustain Chem Eng 3:1155–1163

Shahba H, Sabet M (2020) Two-step and green synthesis of highly fluorescent carbon quantum dots and carbon nanofibers from pine fruit. J Fluoresc 30(4):927–938

Sotiropoulou S, Sierra-Sastre Y, Mark SS, Batt CA (2008) Biotemplated nanostructured materials. Chem Mater 20(3):821–834

Suo F, Xie G, Zhang J, Li J, Li C, Liu X, Ji M (2018) Carbonized sieve-like corn straw cellulose-graphene oxide composite for organophosphorus pesticide removal. RSC Adv 8:7735–7743

Suriani AB, Azira AA, Nik SF, Nor RM, Rusop M (2009) Synthesis of vertically aligned carbon nanotubes using natural palm oil as carbon precursor. Mater Lett 63(30):2704–2706

Sykam N, Madhavi V, Rao GM (2018) Rapid and efficient green reduction of graphene oxide for outstanding supercapacitors and dye adsorption applications. J Environ Chem Eng 6:3223–3232

Tong F, Jia W, Pan Y, Guo J, Ding L, Chen J, Jia D (2019) A green approach to prepare hierarchical porous carbon nanofibers from coal for high-performance supercapacitors. RSC Adv 9:6184–6192

Tuichai W, Karaphun A, Ruttanapun C (2020) Ag nanomaterials deposited reduced graphene oxide nanocomposite as an advanced hybrid electrode material for asymmetric supercapacitor device. J Alloys Compd 849:156516

Ugwuoke CO, Ghali M, Abd El-Moneim A (2024) Green synthesis of carbon dots from Nigella sativa seeds for supercapacitor application. J Energy Storage 95:112634

United States Environmental Protection Agency (2025) Environmental remediation: overview of traditional remediation technologies and their environmental impacts. Wikipedia. Accessed Aug 2025

Vishwakarma PK, Pandey SK, Yadav SK, Shukla P, Srivastava A, Giri R (2022) Multiwalled carbon nanotube-based freestanding filters for efficient removal of fine particulate matters (PM0.3), microplastics (MP0.3), and bioaerosols. ACS Appl Nano Mater 5:9306–9318

Wang Y, Pan C, Chu W, Vipin AK, Sun L (2019) Environmental remediation applications of carbon nanotubes and graphene oxide: adsorption and catalysis. Nanomater 9(3):439

Wang C, Lu L, Liu Z, Lai Y, Li J, Xia Z, Cao J, Wang X, Lu J, Yang W (2023) Green carbon dots illuminate biogenic nanohybrids toward soft, piezo/photoactive, and physically transient nanogenerators. ACS Sustain Chem Eng 11:14646–14658

Zhan W, Gao L, Fu X, Siyal SH, Sui G, Yang X (2019) Green synthesis of amino-functionalized carbon nanotube-graphene hybrid aerogels for high-performance heavy metal ions removal. Appl Surf Sci 467:1122–1133

Zhang Q, Liang J, Zhao L, Wang Y, Zheng Y, Wu Y, Jiang L (2020) Synthesis of novel fluorescent carbon quantum dots from Rosa roxburghii for rapid and highly selective detection of o-nitrophenol and cellular imaging. Front Chem 8:665

Zhang S, Arkin K, Zheng Y, Ma J, Bei Y, Liu D, Shang Q (2022) Preparation of a composite material based on self-assembly of biomass carbon dots and sodium alginate hydrogel and its green, efficient and visual adsorption performance for Pb^{2+}. J Environ Chem Eng 10(1):106921

Zhao Y, Liu Y, Tong C, Ru J, Geng B, Ma Z, Liu H, Wang L (2018) Flexible lignin-derived electrospun carbon nanofiber mats as a highly efficient and binder-free counter electrode for dye-sensitized solar cells. J Mater Sci 53:7637–7647

Zhao S, Song X, Chai X, Zhao P, He H, Liu Z (2020) Green production of fluorescent carbon quantum dots based on pine wood and its application in the detection of Fe^{3+}. J Clean Prod 263:121561

Zhou P, Yang X, He L, Hao Z, Luo W, Xiong B et al (2015) The Young's modulus of high-aspect-ratio carbon/carbon nanotube composite microcantilevers by experimental and modeling validation. Appl Phys Lett 106:111908

Chapter 4
Sustainable Polymeric Materials: Development, Challenges, and Future Benefits for Industrializations

Sriparna De, Dipankar Das, and Dipanjan Bhattacharjee

Abstract Natural renewable resources, waste disposal, and industrial carbon dioxide (CO_2) feedstocks are the sources of sustainable polymers, which may be recycled in an environmentally friendly manner to reach End-of-Life (EoL) through biodegradation and composability. The manufacturing of sustainable polymers from materials that are renewable, and the many sustainability pillars are highlighted in this chapter. A brief discussion is given of the green method, microbial fermentation, and nanotechnological perspectives in the production of polymeric materials. The challenges of recycling and composting, industrial scale-up, data repeatability, and storage via promising results that are still in the early stages of development are some of the barriers to the adoption of sustainable polymers. With the goal to reduce the excessive use of traditional plastics, the development and implementation of customized sustainable polymeric composites in industrial applications such as packaging, electrical, and medical equipment is speeding up and establishing an improved circular economy. In conclusion, the ironic era of synthetic plastics may be ushered in by sustainable polymers through reduce, reuse, and recycle technologies that manage our planet's carbon footprint in a very difficult way.

Keywords Sustainable polymers · Reuse · Reduce · Recycling · End-of-Life (EoL)

4.1 Introduction

Polymers, composed of long chains of repeating molecular units, are essential in a wide range of industries, including packaging, textiles, automotive, electronics, and healthcare. Traditionally, most polymers have been derived from fossil fuels, such

S. De (✉) · D. Das · D. Bhattacharjee
Department of Allied Health Sciences, Brainware University, Kolkata, India
e-mail: sd.ah@brainwareuniversity.ac.in

S. K. Verma et al. (eds.), *Sustainable and Natural Polymers*, Clean Energy Production Technologies, https://doi.org/10.1007/978-981-92-0092-4_4

as polyethylene, polypropylene, and polystyrene, valued for their versatility and cost-effectiveness. In recent years, there has been an increasing focus on multifunctional polymers that are tailored for specific properties, including biodegradability, conductivity, or responsiveness to external stimuli, expanding the potential sustainable applications in diversified field (Banerjee et al. 2025). The concept of sustainable advancement, introduced in 1987 by the World Commission on Environment and Development in their Brundtland Report, is not merely a theoretical idea. As per the report, it is a multifaceted issue that necessitates our profound recognition, throughout environmental science and sustainability. The Commission highlights sustainable development as "a development which meets the needs of the present without compromising the ability of future generations to meet their own needs" (Geir Asheim 1994). The widespread use of petroleum-based polymers has raised significant environmental concerns. Plastic waste accumulation in both land and marine ecosystems, coupled with the environmental footprint of their production and disposal, highlights the pressing need for more sustainable polymer solutions (Singh et al. 2023). With the goal of navigating these barriers, sustainable polymer research emphasises polymers that are recyclable, biodegradable, renewable, or able to function effectively in a circular economy. Many strategies have been developed for rendering polymers sustainably.

Environmental health is pivotal for sustainable development because it influences the potential of future generations, as well as the present, to meet their needs, which depend on the natural environment's life-sustaining functions. A significant tool for evaluating sustainability is life cycle assessment (LCA), which evaluates the impact that has on the environment throughout the course of a product's life cycle. This comprises phases such as the extraction of raw materials, production, packing, transportation, use, and end-of-life (EoL) procedures (Hauschild et al. 2018). These EoL options for goods may include incineration, landfilling, composting (for biodegradable products, either industrial or home composting), and recycling. Sustainability encompasses not only environmental concerns but also social and economic factors (Banerjee and Ray 2022). To ensure a more comprehensive understanding of sustainability, it is important to expand LCA by incorporating social aspects through Social LCA (S-LCA) and financial dimensions via Life Cycle Cost analysis (LCC) (Banerjee et al. 2025) (Fig. 4.1).

Industries are rapidly adopting sustainable polymers to meet growing consumer demand for environmentally friendly products and to comply with stricter regulations. Sectors such as packaging, automotive, electronics, textiles, and healthcare are increasingly using bio-based and biodegradable polymers. For example, materials like polylactic acid (PLA) are gaining traction in the packaging industry due to their compostability and reduced carbon footprint. The automotive sector is also turning to natural fiber composites and recycled polymers to lower vehicle weight and emissions without compromising performance (Dinkar and Kumar 2025) (Fig. 4.2).

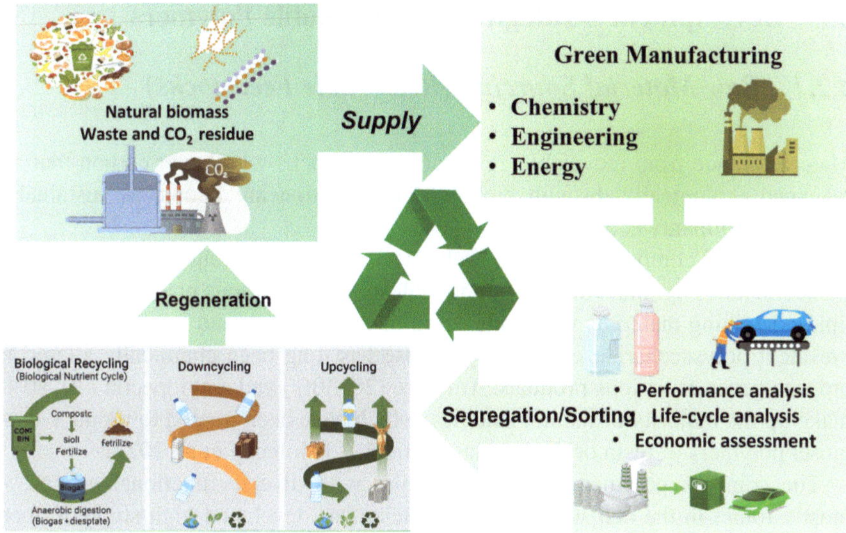

Fig. 4.1 Diagram showing life cycle analysis of sustainable polymers (Mohanty et al. 2022)

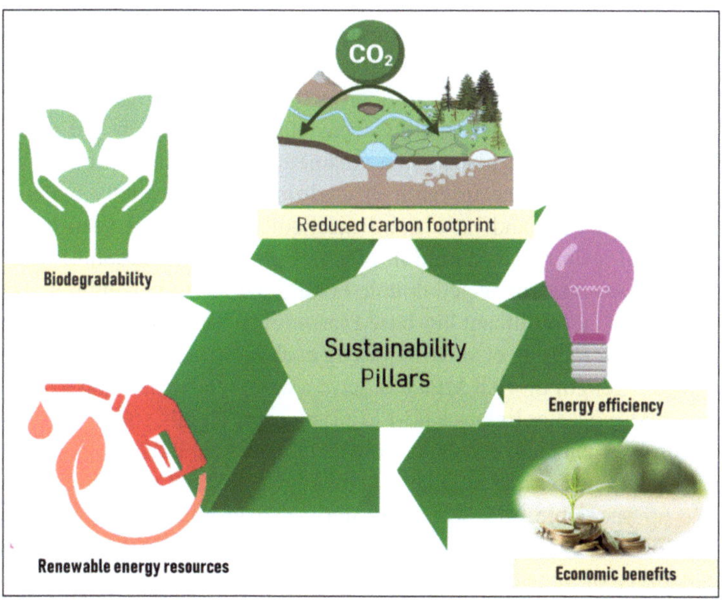

Fig. 4.2 Sustainable polymeric materials: use and importance

4.2 Development Strategies for Sustainable Polymers

4.2.1 Raw Material Sourcing (Renewable Feedstocks)

Materials made from recyclable, renewable, feedstocks with lower carbon that are managed ecologically through biological degradation are known as sustainable polymers (Mohanty et al. 2022).

The typical composition of renewable feedstocks, including plants and marine waste, includes lignin, cellulose, hemicellulose, starch, various proteins, chitin, and lipid-containing materials with wide variations of structure and function. The most prevalent polysaccharide on Earth, cellulose has long been chemically altered in order to generate various products. Amylose (20–30%) and amylopectin (70–80%) make up the majority of natural starches, which have been treated to produce thermoset polymers of resin or thermoplastic starches (Mohanty et al. 2022).

The complex polymer component lignin, with diverse functional groups, is mostly found in the cell walls of plants. Scientists developed high-strength, thermally resistant, and reusable lignin-based polyurethanes (LPU) by combining weak and strong bonds of hydrogen, and dynamic silicone ether covalent bonding pattern with 99 MPa mechanical strength. One of the monoaromatic compounds found in lignin derivatives is vanillin (Luo et al. 2025). Researchers utilized dihydroxy aliphatic monoamine and vanillin to formulate vitrinite with imine structure (VIM), which has a tensile strength of 35.6 MPa. The synthetic glassy structure demonstrated remarkable welding and healing properties (Chen et al. 2023).

Cellulose, which originates mainly from plant cell walls, is the most prolific renewable resource on our planet. The vitrimers made of cellulose and its derivatives are fascinating for the research of environmentally friendly materials since they are biodegradable and have good mechanical properties (Tao et al. 2018). Rosin, a natural resin mostly found in pine trees, is a combination of reactive carboxyl groups, two interconnected double bonds, and an o-phenanthrene backbone. Moreover, rosin is a significant bio-based resource, with 1.2 million tons generated globally each year (Kugler et al. 2019). Zeng et al. constructed epoxy resin matrices with a tensile strength of 40.8 MPa that have a completely rosin-based and include dynamic β-hydroxy ester linkages (C-FPEG) by using rigidly structured rosin derivatives as curing agents and epoxy monomers (Zeng et al. 2022).

Glycerol and unsaturated fatty acids are two examples of the natural compounds that make up vegetable oils, which are derived from plant fruits and seeds. The abundance of soft chain segments, many modification procedures, and potential biodegradability of vegetable oils have led to the production of a variety of bio-based vitrimers (Luo et al. 2025). From the study of Mauro et al., it is observed that using opposing structures of acid and amine or functions of aromatic and aliphatic as hardeners, recyclable resins based on epoxidized linseed capsule oil can be made. A catalyst-free, self-repairing, durable, and totally biobased TOTGE-citric acid vitrimer was produced by Xu et al. by applying renewable tung oil, resulting in methyl esterification, Diels-Alder, and epoxidation to produce TOTGE with terminal epoxy groups. The vitrimer's high density of cross-links and strong reactivity of its terminal epoxy groups allowed it to have good mechanical and thermal stability. (Xu et al. 2021).

Plants high in sugar or starch, such as sugar cane, wheat, or sugar beet, are the source of many commercial sustainable polymers. It is also possible to make polyethylene from sugar cane by dehydrating ethanol to get ethylene monomer. The production process is contentious and requires a robust sugar-cane sector; it is being investigated primarily in Brazil. Plant-based essential oils contain terpenes and terpenoids, which share an isoprene component in their molecular structures. The most commonly identified polyterpene is possibly natural rubber, and more than 10 mega tons are produced annually. A considerably lower extent of research is being done on other terpenes as monomers for the synthesis of polymers. Turpentine, which is collected from pine trees (*Pinus* spp.), is mostly composed of α-pinene and β-pinene. Another example is limonene, which originates from citrus fruit peels. About 0.3 kilotonnes of turpentine and 0.7 kilotonnes of limonene were produced globally in 2013, indicating a small level of production of these monomers. Polymer resins made from these terpenes are already commercially accessible (Zhu et al. 2016). Block polymeric elastomers with thermoplastic properties have been produced by combining terpene monomers from tulips and wild mint with controlled polymerization techniques (Bolton et al. 2014; Zhu et al. 2016).

The most researched polyesters for biomedical uses are polyhydroxyalkanoates, which are naturally synthesized by bacterial cells and other living microorganisms utilising lipids and carbs. The ring opening polymerization or ROP system was invented to produce polyhydroxyalkanoates, and the plant *Cryptocarya massoia* yields δ-Decalactone, a renewable lactone monomer that has been used several times to produce polymers or block copolymers by ROP for drug delivery applications (Bansal and Rosenholm 2020) (Fig. 4.3).

4.2.2 Green Chemistry Approaches in Polymer Synthesis

Environmental challenges have been widely acknowledged since the turn of the twenty-first century as being essential to ensuring a sustainable society. Fossil feedstocks should be used less for both energy generation and material manufacturing in order to reduce carbon dioxide emissions, one of the factors causing global warming. In this regard, research on "green polymer chemistry" has already been carried out and examined in the field of polymer synthesis. It is worth noting that the term "green chemistry" was first introduced in 1998, and the following year, it was expanded to encompass the study of polymers as "green polymer chemistry" (Fig. 4.5). Then, the terms came into play in a significant way (Kobayashi 2017). Recently, Jessop and MacDonald have highlighted the potential of hotspot-driven research to speed up the pace of advancements. By emphasizing the most hazardous phase of a polymer's life cycle, this method, which can be used to identify and select research initiatives, could be employed to "green" polymer chemistry. Synthetic

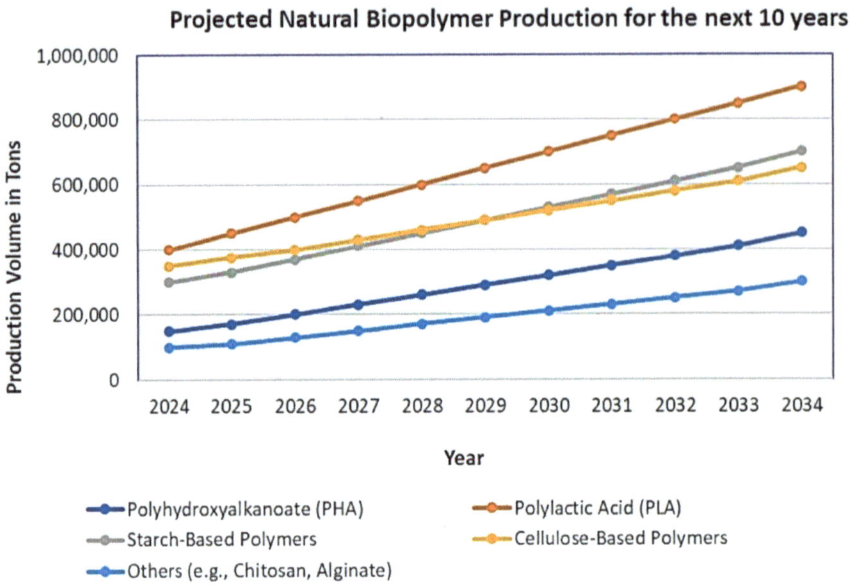

Fig. 4.3 Projected natural biopolymer production. (Picture is taken from an open-access journal (Beena Unni and Muringayil Joseph 2024))

organic chemistry served as the foundation for green chemistry, which significantly altered the production of pharmaceutically active chemicals and lessened their environmental impact (Kerton 2025).

Certain materials, such as those with linkages that are more prone to hydrolysis, are made to biodegrade more easily. Specifically, the degradable polymers that are most frequently discussed—polyhydroxyalkanoates, poly(lactic acid), poly(butylene succinate), poly(butylene terephthalate), poly(e-caprolactone) copolymers—all include ester links. Enzymes found in bacteria and fungi can cause hydrolysis. Additionally, these microorganisms can develop spontaneously or by human intervention to break down some economically important polymers, particularly polyesters (Erickson et al. 2022).

In natural feedstocks, biopolymers such as lignin, chitin, cellulose, hemicellulose, starch, alginic acid, and chitosan can be extracted; some of them are utilized directly, while others can be altered for a variety of uses. Chitosan is generated by deacetylating chitin, and by employing phosphoric acid in a mechanochemical process, the final chitosan may be modified with respect to its molecular weight. The examples showcase the use of environmentally friendly green chemistry methods for biopolymer modification and processing (Yang et al. 2023). The use of carbon dioxide act as a co-monomer for the production of polycarbonates has garnered a lot of interest as it replaces hazardous reagents like phosgene in the manufacturing of polymers (Kerton 2025). Although γ-valerolactone (GVL) has drawn a lot of interest as a renewable fuel, it may also be utilized as an environmentally friendly solvent in a variety of chemical reactions. There are two primary methods for creating GVL: either furfural from hemicellulose or hydroxymethylfurfural from cellulose, which is then converted to levulinic acid. GVL has a boiling point of 207–208 °C and is a colorless, stable liquid that dissolves in water. It has a melting point of −31 °C and does not break down at room temperature. It is a potential green solvent with these qualities that can take the place of many traditional dipolar aprotic solvents in both industry and research circles (Joseph et al. 2021).

4.2.3 An Ionic Liquid-Assisted Method

Ions make up the entirety of ionic liquids. An ionic liquid is, for instance, molten sodium chloride. Notably, they may be liquid at temperatures as low as −96 °C. This solvent is extremely flexible because of its ability to combine several anions and cations to form an ionic liquid (Pereira et al. 2019). In applications that are important to industry, ionic liquids are preferred over traditional solvents primarily because they don't generate any volatile organic molecules. In the presence of functionalized carboxyl ionic liquids, chitosan, nanoclay, and europium (III) complexes were used to produce chitosan films as vapor-responsive luminous sensors. These aqueous free-standing luminous films were made by dissolving chitosan in ionic liquids rather than acetic acid. By decreasing the proton strength on platelets and promoting the creation of high coordination number europium (III) complexes, ionic liquids have the intriguing ability to increase the luminescence with a quantum efficiency of 38% in the film (Joseph et al. 2021). According to various research, cellulose and zeolite were combined to create a recyclable and cell-compatible polymer nanocomposite in the presence of the ionic liquid 1-ethyl-3-methylimidazolium chloride (EMIMCl). The cellulose polymer is dissolved using the EMIMCI ionic liquid. It was reported that regenerated cellulose/sepiolite nanocomposite films were prepared utilizing the ionic liquid 1-butyl-3-methylimidazolium chloride (Soheilmoghaddam et al. 2014). Wang et al. recently reported synthesizing an epoxy nanocomposite with chitin particles that were distributed using 1-ethyl-3-methylimidazolium acetate, an ionic liquid. Ionic liquid promotes cross-linking with epoxy to produce a translucent texture and makes it easier for chitin fibers to disperse as nanofillers in the epoxy matrix (Wang et al. 2020; Joseph et al. 2021).

4.2.4 Method Assisted by Supercritical Carbon Dioxide

For many industrial and chemical applications, carbon dioxide is a desirable solvent due to its affordability and ease of supply. Traditional techniques to produce poly (*N*-isopropylacrylamide)-ferrite nanocomposites (PNC) include the use of volatile and hazardous organic solvents, high temperatures, and expensive materials.

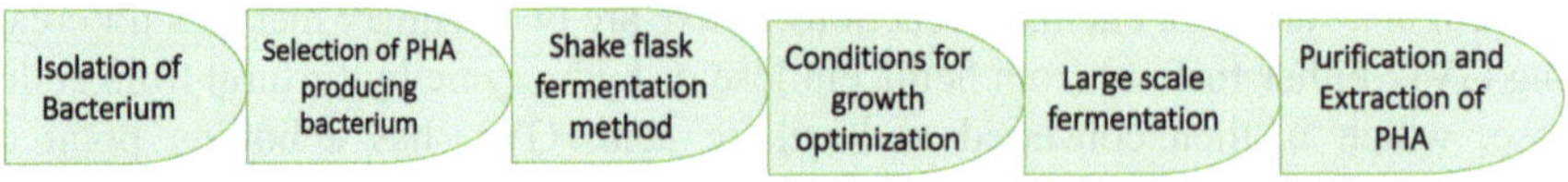

Fig. 4.4 Diagrammatic representation of the stages involved in producing PHA using bacterium (Beena Unni and Muringayil Joseph 2024)

In order to distribute doxorubicin (DOX) for breast cancer in a targeted and controlled manner, Bisht et al. reported developing a biocompatible magnetic nanocomposite made of poly(*N*-isopropylacrylamide) and ferrite utilizing supercritical carbon dioxide (Bisht et al. 2017; Joseph et al. 2021).

4.2.5 Method of Microbial Fermentation

Microbial fermentation technologies contribute to environmentally friendly production methods by using microorganisms to manufacture organic polymers with high yield and purity. In one example, biodegradable polyesters known as polyhydroxyalkanoates (PHAs) are made by microbiologically fermenting renewable substrates such as sugars or lipids (Fig. 4.4) (Beena Unni and Muringayil Joseph 2024).

4.3 Nanotechnology in Sustainable Polymer Development

One of the most difficult problems of our time is to ensure sustainable development for future generations by applying the ideas and principles of green chemistry and green engineering to develop materials and nanoproducts without harmful byproducts. The most significant area of nanotechnology is without a doubt green nanotechnology, which applies the 12 values of green chemistry to a wide range of possible uses. The concepts of green chemistry have excellent prospects in the developing field of nanoscience to establish sustainability and safeguard and advance worker safety and health throughout the production of products and process implementation (Fig. 4.5). Different techniques have been used to create a range of multimodal green nanomaterials with distinct composition, size, shape, and crystallinity, and their possible uses have been investigated. In reality, the bottom-up method of green synthesis of nanostructures utilizing plants uses oxidation and reduction as the primary process (Khan 2020) (Table 4.1).

The demand for these nanoproducts continues to rise every day, and by the end of 2020, the Royal Society estimates that the world's annual output of nanomaterials might surpass 100,000 metric tons. Therefore, developing sustainable substitutes for the innovative synthesis and production processes is imperative in order to mitigate the dangers involved (Khan 2020). In terms of drug administration, therapy, and imaging of integrating molecules on tumor cells, as well as therapeutic results,

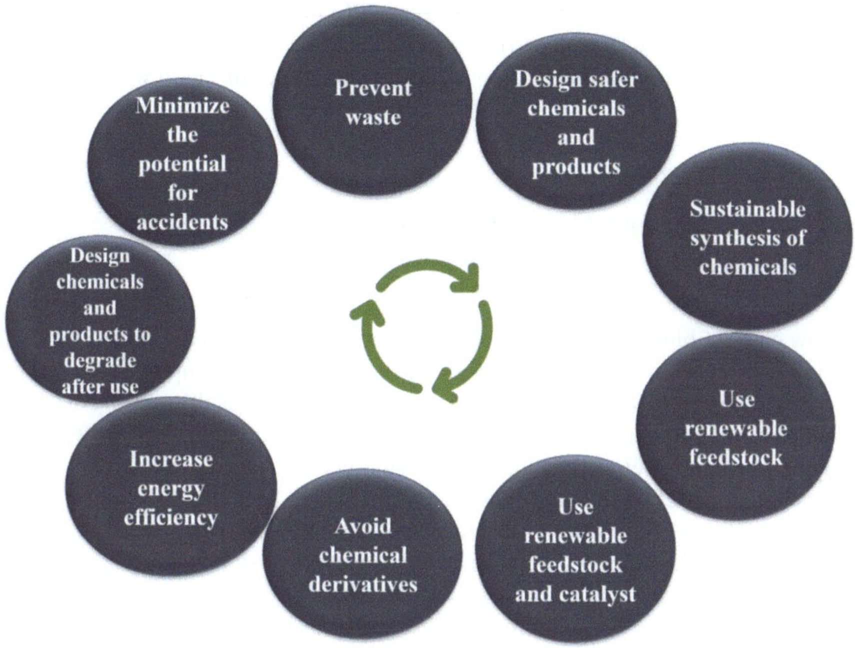

Fig. 4.5 The principles of green chemistry in green synthesis

Table 4.1 Green synthesis of nanostructures utilizing plants

Plant	Nanoparticle	Size (nm)	Part of the plant	References
Alternanthera sessilis	Ag	40–50	Whole	Niraimathi et al. (2013)
Alovera	In_2O_3	5–50	Leaf	Khan (2020)
Azadiricta indica	Ag	10–30	Leaf	
Caria papaya	Ag	15	Fruit	
Cassia fistula	Au	55	Stem	Daisy (2012)
Mirabilis jalapa	Au	100	Flowers	Khan (2020)

maghemite nanoparticles (NPs) (γ-Fe_2O_3-NPs) and magnetite nanoparticles (Fe_3O_4-NPs) have proven to be quite beneficial. Therapeutic substances, for instance, have been affixed to biocompatible, nontoxic super magnetic and magnetic nanoparticles and administered intravenously, whereas magnetic fields are produced externally to concentrate their concentration on certain sites in vivo. Folic acid, which is identified as the targeting agent in combination with poly(ethylene glycol), was used to modify biodegradable and uniformly distributed iron oxide superparamagnetic nanoparticles. According to the biological investigation, altering NPs with poly(ethylene glycol)-Folic acid greatly increased the target cells' intracellular absorption of NPs (Jahangirian et al. 2017). Beta-cyclodextrin (β-CD) functionalized with 3-(3,4-Dihydroxyphenyl) propionic acid was loaded onto hollow mesoporous silica nanoparticles surfaces via boronic acid–catechol ester linkages. The host-guest interaction was then used to place PEG-conjugated adamantine (Ada) onto an HMSNs–β-CD nanocarrier. According to in vivo data, the drug-loaded HMSNs significantly inhibited tumor development while causing few harmful side effects (Liu et al. 2016).

4.4 Additives and Compatibilizers for Eco-Friendly Performance

A new class of bio-composites called eco-composites has been developed by combining waste cellulose with recycled plastics to produce improved materials. Increased aggregate values and possible applications will result from the development of novel eco-composites made of agro-waste reinforcing fillers and recycled polymers with a better understanding of overall performance (Hakamy et al. 2018; Nassar et al. 2021). Adaptive manufacturing, which includes 3D printing eco-composites for new applications that take geometric flexibility into account, is quite interesting. Eco-composites may be able to overcome significant constraints in load-bearing applications because of recent developments in the manufacture of complicated structures. The biomedical items that can be made using 3D printing technology include customized implants, medication delivery devices, organ printing, rehabilitation, and anatomical models for surgical education and planning (Nassar et al. 2021).

Glass fiber-reinforced composites based on petroleum-derived polymers continue to dominate the worldwide polymeric composites market in terms of both value and volume. However, because natural fibers are more readily available, less expensive to produce, renewable, and biodegradable, they are increasingly being employed in composite materials in place of glass fibers. Every component used in the production of completely sustainable and eco-friendly composites must follow a completely green path from the point of origin to the stage of disposal. Given their biobased origins and biodegradability poly-hydroxybutyrate (PHB) and poly-lactic acid (PLA) are viable alternatives to traditional thermoplastic matrices. From Valente BFA et al., GlucoPure® Sense, a sugar-based surfactant, and epoxidized linseed oil are two commercially accessible and environmentally acceptable additives that have been effectively used to PLA and PHB green composites strengthened with cellulose micronized fibers (Valente et al. 2022).

It is seen from various studies that the advantageous outcomes of oxidizing softwood and hardwood nanofibrils in a PVA matrix using 2,2,6,6-tetramethylpiperidi ne1-oxyl. By adding 20% nanofibrils, the PVA film's tensile strength and modulus of elastic were noticeably increased. Atomic force microscopy verified that the nanofibrils were evenly distributed throughout the matrix (Shahzad 2015). Espino-Pérez et al. revealed the beneficial benefits of coupling cellulose nanocrystals (CNC) with *n*-octadecyl isocyanate (ICN) on their interaction with polylactic acid (PLA) matrix. The tensile characteristics of PLA decreased when untreated CNC was added (Haafiz et al. 2013). According to the study by Nipu Dutta et al., it was feasible to successfully examine the role of Epoxidized soybean oil (ESO) as a plasticizer, natural additive, and compatibilizer by including it in the composite system. The system of Polyvinylchloride (PVC) and Rice husk ash (RHA) composite, where all the parameters like thermal, mechanical, water vapor uptake, limiting oxygen index, etc., were enhanced, 7.5 phr ESO was determined to be the ideal amount (Dutta et al. 2021).

4.5 Performance and Application Areas

4.5.1 Packaging Industry

Sustainable polymers have emerged as an essential innovation in the packaging industry, driven by the urgent need to address global plastic pollution, reduce greenhouse gas emissions, and transition toward a circular economy. Unlike conventional petroleum-based plastics such as polyethylene (PE), polypropylene (PP), and polyethylene terephthalate (PET), which persist in the environment for centuries, sustainable polymers are derived from renewable feedstocks, designed for biodegradability, or produced through recycling technologies. Common examples include biodegradable polymers such as polylactic acid (PLA), polyhydroxyalkanoates (PHA), and starch-based blends, as well as bio-based but non-biodegradable options such as bio-PET and bio-PE, alongside recycled plastics (e.g., rPET). These materials have been used increasingly in food trays, liquid bottles, biodegradable baggage, and active packaging films because they provide benefits for consumer-friendly branding, regulatory compliance, and a lower carbon footprint. Nevertheless, they also pose significant challenges, such as increased manufacturing costs, reduced mechanical and thermal durability in comparison to petroleum-based plastics, inadequate industrial composting facilities, and worries about competition with agricultural land usage for feedstocks (Cardoso et al. 2024). Research is moving quickly toward future-oriented polymer systems, such as enzymatically recyclable polymers, hybrid nanocomposites, and intelligent active packaging materials that improve food safety and prolong shelf life, in order to get over these challenges. For instance, Zn-doped TiO_2–reinforced PLA/PBAT films have recently been shown to provide on-demand degradability while acting as ethylene scavengers, thereby both reducing waste and improving fruit preservation (Yuntawattana et al. 2025). Looking forward, the integration of biopolymer engineering, advanced fillers, bio-refinery approaches, and AI-driven waste management systems is expected to significantly improve the functionality, affordability, and scalability of sustainable packaging. Coupled with supportive government policies, industry investments, and increasing consumer environmental awareness, sustainable polymers are poised not just to supplement but eventually redefine the global packaging sector, paving the way for a truly sustainable materials economy.

4.5.2 Automotive and Transportation

Sustainable polymers are playing an increasingly important role in the automotive and transportation sector as industries seek to reduce vehicle weight, enhance fuel efficiency, and minimize environmental footprints. Traditionally, petroleum-based polymers such as polypropylene (PP), polyethylene (PE), and polyurethane (PU) have been used extensively for components like interiors, bumpers, and insulation, but their non-biodegradability and fossil-derived origins pose sustainability concerns. In response, manufacturers are adopting bio-based and recycled polymers such as polylactic acid (PLA), polyhydroxyalkanoates (PHA), bio-based polyamides, recycled PET, and natural fiber-reinforced composites for applications ranging from seat cushions and dashboards to lightweight structural panels (Cardoso et al. 2024). The advantages of these materials include significant weight reduction, which improves fuel economy in conventional vehicles and extends driving range in electric vehicles, alongside recyclability, lower CO_2 emissions, and enhanced consumer acceptance due to their eco-friendly branding. However, their widespread adoption faces challenges such as higher production costs, moisture sensitivity, variable mechanical performance, and limited large-scale recycling and processing infrastructure. Looking ahead, the integration of nanofillers, hybrid composites, and high-performance bio-based polymers is expected to bridge current performance gaps, while advanced manufacturing approaches such as additive manufacturing and Industry 4.0-driven supply chain optimization are likely to accelerate sustainable polymer adoption across transportation systems (Pathak and Dhakate 2022; Ibrahim et al. 2025). Furthermore, coupling sustainable materials with electric vehicles (EVs), autonomous mobility, and lightweight design strategies could enable a new generation of transportation solutions that balance safety, durability, and environmental stewardship (Wazeer et al. 2023). Thus, sustainable polymers represent not just a substitution for traditional plastics but a strategic enabler of low-carbon, high-efficiency mobility systems of the future.

4.5.3 Construction and Infrastructure

The construction and infrastructure sector, one of the largest global consumers of polymeric materials, is under mounting pressure to transition toward sustainable alternatives that reduce environmental impact without compromising durability and functionality. Conventional petroleum-derived plastics, such as PVC, polystyrene, and polyurethane foams, are widely used in pipes, insulation, sealants, adhesives, coatings, and lightweight composites, but they contribute significantly to carbon emissions and long-lived plastic waste. In response, sustainable polymers—including bio-based, biodegradable, and recyclable materials—are gaining traction as greener alternatives with the potential to reshape building practices.

Among the most prominent advances are bio-based insulation foams and composites derived from lignin, starch, and cellulose, which not only lower embodied carbon but also improve indoor air quality by reducing emissions of volatile organic compounds (Raja et al. 2023). Biopolymer blends reinforced with natural fibers are being developed for structural panels and lightweight boards, offering mechanical strength comparable to traditional composites while being renewable and recyclable (Narayanan et al. 2023). Moreover, polymeric binders in sustainable concrete formulations are enabling reduced cement usage, thereby cutting down one of the most carbon-intensive aspects of construction (Golewski 2021).

The advantages of sustainable polymers in this sector include lightweight properties for easier transport, design flexibility for modular construction, and compatibility with circular economy strategies through recycling and biodegradability. They also show promise in enhancing energy efficiency by improving insulation performance and enabling smart building materials such as self-healing coatings and phase-change polymers for thermal regulation. However, limitations persist: weathering resistance, UV degradation, moisture sensitivity, and fire safety standards remain major challenges, as does the higher initial cost of bio-based polymers compared to conventional plastics (Xie et al. 2021).

Looking ahead, the future of sustainable polymers in construction lies in advanced multifunctional materials. Research is increasingly focused on 3D-printed biopolymer composites for modular housing, nanocellulose-reinforced coatings with superior barrier properties, and carbon-sequestering polymers that can actively reduce atmospheric CO_2 levels during their service life (Kumar et al. 2025). Additionally, integrating sustainable polymers with digital construction technologies such as Building Information Modeling (BIM) and smart sensing systems can accelerate adoption by optimizing material selection and lifecycle performance monitoring.

In essence, sustainable polymers represent not only a pathway to lower-carbon and eco-friendly construction materials but also an opportunity to redefine infrastructure resilience, longevity, and adaptability in the face of climate change. By bridging materials science with circular economy principles, these polymers are poised to become central to the next generation of sustainable cities and green infrastructure systems.

4.5.4 Consumer Goods and Electronics

The consumer goods and electronics sector is one of the fastest-growing markets for sustainable polymers, driven by increasing consumer awareness, stricter regulations, and the rapid pace of technological innovation. The reliance on petroleum-derived plastics in electronic housings, packaging, and daily-use products has raised concerns over e-waste and microplastic accumulation. To address these challenges, sustainable polymers—biodegradable, recyclable, or bio-based alternatives—are emerging as viable substitutes.

Recent studies highlight the incorporation of bioplastics and recycled polymers in industrial manufacturing processes for consumer products such as household appliances, furniture, and electronic devices. For instance, olive seed-reinforced acrylonitrile butadiene styrene (ABS) and glass fiber–reinforced recycled polyamide have demonstrated improved mechanical performance while lowering the carbon and water footprint in lock manufacturing and electronic applications (Cardoso et al. 2024). Such innovations reduce dependence on virgin fossil-based polymers and simultaneously contribute to circular economy goals.

In electronics, sustainable materials are increasingly important for wearable devices, sensors, and flexible electronics. Biodegradable conductive films prepared from natural polymers like chitosan and sorbitol, combined with water-soluble conductive polymers, have been shown to yield stretchable, transparent, and conductive substrates for sensors and wearable devices (Lee et al. 2025). These materials not only improve the environmental footprint of production but also ensure safe disposal and minimal ecological damage once devices reach end-of-life. Consumer behavior is also shaping the adoption of sustainable polymers. Surveys conducted in India and other emerging economies show that environmental consciousness and demographic factors strongly influence the demand for eco-friendly products, including electronics and household goods (Mathur 2025). This points to a market transition where producers are explicitly encouraged to use greener packaging and polymers by customer choice.

The development of sustainable polymers in electronics and consumer goods will be determined by developments in additive manufacturing, biodegradability, and recyclability. Microbial isolates that can degrade polyethylene and polyurethane have demonstrated tremendous advancements in biodegradation research, opening the door for more environmentally friendly disposal techniques (Siracusa 2019). Furthermore, employing biodegradable or biosourced feedstocks, 3D printing technologies provide prospects for the fabrication of personalized, lightweight, and sustainable consumer goods based on polymers (Kurkin et al. 2025). Global sustainability goals, improving material efficiency, and lowering e-waste are all made possible by the use of sustainable polymers in electronics and consumer goods. The shift to a circular, low-carbon economy will largely depend on their incorporation into appliances, electronics, and everyday goods.

4.5.5 Medical and Pharmaceutical Applications

Sustainable polymers are increasingly recognized as vital materials in medical and pharmaceutical applications due to their ability to combine therapeutic efficacy with environmental responsibility. Unlike conventional petroleum-based polymers that pose challenges of persistence and waste accumulation, sustainable polymers—such as biopolymers (chitosan, alginate, PLA, PHA), green polymer composites, and bio-based nanoparticles—are derived from renewable resources and offer biodegradability, biocompatibility, and tunable physicochemical properties that make them ideal for healthcare. These materials are being applied in diverse areas, including controlled drug delivery, where they enable targeted and sustained release (Weerarathna et al. 2024a); tissue engineering, where biopolymer scaffolds mimic extracellular matrices to support regeneration (Baranwal et al. 2022); wound healing, where natural polymers exhibit antimicrobial and healing-promoting effects (Weerarathna et al. 2024b); and medical devices, where biocompatible coatings improve functionality and safety (Cháirez-Ramírez et al. 2025). Furthermore, sustainable polymers are transforming pharmaceutical packaging by providing eco-friendly alternatives to conventional plastics, thereby reducing environmental burden while ensuring drug stability (Popović et al. 2024). Recent evidence underscores their potential in addressing chronic disease management, enhancing therapeutic outcomes, and promoting circular economy principles in healthcare, though challenges remain regarding large-scale production, cost-effectiveness, and regulatory integration. Overall, sustainable polymers represent a transformative approach to advancing modern medicine while aligning with global sustainability goals. (Table 4.2; Fig. 4.6).

Table 4.2 A summary of the performance vs. applications of sustainable polymers

Industry	Key sustainable polymers	Performance advantages	Common Applications
Packaging	PLA, PHAs, starch blends, bio-PE	High strength, moisture barrier, process compatibility	Food wraps, beverage bottles, cosmetic containers
Automotive & Transportation	Flax/PP composites, recycled PET, bio-based nylons	Lightweight, impact resistance, thermal stability	Dashboards, bumpers, panels
Construction & Infrastructure	Recycled plastics, PVC alternatives, wood-plastic composites	Durability, weather resistance, insulation	Roofing, piping, cladding
Consumer Goods & Electronics	Biodegradable plastics, recycled ABS, cellulose-based polymers	Customizable properties, chemical resistance	Furniture, electronics casings
Medical & Pharmaceutical	PLA, PGA, PCL, PHAs	Biocompatibility, bioresorbability, sterilization safety	Implants, sutures, sterile packaging

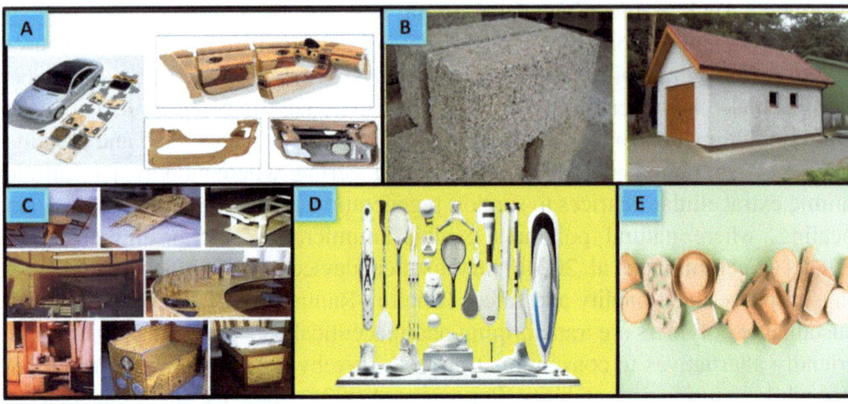

Fig. 4.6 Application of sustainable biopolymers. (**a**) Uses of natural fiber composites in interior automobile components, (**b**) Construction composites composed of flax and hemp shives bound together with Portland cement and lime, (**c**) Composites made of natural fibers for furnishings, (**d**) Composites made of natural fibers for sporting goods, (**e**) Composites made of natural fibers for packaging purposes. (Picture is taken from an open-access journal (Puttegowda 2025))

4.6 Challenges in Sustainable Polymeric Material Development

Sustainable polymeric materials, derived from renewable feedstocks, recycled inputs, or CO_2 utilization pathways, have emerged as promising alternatives to petroleum-based plastics in mitigating the global plastic pollution crisis and enabling circular economy practices. However, despite rapid scientific advancements, their widespread adoption remains constrained by multifaceted challenges that span economic scalability, technical performance, compatibility with existing infrastructure, end-of-life (EoL) management barriers, and regulatory hurdles. From an economic and scalability perspective, sustainable polymers are often more costly to produce than their fossil-based counterparts due to expensive feedstock processing, reliance on specialized catalysts, and energy-intensive manufacturing pathways. For example, biopolymers like polylactic acid (PLA) or cellulose derivatives require additional purification and controlled processing, which inflates production costs. Meanwhile, the fossil-based plastic industry benefits from long-established supply chains and economies of scale that make it more price-competitive (Khandaker et al. 2025). Scaling up also faces barriers from feedstock variability, geographic dependency, and logistical constraints, while solvent- and catalyst-driven recycling methods remain economically prohibitive for widespread adoption (Zhou et al. 2025). Although promising work has explored valorisation of industrial waste streams (e.g., fly ash) to reduce costs (El-kordy et al. 2024). Scaling sustainable polymers for the global market still requires significant cost reduction strategies and infrastructure investments. Equally critical are technical performance limitations, since many bio-based polymers exhibit inferior mechanical, thermal, and barrier properties compared to petroleum-derived plastics. Since these materials are frequently brittle, thermally unstable, and extremely moisture-sensitive, polymers like PLA, starch composites, and chitosan are not suitable for use in industries like construction, biomedicine, or automobiles, where strength and durability are crucial (Iqbal et al. 2023). Even established solutions, such as polymer nanocomposites, have poor interfacial bonding and dispersion issues that limit their ability to improve performance. Low conductivity and poor shape stability are common characteristics of phase change materials and other functional polymers used in energy storage applications (Hu et al. 2025). These drawbacks highlight the need for material design advancements to concentrate on enhancing the inherent qualities of polymers while also guaranteeing their stability in a range of operating conditions. Compatibility with the present framework is another significant challenge. Conventional polymers such as polyethene (PE), polypropylene (PP), and polyethene terephthalate (PET) have essentially been optimized for use in industrial processing lines, moulding equipment, and recycling facilities. It can be challenging to incorporate sustainable alternatives into existing manufacturing processes without incurring expensive retrofitting, since they sometimes call for different processing temperatures, pressures, or extrusion conditions (Han et al. 2021). Although biopolymers degrade differently from traditional plastics, they may contaminate traditional plastic streams in recycling systems, lowering the quality of reclaimed plastics (Yang et al. 2025). Furthermore, safety and regulatory requirements for building and packaging are still mostly geared toward fossil-based plastics, providing deficiencies that sustainable polymers find difficult to bridge (Soussi et al. 2025). Furthermore, bio-based polymer matrices frequently interact poorly with adhesives, coatings, and additives, making it difficult to integrate them into multi-material systems such as multilayer packaging. Barriers to recycling and end-of-life care are equally important, since they provide a major hindrance to attaining circularity. Chemical recycling techniques, while promising, are still expensive and energy-intensive, and mechanical recycling is sometimes inappropriate for bio-based polymers because of heat deterioration and incompatibility with current recycling processes (Saitta et al. 2025). Isolating sustainable polymers from mixed plastic waste sources is challenging due to inadequate waste segregation procedures and a lack of consistent labelling. The environmental promise of biodegradable plastics is undermined when they regularly wind up in incinerators or landfills because of inadequate composting infrastructure and low consumer awareness. Moreover, recyclability and biodegradability have not been sufficiently incorporated into material design specifications by policy and regulatory frameworks, which perceive them as optional rather than necessary characteristics (Machala et al. 2025). These obstacles will keep preventing sustainable polymers from achieving full lifetime sustainability unless they are removed.

Finally, the lack of standardised international standards sometimes causes ambiguity for enterprises trying to incorporate these materials into common applications, making regulatory and certification barriers a significant obstacle. Bio-based and biodegradable polymers confront fragmented certification schemes across areas, especially in biodegradability, composability, and food-contact safety requirements, whereas conventional plastics have the advantages of long-standing regulatory frameworks. Because manufacturers must go through expensive and time-consuming certification procedures specific to each country, this regulatory uncertainty slows down market acceptance and deters investment (Samir et al. 2022). Further, postponements in developing review processes for innovative polymers impede commercialization timeframes, and discrepancies between testing in laboratory standards and real-world settings make compliance issues worse (Sharma et al. 2025). In addition, uneven certification logos and labeling confuse end users, which reduces consumer confidence and the adoption of sustainable polymers in society. In order to enable the quicker, safer, and more dependable integration of sustainable polymeric materials, these obstacles underscore the critical need for global standardization, expedited certification procedures, and increased cooperation between industry players and regulatory agencies (Autenrieth et al. 2015).

A multifaceted, forward-thinking strategy that connects scientific innovation, industry adaptation, and regulatory change is needed to overcome the aforementioned obstacles. Developments in synthetic biology and green chemistry are making it possible to synthesize customized polymers with improved performance while reducing the need for hazardous solvents and energy-intensive procedures. Similarly, computational modeling and AI-driven material discovery can speed up the process of finding bio-based polymer compositions with better mechanical and thermal stability for filling deficiencies in performance. On the industrial front, investment in retrofit-friendly technologies and hybrid processing lines can reduce compatibility barriers, allowing gradual integration of sustainable polymers into existing infrastructure (Fig. 4.7).

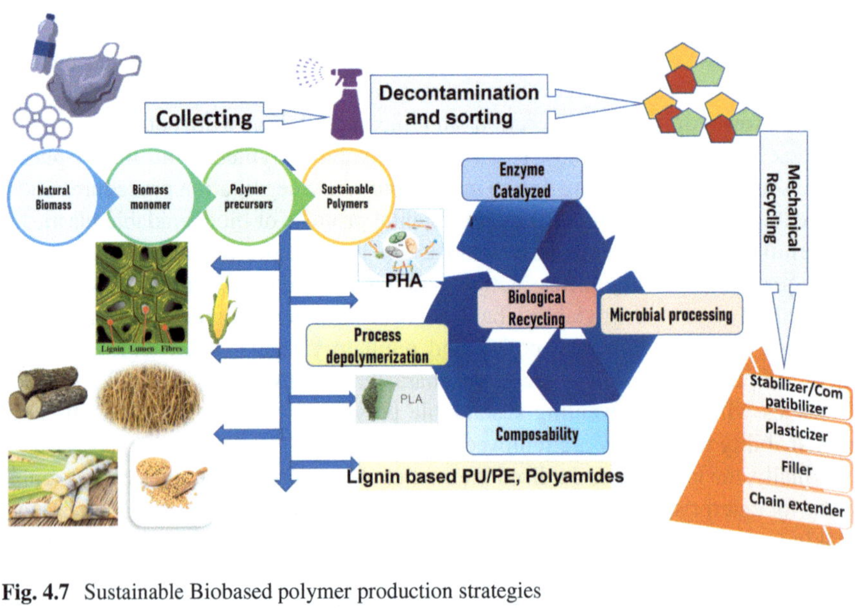

Fig. 4.7 Sustainable Biobased polymer production strategies

4.7 Conclusion and Future Outlook

Tremendous efforts have been devoted to the advanced research and development of sustainable polymer, via integrating biobased renewable monomers into a multifunctional highly tuned polymers for formulating innovative solutions for their biodegradation and recycling. Recent advances in polymer chemistry, biotechnology, computational tools like AI, and life-cycle assessment have significantly expanded our understanding of sustainable materials. However, the complete replacement of petroleum-based polymers with green alternatives across all sectors remains a long-term goal. Future research is expected to focus on designing affordable, multifunctional, recyclable, or compostable polymers, as well as developing high-performance composites from agricultural residues and non-edible resources using energy-efficient, waste-free processes. Achieving this vision will require strong interdisciplinary collaborations among scientists, industries, policymakers, and social groups, along with greater public awareness of sustainability goals.

For researchers to develop effective, adaptable, and sustainable products, major bottlenecks depend on the initial screening of biomass; for this reason, next-generation research focuses on molecular design, recycling techniques, catalytic technology, and waste reduction, followed by appropriate databanking for plastic utilization and applications. Modifying the chemical or mechanical recycling method in order to scale up technology with the highest intensity is a further major issue. There remains scope for enhancing the development of sustainable polymers by using waste materials like lignin, CO_2, and non-edible biomass; implementing green chemistry processes that use less energy and water; creating materials with eco-friendly end-of-life pathways; and upholding the 3R policy, which prioritizes the widespread acceptance of sustainable polymers in our daily lives.

Conflict of Interest Author declare no conflict of interest.

References

Autenrieth B, Jeong H, Forrest WP, Axtell JC, Ota A, Lehr T, Buchmeiser MR, Schrock RR (2015) Stereospecific ring-opening metathesis polymerization (ROMP) of *endo*-dicyclopentadiene by molybdenum and tungsten catalysts. Macromolecules 48(8):2480–2492. https://doi.org/10.1021/acs.macromol.5b00123

Banerjee R, Ray SS (2022) Sustainability and life cycle assessment of thermoplastic polymers for packaging: a review on fundamental principles and applications. Macromol Mater Eng 307(6). https://doi.org/10.1002/mame.202100794

Banerjee R, Bandyopadhyay J, Ray SS (2025) Functional polymers and their nanocomposites for sustainable packaging applications. Macromol Mater Eng. https://doi.org/10.1002/mame.202500130

Bansal KK, Rosenholm JM (2020) Synthetic polymers from renewable feedstocks: an alternative to fossil-based materials in biomedical applications. Ther Deliv 11(5):297–300

Baranwal J, Barse B, Fais A, Delogu GL, Kumar A (2022) Biopolymer: a sustainable material for food and medical applications. Polymers 14(5):983. https://doi.org/10.3390/polym14050983

Beena Unni A, Muringayil Joseph T (2024) Enhancing polymer sustainability: eco-conscious strategies. Polymers 16(13):1769. https://doi.org/10.3390/polym16131769

Bisht G, Zaidi MGH, Rayamajhi S (2017) Supercritical carbon dioxide–assisted synthesis of stimuli-responsive magnetic poly(*N*-isopropylacrylamide)–ferrite biocompatible nano-composites for targeted and controlled drug delivery. Int J Polym Mater Polym Biomater 66(14):708–716. https://doi.org/10.1080/00914037.2016.1263949

Bolton JM, Hillmyer MA, Hoye TR (2014) Sustainable thermoplastic elastomers from terpene-derived monomers. ACS Macro Lett 3(8):717–720. https://doi.org/10.1021/mz500339h

Cardoso V d F, Batista CN d S, Moura ML, da Rocha LVM, Gomes AC d O (2024) Sustainability in polymeric materials: a review on basic concepts, socioeconomic challenges, and innovations in the field of nanotechnology. In: Academic education navigating the path of knowledge. Seven Editora. https://doi.org/10.56238/sevened2023.008-024

Cháirez-Ramírez MH, Díaz-Rivas JO, Contreras-Ramírez JI, González-Laredo RF, Gallegos-Infante JA (2025) Green polymer-based materials as promising therapeutical agents for non-communicable diseases. Polym Int 74(6):483–493. https://doi.org/10.1002/pi.6708

Chen P, Ding Y, Wang Y, Zhao H, Li P, Liu Y, Gao C (2023) Functional bio-based vitrimer with excellent healing and recyclability based on conjugated deflection self-toughening. Chem Eng J 474:145680. https://doi.org/10.1016/j.cej.2023.145680

Daisy P, Saipriya (2012) Biochemical analysis of Cassia fistula aqueous extract and phytochemically synthesized gold nanoparticles as hypoglycemic treatment for diabetes mellitus. Int J Nanomedicine:1189. https://doi.org/10.2147/IJN.S26650

Dinkar VC, Kumar V (2025) A review study on the mechanical behaviour of natural fibre-reinforced epoxy composite. Polym Bull. https://doi.org/10.1007/s00289-025-05902-4

Dutta N, Bhadra B, Gogoi G, Kumar Maji T (2021) Development of polyvinyl chloride/waste rice husk ash/modified montmorillonite nanocomposite using epoxidized soybean oil as green additive substituting synthetic plasticizer and compatibiliser. Cleaner Mater 2:100033. https://doi.org/10.1016/j.clema.2021.100033

El-kordy A, Elgamouz A, Khalil AKA, Tijani N, Kawde A-N, Laoui T (2024) Synthesis of clay/Fe/ZSM-5 composite membrane for chromium (vi) removal from aqueous media and its electrochemical performance. J Environ Chem Eng 12(4):113235. https://doi.org/10.1016/j.jece.2024.113235

Erickson E, Shakespeare TJ, Bratti F, Buss BL, Graham R, Hawkins MA, König G, Michener WE, Miscall J, Ramirez KJ, Rorrer NA, Zahn M, Pickford AR, McGeehan JE, Beckham GT (2022) Comparative performance of PETase as a function of reaction conditions, substrate properties, and product accumulation. ChemSusChem 15(1). https://doi.org/10.1002/cssc.202101932

Geir Asheim (1994) Sustainability: ethical foundations and economic properties. The World Bank

Golewski GL (2021) Green concrete based on quaternary binders with significant reduced of CO_2 emissions. Energies 14(15):4558. https://doi.org/10.3390/en14154558

Haafiz MKM, Hassan A, Zakaria Z, Inuwa IM, Islam MS, Jawaid M (2013) Properties of polylactic acid composites reinforced with oil palm biomass microcrystalline cellulose. Carbohydr Polym 98(1):139–145. https://doi.org/10.1016/j.carbpol.2013.05.069

Hakamy A, Shaikh FUA, Low IM (2018) High-performance natural fiber–reinforced cement composites. In: Advances in ceramic matrix composites. Elsevier, pp 277–305. https://doi.org/10.1016/B978-0-08-102166-8.00012-8

Han R, Liu Y, Shi J, Chen G-X, Li Q (2021) Enhanced dielectric properties and breakdown strength of polymer/carbon nanotube composites by coating an $SrTiO_3$ layer. E-Polymers 21(1):272–278. https://doi.org/10.1515/epoly-2021-0028

Hauschild MZ, Rosenbaum RK, Olsen SI (2018) Life cycle assessment: theory and practice. Springer

Hu L, Zhang L, Cui W, An Q, Ma T, Wang Q, Mai L (2025) Carbon-based porous materials for performance-enhanced composite phase change materials in thermal energy storage: materials, fabrication and applications. J Mater Sci Technol 210:204–226. https://doi.org/10.1016/j.jmst.2024.05.035

Ibrahim MI, Rejab MRM, Alang NA, Rani MF (2025) Lightweight hybrid composite B-pillars for automotive applications; flexural performance evaluation. J Phys Conf Ser 2933(1):012002. https://doi.org/10.1088/1742-6596/2933/1/012002

Iqbal Y, Ahmed I, Irfan MF, Chatha SAS, Zubair M, Ullah A (2023) Recent advances in chitosan-based materials; the synthesis, modifications and biomedical applications. Carbohydr Polym 321:121318. https://doi.org/10.1016/j.carbpol.2023.121318

Jahangirian H, Ghasemian lemraski E, Webster TJ, Rafiee-Moghaddam R, Abdollahi Y (2017) A review of drug delivery systems based on nanotechnology and green chemistry: green nanomedicine. Int J Nanomedicine 12:2957–2978. https://doi.org/10.2147/IJN.S127683

Joseph B, Krishnan S, Kavil SV, Pai AR, James J, Kalarikkal N, Thomas S (2021) Green chemistry approach for fabrication of polymer composites. Sustain Chem 2(2):254–270. https://doi.org/10.3390/suschem2020015

Kerton FM (2025) Applying the principles of green chemistry to achieve a more sustainable polymer life cycle. Curr Opin Green Sustain Chem 51:100996. https://doi.org/10.1016/j.cogsc.2024.100996

Khan SH (2020) Green nanotechnology for the environment and sustainable development, pp 13–46. https://doi.org/10.1007/978-3-030-17724-9_2

Khandaker T, Islam T, Nandi A, Anik MAAM, Hossain MS, Hasan MK, Hossain MS (2025) Biomass-derived carbon materials for sustainable energy applications: a comprehensive review. Sustain Energy Fuels 9(3):693–723. https://doi.org/10.1039/D4SE01393J

Kobayashi S (2017) Green polymer chemistry: new methods of polymer synthesis using renewable starting materials. Struct Chem 28(2):461–474. https://doi.org/10.1007/s11224-016-0861-3

Kugler S, Ossowicz P, Malarczyk-Matusiak K, Wierzbicka E (2019) Advances in rosin-based chemicals: the latest recipes, applications and future trends. Molecules 24(9):1651. https://doi.org/10.3390/molecules24091651

Kumar A, Moolemane R, Rajendran T, Krishna SBN (2025) Nanocellulose: a comprehensive review of sustainable applications and innovations. J Renew Mater 13(7):1315–1346. https://doi.org/10.32604/jrm.2025.02024-0050

Kurkin E, Kishov E, Chertykovtseva V, Alekseev V (2025) Bimaterial topology optimization for Ti-6Al-4V lattice bushing design in PA6 injection molded structures. J Manuf Process 141:1368–1384. https://doi.org/10.1016/j.jmapro.2025.03.012

Lee M-H, Teng K-H, Liang Y-Y, Ding C-F, Chen Y-C (2025) Flexible biodegradable wearables based on conductive leaf networks. Sustain Mater Technol 43:e01263. https://doi.org/10.1016/j.susmat.2025.e01263

Liu J, Luo Z, Zhang J, Luo T, Zhou J, Zhao X, Cai K (2016) Hollow mesoporous silica nanoparticles facilitated drug delivery via cascade pH stimuli in tumor microenvironment for tumor therapy. Biomaterials 83:51–65. https://doi.org/10.1016/j.biomaterials.2016.01.008

Luo T, Hu Y, Zhang M, Jia P, Zhou Y (2025) Recent advances of sustainable and recyclable polymer materials from renewable resources. Resour Chem Mater 4(2):100085. https://doi.org/10.1016/j.recm.2024.10.004

Machala ML, Chen X, Bunke SP, Forbes G, Yegizbay A, de Chalendar JA, Azevedo IL, Benson S, Tarpeh WA (2025) Life cycle comparison of industrial-scale lithium-ion battery recycling and mining supply chains. Nat Commun 16(1):988. https://doi.org/10.1038/s41467-025-56063-x

Mathur M (2025) Sustainable consumer behaviour in emerging markets. In: Marketing to the base of the pyramid. Springer Nature, Singapore, pp 171–187. https://doi.org/10.1007/978-981-96-7105-2_9

Mohanty AK, Wu F, Mincheva R, Hakkarainen M, Raquez J-M, Mielewski DF, Narayan R, Netravali AN, Misra M (2022) Sustainable polymers. Nat Rev Methods Primers 2(1):46. https://doi.org/10.1038/s43586-022-00124-8

Narayanan D, Giffin MT, Neetha PS, Jose C, Thomas S (2023) Natural fiber-reinforced biopolymers as construction materials. In: Handbook of biopolymers. Springer Nature, Singapore, pp 1173–1193. https://doi.org/10.1007/978-981-19-0710-4_43
Nassar MMA, Alzebdeh KI, Pervez T, Al-Hinai N, Munam A (2021) Progress and challenges in sustainability, compatibility, and production of eco-composites: a state-of-art review. J Appl Polym Sci 138(43). https://doi.org/10.1002/app.51284
Niraimathi KL, Sudha V, Lavanya R, Brindha P (2013) Biosynthesis of silver nanoparticles using Alternanthera sessilis (Linn.) extract and their antimicrobial, antioxidant activities. Colloids Surf B: Biointerfaces 102:288–291. https://doi.org/10.1016/j.colsurfb.2012.08.041
Pathak AK, Dhakate SR (2022) Carbon nanomaterial-carbon fiber hybrid composite for lightweight structural composites in the aerospace industry: synthesis, processing, and properties. In: Advanced composites in aerospace engineering applications. Springer, pp 445–470. https://doi.org/10.1007/978-3-030-88192-4_23
Pereira VA, Mendonça PV, Coelho JFJ, Serra AC (2019) Liquid salts as eco-friendly solvents for atom transfer radical polymerization: a review. Polym Chem 10(36):4904–4913. https://doi.org/10.1039/C9PY00865A
Popović K, Živanović S, Jevtić I (2024) Biopolymer packaging materials in the pharmaceutical industry. AIDASCO Rev 2(1):46–56. https://doi.org/10.59783/aire.2024.43
Puttegowda M (2025) Eco-friendly composites: exploring the potential of natural fiber reinforcement. Discov Appl Sci 7(5):401. https://doi.org/10.1007/s42452-025-06981-8
Raja P, Murugan V, Ravichandran S, Behera L, Mensah RA, Mani S, Kasi A, Balasubramanian KBN, Sas G, Vahabi H, Das O (2023) A review of sustainable bio-based insulation materials for energy-efficient buildings. Macromol Mater Eng 308(10). https://doi.org/10.1002/mame.202300086
Saitta L, Dattilo S, Rizzo G, Tosto C, Blanco I, Ferrari F, Carallo GA, Cafaro F, Greco A, Cicala G (2025) Chemical recycling of bio-based epoxy matrices based on precursors derived from waste flour: recycled polymers characterization. Polymers 17(3):335. https://doi.org/10.3390/polym17030335
Samir A, Ashour FH, Hakim AAA, Bassyouni M (2022) Recent advances in biodegradable polymers for sustainable applications. Npj Mater Degrad 6(1):68. https://doi.org/10.1038/s41529-022-00277-7
Shahzad A (2015) Mechanical properties of eco-friendly polymer nanocomposites, pp 527–559. https://doi.org/10.1007/978-81-322-2470-9_18
Sharma R, Verma N, Dewangan R, Sharma AK, Tripathi M, Pani B (2025) Delineating the techniques employed for the fabrication of sustainable polymers *via* 3D printing phenomena. In: Sustainable 3D printing for innovative biopolymer production and applications. Wiley, pp 25–47. https://doi.org/10.1002/9781119792314.ch2
Singh A, Srivastava AK, Singh G, Singh AD, Singh HK, Kumar A, Singh GK (2023) Utilization of plastic waste for developing composite bricks and enhancing mechanical properties: a review on challenges and opportunities. Adv Polym Technol 2023:1–24. https://doi.org/10.1155/2023/6867755
Siracusa V (2019) Microbial degradation of synthetic biopolymers waste. Polymers 11(6):1066. https://doi.org/10.3390/polym11061066
Soheilmoghaddam M, Wahit MU, Tuck Whye W, Ibrahim Akos N, Heidar Pour R, Ali Yussuf A (2014) Bionanocomposites of regenerated cellulose/zeolite prepared using environmentally benign ionic liquid solvent. Carbohydr Polym 106:326–334. https://doi.org/10.1016/j.carbpol.2014.02.085
Soussi MAAB, El Madsia N, Zaki C, Ramadan A, Saker L, Ibrahim M (2025) Analytic hierarchy process geographic information system based model for sustainable construction and demolition waste landfill site selection. Int J Electr Comput Eng (IJECE) 15(1):803. https://doi.org/10.11591/ijece.v15i1.pp803-816
Tao P, Ni G, Song C, Shang W, Wu J, Zhu J, Chen G, Deng T (2018) Solar-driven interfacial evaporation. Nat Energy 3(12):1031–1041. https://doi.org/10.1038/s41560-018-0260-7

Valente BFA, Silvestre AJD, Neto CP, Vilela C, Freire CSR (2022) Improving the processability and performance of micronized fiber-reinforced green composites through the use of biobased additives. Polymers 14(17):3451. https://doi.org/10.3390/polym14173451
Wang J, Chen Z, Guan AQ, Demarquette NR, Naguib HE (2020) Ionic liquids facilitated dispersion of chitin nanowhiskers for reinforced epoxy composites. Carbohydr Polym 247:116746. https://doi.org/10.1016/j.carbpol.2020.116746
Wazeer A, Das A, Abeykoon C, Sinha A, Karmakar A (2023) Composites for electric vehicles and automotive sector: a review. Green Energy Intell Transp 2(1):100043. https://doi.org/10.1016/j.geits.2022.100043
Weerarathna IN, Kumar P, Luharia A, Mishra G (2024a) Sustainable biomaterials for pharmaceutical and medical applications. Multidiscip Rev 8(5):2025141. https://doi.org/10.31893/multirev.2025141
Weerarathna IN, Kumar P, Luharia A, Mishra G (2024b) Sustainable biomaterials for pharmaceutical and medical applications. https://doi.org/10.20944/preprints202409.2108.v1
Xie C, Cao M, Khan M, Yin H, Guan J (2021) Review on different testing methods and factors affecting fracture properties of fiber reinforced cementitious composites. Constr Build Mater 273:121766. https://doi.org/10.1016/j.conbuildmat.2020.121766
Xu Y, Fu P, Dai S, Zhang H, Bi L, Jiang J, Chen Y (2021) Catalyst-free self-healing fully bio-based vitrimers derived from tung oil: strong mechanical properties, shape memory, and recyclability. Ind Crop Prod 171:113978. https://doi.org/10.1016/j.indcrop.2021.113978
Yang G, Lam E, Moores A (2023) Controlled chitosan molecular weight reduction by mechanochemical and aging-based phosphoric acid hydrolysis. ACS Sustain Chem Eng 11(20):7765–7774. https://doi.org/10.1021/acssuschemeng.3c00367
Yang N, Zhou Z, Zhu X, Wu J, Zhang Y, Wang T, Wu X, Tian C, Jiang X, Dai S (2025) Sulfur conversion to donor-acceptor ladder polymer networks through mechanochemical nucleophilic aromatic substitution for efficient CO_2 photoreduction. Angew Chem 137(20). https://doi.org/10.1002/ange.202419108
Yuntawattana N, Buaban T, Siri T (2025) Sustainable active packaging from on-demand degradable PLA/PBAT and Zn-doped TiO_2 composites. ACS Omega 10(3):2931–2939. https://doi.org/10.1021/acsomega.4c09138
Zeng Y, Yang B, Luo Z, Pan X, Ning Z (2022) Fully rosin-based epoxy vitrimers with high mechanical and thermostability properties, thermo-healing and closed-loop recycling. Eur Polym J 181:111643. https://doi.org/10.1016/j.eurpolymj.2022.111643
Zhou B, Su H, Liu W, Zhu Z, Wang L, Qi T (2025) Solvent extraction of metal ions from the leaching solutions of waste lithium-ion battery materials: a review. Sep Purif Technol 354:129173. https://doi.org/10.1016/j.seppur.2024.129173
Zhu Y, Romain C, Williams CK (2016) Sustainable polymers from renewable resources. Nature 540(7633):354–362. https://doi.org/10.1038/nature21001

Chapter 5
Thermoelectric Materials for Sustainable Energy Conversion: Insights from Defect Engineering Approaches

Savita, Sarva Shakti Singh, Punit K. Dhawan, Dhirendra K. Chaudhary, Ranjit Kumar, Nilesh Kumar Rai, and Hari. P. Bhasker

Abstract Thermoelectric (TE) materials convert heat directly into electricity, and the efficiency of TE materials is measured by the figure of merit (ZT). This chapter explores how defect engineering improves thermoelectric performance by tuning electrical conductivity, Seebeck coefficient, and lattice thermal conductivity across inorganic, polymers, perovskite, and hybrid materials, etc. Defects such as vacancies, dislocations, grain boundaries, and nano-precipitates play a crucial role in tailoring the electronic band structure, carrier mobility, and phonon scattering mechanisms, thereby enhancing both the thermoelectric performance (high ZT) and mechanical stability of materials. This chapter also highlights recent advances in

Savita
Department of Physics, Prof. Rajendra Singh (Rajju Bhaiya) Institute of Physical Sciences for Study and Research, V. B. S. Purvanchal University, Jaunpur, India

Centre for Renewable Energy, Prof.Rajendra Singh (RajjuBhaiya) Institute of Physical Sciences for Study and Research, V. B. S. Purvanchal University, Jaunpur, India

S. S. Singh · N. K. Rai · H. P. Bhasker (✉)
Department of Physics, Chaudhary Mahadeo Prasad Degree College, University of Allahabad, Allahabad, India

P. K. Dhawan
Department of Physics, Prof. Rajendra Singh (Rajju Bhaiya) Institute of Physical Sciences for Study and Research, V. B. S. Purvanchal University, Jaunpur, India

D. K. Chaudhary (✉)
Centre for Renewable Energy, Prof.Rajendra Singh (RajjuBhaiya) Institute of Physical Sciences for Study and Research, V. B. S. Purvanchal University, Jaunpur, India

Interdisciplinary Centre for Water and Energy Research, V. B. S. Purvanchal University, Jaunpur, India

R. Kumar
Department of Physics, Arignar Anna Government Arts and Science College, Karaikal, Puducherry, India

Department of Physics, Dr. Kalaignar M. Karunanidhi Government Institute for Postgraduate Studies and Research, Karaikal (UT of Puducherry), India

S. K. Verma et al. (eds.), *Sustainable and Natural Polymers*, Clean Energy Production Technologies, https://doi.org/10.1007/978-981-92-0092-4_5

computational modeling and high-resolution characterization techniques—such as Transmission Electron Microscopy (TEM), synchrotron-based X-ray analysis, and positron annihilation spectroscopy—which provide valuable insights into how defects alter local atomic structures and electronic states. The discussion concludes with future perspectives that emphasize the integration of multiscale modeling, in-situ experimental studies, and data-driven materials design to develop efficient, durable, and sustainable thermoelectric materials for waste-heat recovery applications.

5.1 Introduction

The rapid rise in energy consumption and growing demand have compelled us to explore alternative energy sources that are affordable, renewable, and environmentally clean, thereby reducing our reliance on carbon-based fuels. Thermoelectric (TE) materials are operated on the principle of the Seebeck effect (Bell 2008; Champier 2017; Chandra et al. 2021; Mahan and Sofo 1996; Rowe 2018). TE materials provide an eco-friendly and efficient means of converting waste heat or natural thermal energy into electricity, leveraging the Seebeck effect to transform temperature gradients into electrical voltage. The performance of TE materials is quantified by the dimensionless figure of merit ($ZT = (\sigma S^2/k)T$), where σ is electrical conductivity, k is thermal conductivity, S-Seebeck coefficient, and T is temperature. High TE efficiency requires materials of unique combination with large Seebeck coefficient, superior electrical, and low thermal conductivity (Kumar et al. 2021; Liu et al. 2015; Somdock et al. 2019; Tran et al. 2017). Therefore, to enhance the figure of merit (ZT) of thermoelectric materials, an effective strategy is highly desirable to be adopted. In this regard, numerous approaches have been proposed to improve the power factor ($S^2\sigma$) of thermoelectric materials (Mahan and Sofo 1996; Kumar et al. 2021; Liu et al. 2015; Somdock et al. 2019; Tran et al. 2017; Snyder and Toberer 2008; Sootsman et al. 2009; Tabaie and Omidvar 2023; Tritt 2011).

Bismuth telluride (Bi_2Te_3) and its alloys are among the most well-established thermoelectric materials (Rowe 2018; Tritt 2011; Goldsmid 2018). Their heavy constituent elements—bismuth and tellurium—contribute to low phonon velocity, low thermal conductivity, small band gaps, and high charge mobility, making them ideal for thermoelectric applications. Other materials, such as conducting oxides and skutterudites, have also been explored but face challenges in growth and power conversion (Nolas et al. 1999). Although inorganic materials like bismuth telluride Bi_2Te_3 offer high performance, they are costly and brittle, limiting large-scale use. In contrast, organic materials are flexible, inexpensive, and possess low intrinsic thermal conductivity, making them promising candidates for thermoelectric generators (Tritt 2011; Goldsmid 2018; Rowe 2018).

Recent advances in organic–inorganic halides have opened opportunities for flexible electronic devices that can harvest waste heat (Xu et al. 2020). Conducting polymers such as PEDOT: PSS are particularly attractive due to their tunable electrical and Seebeck properties, low thermal conductivity, and water solubility—key factors driving their use in hybrid flexible thermoelectric systems. Recently, flexible polymer -based thermoelectric materials such as PEDOT:PSS (poly(3,4-ethylenedioxythiophene):poly(styrene sulfonate)) (Zhang et al. 2023; Du et al. 2017), PANI (polyaniline) (Panigrahy and Kandasubramanian 2020), PEI (polyethyleneimine), PTH (polythiophene), and P_3HT (poly(3-hexylthiophene)) (Hong et al. 2015) have attracted significant attention due to their excellent flexibility, lightweight nature, low thermal conductivity, and environmental friendliness. However, these materials typically suffer from high electrical resistivity (ρ) and a low power factor (PF), which limits their overall thermoelectric efficiency.

Currently, a diverse range of inorganic compounds—such as Zintl phases (Zhou et al. 2020), clathrates (Kishimoto et al. 2021), Half-Heusler alloys (Yan et al. 2022), etc., have exhibited promising thermoelectric performance, achieving high figures of merit (ZT). Among them, bismuth telluride (Bi_2Te_3), tin selenide (SnSe) (Chandra et al. 2021), lead telluride (PbTe) (Heremans et al. 2008), etc., have been extensively studied and widely employed due to their superior *ZT* values, often exceeding unity. Despite these advantages, conventional inorganic thermoelectric materials face several challenges, including high production cost, toxicity, elevated intrinsic thermal conductivity, limited thermal stability at high temperatures, and complex synthesis routes. In contrast, organic thermoelectric materials offer a cost-effective and environmentally sustainable alternative. Their low toxicity, lightweight nature, mechanical flexibility, and compatibility with low-temperature, solution-based fabrication techniques (such as printing and coating) make them particularly attractive for flexible and wearable applications. Moreover, their inherently low thermal conductivity (~0.9 Wm^{-1} K^{-1}) further enhances their thermoelectric potential. However, to compete effectively with other renewable energy technologies, thermoelectric devices must achieve *ZT* values exceeding 3. To date, even with advanced material processing, the room-temperature *ZT* has only modestly improved from about 0.6 to 1.0 over the past five decades. Defects such as vacancies, dislocations, grain boundaries, and nanoprecipitates modify band structures, carrier mobility, and phonon scattering, helping achieve high ZT and mechanical stability (Hu et al. 2014; Hu 2023; Kalbacova et al. 2019; Wu et al. 2024; Zheng et al. 2021; Mukherjee et al. 2022).

This chapter examines how defect engineering can be employed to tailor and enhance the thermoelectric performance of advanced materials for practical thermoelectric applications.

5.2 Role of Structural Defects in Shaping Thermoelectric Performance

Structural defects play a pivotal role in determining the overall thermoelectric efficiency of materials by simultaneously influencing electronic and phononic transport processes (Hu et al. 2014; Hu 2023; Kalbacova et al. 2019; Wu et al. 2024; Zheng et al. 2021; Mukherjee et al. 2022; Li et al. 2016; Liu et al. 2016; Zhao et al. 2020). Defects such as vacancies, interstitials, antisite substitutions, and dislocations disrupt the ideal periodicity of the crystal lattice, thereby modifying the electronic band structure, carrier concentration and mobility, and phonon scattering mechanisms (Hu et al. 2014; Hu 2023; Kalbacova et al. 2019; Wu et al. 2024; Zheng et al. 2021; Mukherjee et al. 2022; Li et al. 2016; Liu et al. 2016; Zhao et al. 2020). Through careful engineering of these imperfections, it becomes possible to achieve an optimal balance among the three interdependent thermoelectric parameters—high electrical conductivity (σ), large Seebeck coefficient (S), and low lattice thermal conductivity (k_l) which collectively define the figure of merit, $ZT = (\sigma S^2/k)T$.

Figure 5.1 illustrates the multifaceted role of defects in governing thermoelectric performance through the introduction of new degrees of freedom in materials. Traditionally, thermoelectric efficiency—expressed through parameters such as Seebeck coefficient (S), electrical conductivity (σ), and thermal conductivity (k)

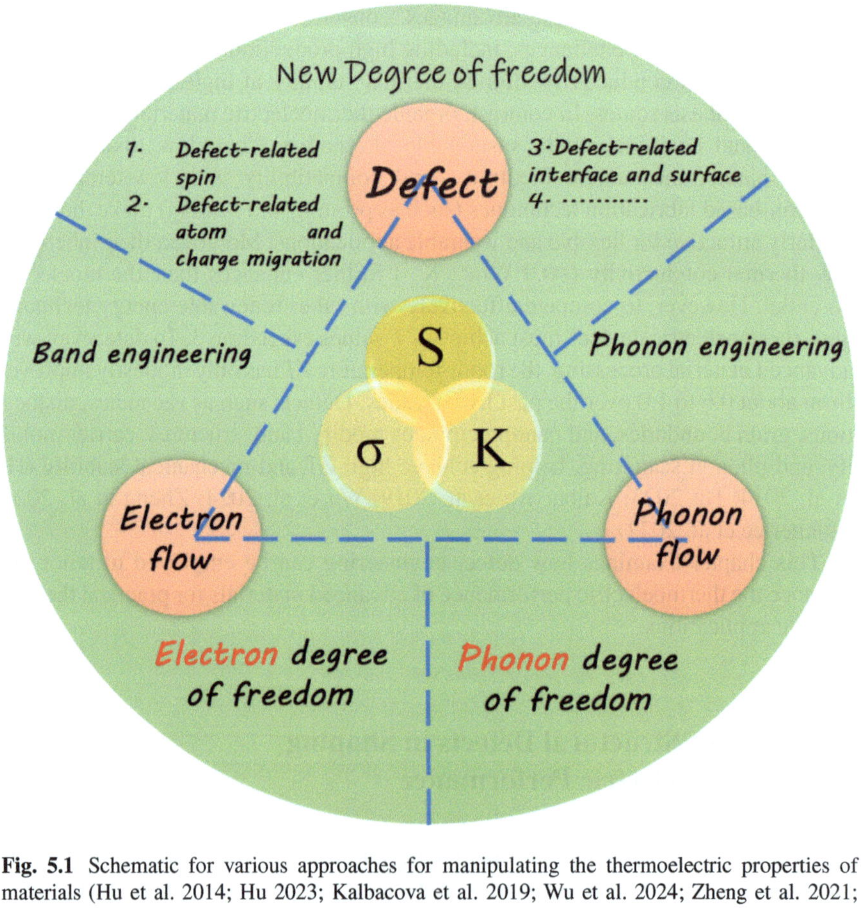

Fig. 5.1 Schematic for various approaches for manipulating the thermoelectric properties of materials (Hu et al. 2014; Hu 2023; Kalbacova et al. 2019; Wu et al. 2024; Zheng et al. 2021; Mukherjee et al. 2022; Li et al. 2016; Liu et al. 2016; Zhao et al. 2020)

depends on the coupled transport of electrons and phonons. Here, defects emerge as a central controlling factor that bridges electron flow and phonon flow, enabling both band engineering and phonon engineering strategies. Defects influence electronic and lattice transport by modifying band structures, scattering mechanisms, and local atomic arrangements. The above diagram categorizes these effects into distinct contributions (Hu et al. 2014; Hu 2023; Kalbacova et al. 2019; Wu et al. 2024; Zheng et al. 2021; Mukherjee et al. 2022; Li et al. 2016; Liu et al. 2016; Zhao et al. 2020):

(i) Defect-related spin degrees of freedom
(ii) Defect-related atomic and charge migration
(iii) Defect-related interfaces and surfaces

Through such defect-mediated interactions, materials gain new control parameters (degrees of freedom) that can be tailored to optimize transport processes, reduce lattice thermal conductivity, and enhance power factor ($S^2\sigma$) (Hu et al. 2014; Hu 2023; Kalbacova et al. 2019; Wu et al. 2024; Zheng et al. 2021; Mukherjee et al. 2022; Li et al. 2016; Liu et al. 2016; Zhao et al. 2020).

By deliberately introducing and controlling defects, researchers can fine-tune carrier concentration, enhance energy filtering, and suppress heat flow—achieving materials with high power factors and ultralow thermal conductivity (Hu et al. 2014; Hu 2023; Kalbacova et al. 2019; Wu et al. 2024; Zheng et al. 2021; Mukherjee et al. 2022; Li et al. 2016; Liu et al. 2016; Zhao et al. 2020). Defect engineering thus forms the foundation for next-generation thermoelectric materials offering superior efficiency and structural stability. Defects (i.e. point, extended, or interfacial, etc.) modify local potentials, mass distribution, and strain, thereby influencing carrier mobility, concentration, and phonon transport. Their characteristics are often described using Kröger–Vink notation (Hu et al. 2014; Hu 2023). Since electrical conductivity ($\sigma = ne\mu$) depends on carrier density (n) and mobility (μ), precise optimization of defect type and concentration is essential; excessive or compensating defects can trap carriers and degrade mobility. Properly balanced defect control ensures the ideal interplay between electrical and thermal properties for enhanced thermoelectric performance. The Seebeck coefficient, another essential parameter, is related to the position of the Fermi level and can be expressed as (Hu et al. 2014; Hu 2023; Kalbacova et al. 2019; Wu et al. 2024; Zheng et al. 2021)

$$S = \frac{8\pi^2 k_B^2 T}{3eh^2} m_{\mathrm{DOS}}^* \left(\frac{\pi}{3n}\right)^{2/3} \tag{5.1}$$

where m_{DOS}^* is the density-of-states effective mass. By controlling the defect concentration, it is possible to position the Fermi level optimally within the electronic bands, thereby maximizing the power factor ($S^2\sigma$).

In addition to their influence on electronic properties, defects play a dominant role in suppressing the lattice component of thermal conductivity, κ_L, by scattering phonons of various wavelengths. According to the Callaway–Klemens model, the phonon relaxation rate for point-defect scattering is given by

$$\frac{1}{\tau_{\mathrm{pd}}} = A\omega^4, \quad A = \frac{V_0}{4\pi v^3}\Gamma \tag{5.2}$$

where V_0 is the atomic volume, v is the sound velocity, and Γ is the scattering parameter defined as

$$\Gamma = \sum_i c_i \left[\left(\frac{\Delta M}{M} \right)^2 + \varepsilon \left(\frac{\Delta r}{r} \right)^2 \right] \tag{5.3}$$

Here c_i is the concentration of defect species, and ΔM and Δr denote the differences in atomic mass and ionic radius between defect atoms and host atoms. This expression shows that even a small concentration of substitutional defects with large mass or strain contrast can substantially reduce κ_L.

Defect engineering strategies have been widely explored in various classes of thermoelectric materials (Hu et al. 2014; Hu 2023; Kalbacova et al. 2019; Wu et al. 2024; Zheng et al. 2021; Mukherjee et al. 2022; Li et al. 2016; Liu et al. 2016; Zhao et al. 2020). In inorganic semiconductors such as PbTe, SnTe, and SnSe, intentional alloying and resonant doping create strong phonon scattering centers while enhancing the Seebeck response. Sodium-doped SnSe, for example, exhibits a remarkably high figure of merit of about 2.6 at 923 K due to optimized carrier concentration and suppressed thermal conductivity. In Mg_3Sb_2, careful control of the Mg-rich synthesis environment minimizes compensating magnesium vacancies, while tellurium substitution provides additional electron donors, yielding a ZT value exceeding 1.5 at 773 K. In half-Heusler compounds like TiNiSn or (Ti, Zr, Hf)NiSn, antisite disorder and multielement alloying generate complex mass fluctuations that scatter phonons effectively and promote band convergence (Hu et al. 2014; Hu 2023; Kalbacova et al. 2019; Wu et al. 2024; Zheng et al. 2021; Mukherjee et al. 2022; Li et al. 2016; Liu et al. 2016; Zhao et al. 2020).

5.3 Defect-Related Thermoelectric Transport Mechanisms

Understanding carrier and phonon transport is key to explaining how defects govern thermoelectric performance. Carrier transport determines electrical conductivity and the Seebeck coefficient, while phonon transport controls heat conduction—both strongly influenced by defects that modify carrier concentration, mobility, and scattering (Hu et al. 2014; Hu 2023; Kalbacova et al. 2019; Wu et al. 2024; Zheng et al. 2021; Mukherjee et al. 2022; Li et al. 2016; Liu et al. 2016; Zhao et al. 2020). Models like the Single Parabolic Band and Kane Band describe how vacancies, dopants, or dilocations tune these effects (Sun et al. 2019, 2021; Hu et al. 2014; Hu

2023; Kalbacova et al. 2019; Wu et al. 2024; Zheng et al. 2021; Mukherjee et al. 2022; Li et al. 2016; Liu et al. 2016; Zhao et al. 2020). Ultimately, defect engineering offers a powerful means to balance electron flow and phonon scattering, enabling optimized electrical and thermal transport for high thermoelectric efficiency (Zheng et al. 2021). Carrier transport behavior is closely linked to the electronic band structure, and models such as the Single Parabolic Band (SPB) and Kane Band (KB) models are often used to describe it (Hu et al. 2014; Hu 2023; Kalbacova et al. 2019; Wu et al. 2024; Zheng et al. 2021; Mukherjee et al. 2022; Li et al. 2016; Liu et al. 2016; Zhao et al. 2020). Within the SPB framework, carrier transport parameters can be expressed as:

$$S = \pm \frac{k_B}{e}\left[(2+\lambda)\frac{F_{1+\lambda}(\eta)}{(1+\lambda)F_\lambda(\eta)} - \eta\right] \tag{5.4}$$

$$n = 4\pi\left(\frac{2m^* k_B T}{h^2}\right)^{3/2} F_\lambda(\eta) \tag{5.5}$$

$$= \frac{2e^3\tau_0(1+\lambda)}{m^*}(k_B T)^{\lambda-1/2}\frac{F_\lambda(\eta)}{F_{1/2}(\eta)}, \quad \sigma = ne\mu \tag{5.6}$$

$$\mu = \frac{2e^3\tau_0(1+\lambda)}{m^*}(k_B T)^{\lambda-1/2}\frac{F_\lambda(\eta)}{F_{1/2}(\eta)}, \quad \sigma = ne\mu \tag{5.7}$$

$$L = \left(\frac{k_B}{e}\right)^2\left[(3+\lambda)\frac{F_{\lambda+2}(\eta)}{(1+\lambda)F_\lambda(\eta)} - \left((2+\lambda)\frac{F_{1+\lambda}(\eta)}{(1+\lambda)F_\lambda(\eta)}\right)^2\right] \tag{5.8}$$

where $F_i(\eta)$ is the Fermi integral, m^* is the effective mass, λ is the scattering parameter, τ_0 is the relaxation time, and L is the Lorenz number. These parameters link fundamental material characteristics, such as effective mass and scattering mechanisms, to measurable thermoelectric properties.

5.4 Effect on Carrier Scattering and Transport

Carrier scattering in thermoelectric materials originates from interactions with lattice vibrations, impurities, and structural defects (Hu et al. 2014; Hu 2023; Kalbacova et al. 2019; Wu et al. 2024; Zheng et al. 2021; Mukherjee et al. 2022; Li et al. 2016; Liu et al. 2016; Zhao et al. 2020). Through deliberate defect engineering, these scattering mechanisms can be precisely tuned to control carrier concentration and mobility. Appropriate doping introduces additional charge carriers, enhancing electrical conductivity, while engineered point and interfacial defects can minimize boundary scattering and improve carrier transport. Achieving an optimal balance between carrier generation and scattering is therefore critical for maximizing electrical conductivity and overall thermoelectric efficiency.

Thermal transport in solids is primarily governed by electrons and phonons, with photon-mediated contributions being negligible within the typical operating temperature range of thermoelectric devices (Hu et al. 2014; Hu 2023; Kalbacova et al. 2019; Wu et al. 2024; Zheng et al. 2021; Mukherjee et al. 2022; Li et al. 2016; Liu et al. 2016; Zhao et al. 2020). Accordingly, the total thermal conductivity (κ) of a thermoelectric material is expressed as the sum of the electronic (κ_e) and lattice thermal conductivity (κ_L) components. Controlled phonon scattering through defect engineering thus provides an effective means to tailor heat transport while maintaining favourable electronic conduction, ultimately enhancing thermoelectric performance. The thermal conductivity can be expressed as

$$\kappa = \kappa_e + \kappa_L, \quad \kappa_e = L\sigma T \tag{5.9}$$

At high temperatures, intrinsic excitation generates electron–hole pairs, leading to bipolar diffusion, which increases thermal conductivity and reduces the Seebeck coefficient. The bipolar contribution (κ_B) is expressed as:

$$\kappa_B = \frac{\sigma_h \sigma_e}{\sigma_h + \sigma_e}\left(S_h - S_e\right)^2 T \tag{5.10}$$

where σ_h and σ_e are hole and electron conductivities, respectively, and S_h, S_e are their corresponding Seebeck coefficients. To suppress bipolar diffusion, heavy doping is often employed to maintain single-carrier conduction (Hu et al. 2014; Hu 2023; Kalbacova et al. 2019; Wu et al. 2024; Zheng et al. 2021; Mukherjee et al. 2022; Li et al. 2016; Liu et al. 2016; Zhao et al. 2020).

The lattice thermal conductivity is primarily governed by phonon transport and can be expressed as:

$$\kappa_{\mathrm{L}} = \frac{1}{3} C_V v l = \frac{1}{3} C_{\mathrm{V}} v^2 \tau \tag{5.11}$$

where C_V is the specific heat, v is the phonon velocity, l is the mean free path, and τ is the phonon relaxation time. A shorter τimplies stronger phonon scattering and consequently lower κ_L. A more rigorous description of lattice heat conduction is given by the Debye–Callaway model:

$$\kappa_L = \frac{k_B^4 T^3}{2\pi^2 v \hbar^3} \int_0^{\theta_D/T} \frac{\zeta^4 e^{\zeta}}{\left(e^{\zeta} - 1\right)^2} \tau_t\left(\zeta\right) \mathrm{d}\zeta \tag{5.12}$$

where θ_D is the Debye temperature, $\zeta = \hbar\omega/k_BT$, and τ_t represents the total phonon relaxation time. The total relaxation rate follows Matthiessen's rule:

$$\tau_t^{-1} = \tau_U^{-1} + \tau_{PD}^{-1} + \tau_B^{-1} + \tau_D^{-1} + \tau_{NP}^{-1} + \tau_r^{-1} + \cdots \tau_t^{-1} \quad (5.13)$$

This include contributions from Umklapp processes (U), point defects (PD), grain boundaries (B), dislocations (D), nanoprecipitates (NP), and resonance scattering (r) (Hu et al. 2014; Hu 2023; Kalbacova et al. 2019; Wu et al. 2024; Zheng et al. 2021; Mukherjee et al. 2022; Li et al. 2016; Liu et al. 2016; Zhao et al. 2020).

In thermoelectric materials, various defect types influence phonon transport across different frequency regimes. Grain boundaries predominantly scatter low-frequency phonons, while dislocations and planar defects interact with phonons of intermediate frequencies. In contrast, point defects are more effective in scattering high-frequency phonons. A synergistic combination of these scattering mechanisms—achieved through deliberate defect engineering—facilitates comprehensive phonon suppression over a wide spectral range, thereby substantially reducing the lattice thermal conductivity. At lower temperatures, phonon scattering is primarily governed by grain boundaries, whereas at intermediate and elevated temperatures, point-defect and phonon–phonon (Umklapp) interactions become the dominant processes. Consequently, careful optimization of defect type, density, and spatial distribution is essential for achieving ultralow k_L while preserving high electrical conductivity—an indispensable requirement for the design of high-efficiency thermoelectric materials.

5.5 Thermoelectric Performance Optimization

Achieving a high thermoelectric figure of merit (ZT) requires balancing and optimizing the fundamental transport parameters that govern carrier and phonon dynamics. One of the most critical factors in this optimization is the adjustment of the carrier concentration (n), which directly influences the Seebeck coefficient (S), electrical conductivity (σ), and electronic thermal conductivity (κ_e) (Hu et al. 2014; Hu 2023; Kalbacova et al. 2019; Wu et al. 2024; Zheng et al. 2021; Mukherjee et al. 2022; Li et al. 2016; Liu et al. 2016; Zhao et al. 2020). In thermoelectric materials, electrical conductivity and the Seebeck coefficient are inversely related—higher carrier concentration improves conductivity but reduces the Seebeck coefficient (Hu et al. 2014; Hu 2023; Kalbacova et al. 2019; Wu et al. 2024; Zheng et al. 2021; Mukherjee et al. 2022; Li et al. 2016; Liu et al. 2016; Zhao et al. 2020). Optimal performance typically occurs at carrier densities of 10^{19}–10^{20} cm^{-3}. Excessive doping increases electronic thermal conductivity, so decoupling charge and heat transport is essential for maximizing ZT (Hu et al. 2014; Hu 2023; Kalbacova et al. 2019; Wu et al. 2024; Zheng et al. 2021; Mukherjee et al. 2022; Li et al. 2016; Liu et al. 2016; Zhao et al. 2020). Theoretical models such as the Single Parabolic Band and

Kane Band, supported by DFT studies, help identify ideal carrier concentrations, while experimental deviations often stem from imperfect doping or thermal effects. Defect and dopant engineering can finely tune the band structure—for example, Eu-doped PbTe achieves enhanced Seebeck coefficients through band convergence without sacrificing conductivity. After optimizing electrical transport, reducing lattice thermal conductivity via phonon scattering becomes crucial. Different defects scatter phonons across distinct frequency ranges: point defects target high-frequency modes, dislocations affect midrange phonons, and grain boundaries suppress long-wavelength vibrations. Temperature-dependent studies using in situ XRD and TEM now reveal how these dynamic defects evolve, providing valuable insights for designing thermally stable, high-efficiency thermoelectric materials.

5.6 Characterization of Defects in Thermoelectric Materials

Defect characterization is key to improving thermoelectric performance because defects directly control both electrical and thermal transport. Advanced tools such as synchrotron X-ray diffraction, Cs-corrected electron microscopy, STM/STS, atom probe tomography, and positron annihilation spectroscopy now allow atomic-level visualization of lattice strain, vacancies, dopants, and electronic inhomogeneities. Combining these methods offers a comprehensive view of how defects shape transport behaviour, guiding the design of stable, high-efficiency thermoelectrics.

Point Defects Defects such as vacancies, interstitials, substitutions, and antisites—critically affect charge and phonon transport by modifying lattice symmetry, scattering centers, and carrier mobility. Controlled introduction of these defects through doping or synthesis enables precise tuning of the Seebeck coefficient, electrical conductivity, and lattice thermal conductivity. Complementary DFT studies reveal defect energetics and stability; for example, Co in Mg_3Sb_2 and Cu in SnSe adjust vacancy formation to improve conductivity (Hu et al. 2014; Hu 2023; Wu et al. 2024; Mukherjee et al. 2022; Li et al. 2016).

Hu et al. demonstrated the power of point defect engineering in $(Bi,Sb)_2(Te,Se)_3$ thermoelectrics by optimizing antisite and donor-like defects via controlled formation energy and hot deformation. This approach enhanced both electrical and thermal transport, yielding record *ZT* values of ≈1.2 at 445 K for *n-type* $Bi_2Te_{2.3}Se_{0.7}$ and ≈1.3 at 380 K for *p-type* $Bi_{0.3}Sb_{1.7}Te_3$ surpassing commercial standards (Hu et al. 2014; Hu 2023; Wu et al. 2024; Mukherjee et al. 2022; Li et al. 2016). These findings establish atomic-scale defect engineering as a central strategy for achieving high-efficiency thermoelectric energy conversion.

Vacancies break lattice periodicity and serve as scattering centers for both charge carriers and phonons, reducing lattice thermal conductivity (k_L) and tuning carrier density. While, Substitutional and interstitial defects alter local bonding and electronic structure, influencing carrier type and phonon scattering. These defects form the basis of entropy engineering, which enhances phonon scattering without significantly affecting electrical conductivity (Ghosh et al. 2024; Tang et al. 2024; Chang and Kanatzidis 2023).

Linear Defects The dislocations (Zheng et al. 2021; Sun et al. 2021; Sun et al. 2019) or linear defects serve as strong phonon-scattering centers that effectively reduce lattice thermal conductivity (κ_L) while maintaining reasonable carrier mobility (Zhao et al. 2020). Controlled formation through vacancy diffusion, dopant-induced strain, or thermal annealing—observed in materials like PbSe, Zn_4Sb_3, and Bi_2Te_3 provides an efficient route to enhance the thermoelectric figure of merit (ZT). Thus, balancing dislocation-induced phonon scattering with minimal electron scattering, guided by microscopy and DFT modeling, enables precise defect control and the design of robust, high-efficiency thermoelectric materials through multiscale defect engineering. According to this model, μ varies with dislocation density (ρ_d) as $\mu \propto 1/\rho_d$, and its temperature dependence can be expressed as:

$$\mu(T) \propto \frac{T^{3/2}}{\rho_d L_D^2 f(\xi)}$$

where L_D is the Debye screening length and $f(\xi)$ represents the occupation probability of acceptor centers. High dislocation densities generally reduce carrier mobility (μ) by intensifying charge-carrier scattering—a phenomenon confirmed experimentally in materials such as GaN and Ge at low temperatures. Interestingly, in $Pb_{1-x}Sb_{2/3}Se$ systems with dense dislocations, μ shows little temperature dependence, likely because the high dielectric constant of PbSe screens the electrostatic fields generated by charged dislocation cores. Experiments and simulations on $Pb_{1-x}Sb_xSe$ reveal that accounting for dislocation scattering greatly reduces lattice thermal conductivity (κ_L) compared to models with only Umklapp and point-defect scattering. This reduction occurs because atomic disorder and strain fields around dislocations effectively scatter phonons across a broad frequency range.

Planar Defects These defects are two-dimensional imperfections—such as stacking faults, twin boundaries, and grain boundaries—that critically influence thermoelectric properties (Hu et al. 2014; Hu 2023; Kalbacova et al. 2019; Wu et al. 2024; Zheng et al. 2021; Mukherjee et al. 2022; Li et al. 2016; Liu et al. 2016; Zhao et al. 2020). Grain boundaries scatter charge carriers and lower mobility (μ) but also enhance phonon scattering, reducing lattice thermal conductivity (k_L) and improving heat resistance. Experiments on materials like Bi_2Te_3, PbTe, and Mg_2Si show that nanostructuring can halve k_L without major loss of electrical conductivity. Twin boundaries are particularly effective, suppressing phonon transport while maintaining electron flow. Therefore, optimized planar defects provide an efficient route to control heat transport and boost thermoelectric efficiency. Twin boundaries are particularly beneficial—they scatter long-wavelength phonons while maintaining electron flow—whereas stacking faults enhance phonon localization and lattice anharmonicity. Therefore, planar defects provide a powerful means of controlling heat transport while maintaining electrical performance, leading to significant gains in thermoelectric efficiency.

Hierarchical microstructuring—combining point defects, twin boundaries, and grain boundaries—has proven highly effective in scattering phonons over multiple length scales. In thermoelectric systems such as Mg_2Si–Sb and Zn_4Sb_3, these multiscale defect architectures have driven (κ_L) values near the amorphous limit while maintaining good electrical conductivity. For example, nanostructured Zn_4Sb_3 containing both planar and linear defects forms a percolative strain network that scatters phonons of all frequencies, achieving $k_l \approx 0.6\ Wm^{-1}\ K^{-1}$ and a high ZT ≈ 1.4 at 650 K. Optimizing their density, coherence, and spatial distribution enables an ideal balance between electrical and thermal performance. Future progress in in-situ TEM analysis and atomistic simulations will deepen understanding of phonon–defect interactions, guiding the design of next-generation low-k_l thermoelectric materials.

Bulk Defects Bulk or 3D defects—like secondary phases, pores, and cracks—are large-scale imperfections that strongly influence thermoelectric transport. When engineered properly, they can enhance performance through interfacial effects and energy filtering, which selectively scatter low-energy carriers while reducing lattice thermal conductivity (k_L). Guest-phase inclusions form heterojunctions that improve the Seebeck coefficient (S) and maintain good conductivity, as seen in systems like Yb–$CoSb_3$/GaSb and PbTe-based composites. Conductive or magnetic inclusions can trigger percolation conduction or spin-dependent scattering, further enhancing ZT values up to ~1.8 at high temperatures. Overall, controlled bulk defects and inclusions provide a versatile strategy to optimize both electrical and thermal transport, significantly boosting thermoelectric efficiency.

5.7 Defect-Related Spin Effects

This introduces a new frontier in thermoelectric design. Doping with magnetic ions carrying unpaired 3d or 4f electrons introduces localized moments that interact with conduction electrons via the *Kondo effect*, leading to giant Seebeck coefficients. Materials like $Ca_3Co_4O_9$ demonstrate spin-state transitions under aliovalent doping, altering carrier transport and improving efficiency.

In strongly correlated materials like $Na_xCo_2O_4$ and $Ca_3Co_4O_9$, spin entropy from Co^{3+}/Co^{4+} ions yields exceptionally high S values and $ZT \approx 0.8$. Magnetic doping, as in Ni- or Co-doped Cu_2ZnSnS_4, further enhances spin entropy, improves conductivity, and lowers lattice thermal conductivity (κ_L), achieving up to seven to nine times higher ZT. The Kondo effect in Ce-, Yb-, and U-based intermetallics also boosts S via strong electron–spin interactions. Spin entropy and Kondo mechanisms together define a new direction in spin-coupled defect engineering, enabling simultaneous control over charge, lattice, and spin interactions for improved thermoelectric performance. Additionally, perovskite oxides ($SrTiO_3$, $CaMnO_3$) and lead-free halide perovskites ($Cs_2AgBiBr_6$, $Cs_3Bi_2I_9$) demonstrate how tailored defects—such as oxygen or halide vacancies and cation substitutions—can tune conductivity, carrier density, and thermal transport (Wu and Gao 2018; Kumar and Chauhan 2019; Sharma et al. 2024; Yadav et al. 2025). Hybrid composites like SnSe–PEDOT:PSS also leverage interfacial defect states for energy filtering, enhancing Seebeck performance. Overall, a multiscale defect-engineering strategy—combining atomic, nanoscale, and mesoscale control—enables broad phonon scattering while maintaining good electronic transport, offering a holistic path toward next-generation, high-efficiency thermoelectric materials.

5.8 Summary

Defect engineering plays a pivotal role in optimizing thermoelectric materials by controlling charge and heat transport at the atomic level. Through precise manipulation of defects, properties like electrical conductivity, Seebeck coefficient, and thermal conductivity can be balanced for maximum efficiency and stability. Future advancements using first-principles modeling, machine learning, and in-situ characterization will enable atomic-level defect control, driving the design of high-performance, durable, and sustainable thermoelectric materials for efficient waste-heat recovery and clean energy conversion.

Acknowledgments H. P. Bhasker gratefully acknowledges the financial support from the Council of Science and Technology, Uttar Pradesh (Project Ref. No. CST/D-1302), and the University Grants Commission (UGC) (Project Ref. No. F-30-461/2019 (BSR)).

References

Bell LE (2008) Cooling, heating, generating power, and recovering waste heat with thermoelectric systems. Science 321:1457–1461

Champier D (2017) Thermoelectric generators: a review of applications. Energy Convers Manag 140:167–181

Chandra S, Dutta P, Biswas K (2021) High-performance thermoelectrics based on solution-grown SnSe nanostructures. ACS Nano 16:7–14

Chang C, Kanatzidis MG (2023) High-entropy thermoelectric materials emerging. Mater Lab 2:220048

Du Y, Cai K, Shen SZ, Donelsonand R, Xu J, Wang H et al (2017) Multifold enhancement of the output power of flexible thermoelectric generators made from cotton fabrics coated with conducting polymer. RSC Adv 7:43737–43742

Ghosh S, Raman L, Sridar S, Li W (2024) High-entropy engineering in thermoelectric materials: a review. Crystals 14(5)

Goldsmid H (2018) Conversion efficiency and figure-of-merit. In: CRC handbook of thermoelectrics. CRC Press, pp 19–26

Heremans JP, Jovovic V, Toberer ES, Saramat A, Kurosaki K, Charoenphakdee A et al (2008) Enhancement of thermoelectric efficiency in PbTe by distortion of the electronic density of states. Science 321:554–557

Hong CT, Kang YH, Ryu J, Cho SY, Jang K-S (2015) Spray-printed CNT/P3HT organic thermoelectric films and power generators. J Mater Chem A 3:21428–21433
Hu Y-J (2023) First-principles approaches and models for crystal defect energetics in metallic alloys. Comput Mater Sci 216:111831
Hu L, Zhu T, Liu X, Zhao X (2014) Point defect engineering of high-performance bismuth-telluride-based thermoelectric materials. Adv Funct Mater 24:5211–5218
Kalbacova J, Garratt E, Rodriguez RD, Hight Walker AR, Twedt KA, Fagan JA et al (2019) Defect evolution of ion-exposed single-wall carbon nanotubes. J Phys Chem C 123:2496–2505
Kishimoto K, Hayashi M, Akai K (2021) Thermoelectric properties and band structures of vacancy-containing Sn-based clathrates $K_8Sn_{44-x}Ge_x$. J Solid State Chem 303:122471
Kumar P, Chauhan AK (2019) Highly efficient flexible perovskite solar cells and their photostability. J Phys D: Appl Phys 53:035101
Kumar A, Bano S, Govind B, Bhardwaj A, Bhatt K, Misra DK (2021) A review on fundamentals, design and optimization to high ZT of thermoelectric materials for application to thermoelectric technology. J Electron Mater 50:6037–6059
Li Z, Xiao C, Zhu H, Xie Y (2016) Defect chemistry for thermoelectric materials. J Am Chem Soc 138:14810–14819
Liu S, Deng H, Zhao Y, Ren S, Fu Q (2015) The optimization of thermoelectric properties in a PEDOT: PSS thin film through post-treatment. RSC Adv 5:1910–1917
Liu X, Xi L, Qiu W, Yang J, Zhu T, Zhao X et al (2016) Significant roles of intrinsic point defects in Mg_2X (X = Si, Ge, Sn) thermoelectric materials. Adv Electron Mater 2:1500284
Mahan GD, Sofo JO (1996) The best thermoelectric. Proc Natl Acad Sci U S A 93:7436–7439
Mukherjee M, Srivastava A, Singh AK (2022) Recent advances in designing thermoelectric materials. J Mater Chem C 10:12524–12555
Nolas GS, Morelli DT, Tritt TM (1999) Skutterudites: a phonon-glass-electron crystal approach to advanced thermoelectric energy conversion applications. Annu Rev Mater Sci 29:89–116
Panigrahy S, Kandasubramanian B (2020) Polymeric thermoelectric PEDOT: PSS & composites: synthesis, progress, and applications. Eur Polym J 132:109726
Rowe DM (2018) CRC handbook of thermoelectrics. CRC Press
Sharma A, Kumar A, Yadav J, Chauhan U, Mathur M, Bhasker HP et al (2024) 2-Methoxy ethanol-driven solvent engineering of $Cs_3Bi_2I_9$ perovskite films and its impact on lead free perovskite solar cell devices. Indian J Pure Appl Phys (IJPAP) 62:918–924
Snyder GJ, Toberer ES (2008) Complex thermoelectric materials. Nat Mater 7:105–114
Somdock N, Kianwimol S, Harnwunggmoung A, Sakulkalavek A, Sakdanuphab R (2019) Simultaneous stoichiometric composition and highly (00l) orientation of flexible Bi_2Te_3 thin films via optimising the DC magnetron sputter-deposition process. J Alloys Compd 773:78–85
Sootsman JR, Chung DY, Kanatzidis MG (2009) New and old concepts in thermoelectric materials. Angew Chem Int Ed 48:8616–8639
Sun Y, Zhou Y, Han J, Liu W, Nan C, Lin Y et al (2019) Strong phonon localization in PbTe with dislocations and large deviation to Matthiessen's rule. npj Comput Mater 5:97
Sun Y, Zhou Y, Gurunathan R, Zhang J-Y, Hu M, Liu W et al (2021) Phonon scattering in the complex strain field of a dislocation in PbTe. J Mater Chem C 9:8506–8514
Tabaie Z, Omidvar A (2023) Human body heat-driven thermoelectric generators as a sustainable power supply for wearable electronic devices: recent advances, challenges, and future perspectives. Heliyon
Tang Q, Jiang B, Wang K, Wang W, Jia B, Ding T et al (2024) High-entropy thermoelectric materials. Joule 8:1641–1666
Tran V-T, Saint-Martin J, Dollfus P, Volz S (2017) Optimizing the thermoelectric performance of graphene nano-ribbons without degrading the electronic properties. Sci Rep 7:1–11
Tritt TM (2011) Thermoelectric phenomena, materials, and applications. Annu Rev Mater Res 41:433–448
Wu T, Gao P (2018) Development of perovskite-type materials for thermoelectric application. Materials 11(6)

Wu C, Shi X-L, Wang L, Lyu W, Yuan P, Cheng L et al (2024) Defect engineering advances thermoelectric materials. ACS Nano 18:31660–31712

Xu Q, Qu S, Ming C, Qiu P, Yao Q, Zhu C et al (2020) Conformal organic–inorganic semiconductor composites for flexible thermoelectrics. Energy Environ Sci 13:511–518

Yadav J, Sharma AK, Kumar M, Rajput P, Patel SP, Bhasker HP et al (2025) Synthesis and analysis of temperature-driven charge transport and dielectric relaxation in $Cs_2AgFeCl_6$ double perovskite single crystal. Trans Electr Electron Mater:1–11

Yan R, Xie W, Weidenkaff A (2022) Improving thermoelectric performance of half-Heusler $Ti_{0.2}Hf_{0.8}CoSb_{0.8}Sn_{0.2}$ compounds via the introduction of excessive Ga and Co-deficiencies. Ceram Int

Zhang M, Cao X, Wen M, Chen C, Wen Q, Fu Q et al (2023) Highly electrical conductive PEDOT:PSS/SWCNT flexible thermoelectric films fabricated by a high-velocity non-solvent turbulent secondary doping approach. ACS Appl Mater Interfaces 15:10947–10957

Zhao C, Li Z, Fan T, Xiao C, Xie Y (2020) Defects engineering with multiple dimensions in thermoelectric materials. Research

Zheng Y, Slade TJ, Hu L, Tan XY, Luo Y, Luo Z-Z et al (2021) Defect engineering in thermoelectric materials: what have we learned? Chem Soc Rev 50:9022–9054

Zhou T, Feng Z, Mao J, Jiang J, Zhu H, Singh DJ et al (2020) Thermoelectric properties of Zintl phase $YbMg_2Sb_2$. Chem Mater 32:776–784

Chapter 6
Development of Sustainable Materials for Energy Harvesting: Views and Prospects

Jimli Sarma, Sonika, and Akash Ranjan

Abstract This chapter reviews the recent emergence and evolution of sustainable materials designed to convert existing energy sources—like solar, thermal, mechanical, and electromagnetic—into functional electrical power. Particular attention is paid to eco-friendly, biodegradable, and affordable materials, like organic semiconductors, perovskites, piezoelectric polymers, carbon-based nanostructures, bioinspired systems, which not only ensure energy conversion efficiency additionally align with environmental goals due to their reduced ecological footprint and potential for recyclability. This chapter reviews at the underlying mechanisms of energy harvesting in these materials and highlights recent advancements in material synthesis, structural design, and device integration. Important issues are investigated critically, including compatibility with current technologies, scalability, and long--term stability. To maximize material performance and sustainability, the chapter also explores interdisciplinary trends that combine materials science with artificial intelligence, additive manufacturing, and green chemistry. The chapter expects future developments, such as the emergence of multipurpose hybrid materials, the creation of self-powered and self-healing systems, and the incorporation of energy harvesters into implantable and wearable electronics. These developments represent a paradigm shift away from traditional, resource-intensive systems and toward decentralized, intelligent, and sustainable energy solutions. This chapter attempts to give researchers, engineers, and legislators a thorough understanding of the situation and a glimpse into the future of technology to fully utilize sustainable materials in the pursuit of persistent and clean power systems.

J. Sarma (✉)
Research Scholar, Centre for Multidisciplinary Research, Tezpur University, Napaam, Sonitpur, Assam, India

Sonika
Department of Physics, Rajiv Gandhi University, Rono Hills, Doimukh, Papumpare, Arunachal Pradesh, India
e-mail: Sonika.gupta@rgu.ac.in

A. Ranjan
Faculty of Education, Banaras Hindu University, Kamachha, Varanasi, Uttar Pradesh, India

S. K. Verma et al. (eds.), *Sustainable and Natural Polymers*, Clean Energy Production Technologies, https://doi.org/10.1007/978-981-92-0092-4_6

Keywords Sustainable materials · Eco-friendly · Biodegradable · Energy harvesting · Material synthesis · Structural design · Device integration · Multipurpose hybrid materials · Clean energy systems · Green chemistry

6.1 Introduction

Due to rapid industrialisation, population growth, and an ever-growing reliance on electronic devices, the twenty-first century is characterised by an unprecedented global energy demand. Fossil fuels have been the primary source of this growing energy consumption, which has made serious environmental problems like pollution, resource depletion, and climate change worse. It is more important than ever to minimize dependency on fossil fuels and develop renewable energy sources like hydropower, wind, and solar energy. Energy harvesting, which provides capacity to convert ambient energy that is present everywhere—for example, solar radiation, thermal gradients, mechanical vibrations, and electromagnetic waves—into usable electrical power, has emerged as a highly promising solution within this changing environment (Yeatman 2009; Yildiz 2009; Xu et al. 2015). This innate ability to use otherwise squandered energy sources offers a strong argument for lowering carbon emissions and promoting energy independence. However, the materials used to build energy harvesting technologies are inextricably linked to their effectiveness and actual sustainability. In the past, a large number of advanced materials used in energy conversion devices have been resource-intensive, challenging to process, or, at the end of their useful lives, presented environmental challenges. A true paradigm shift towards a sustainable energy future requires a revolution in materials science at the same time, with an emphasis on the creation and application of inexpensive, biodegradable, and environmentally friendly materials. These next-generation materials are evaluated not only on the basis of their energy conversion efficiency but also on the basis of their decreased environmental impact, potential for recycling, and overall compliance with environmental standards over the course of their whole lifecycle (Greenman et al. 2024; Das 2020; Cordaro et al. 2023) (Table 6.1). Building genuinely robust and clean energy systems that can sustain a sustainable global society requires a dedication to green materials.

This chapter provides a thorough analysis of the development, recent appearance of these environmentally friendly materials created especially for energy harvesting applications. We will explore the new developments in a wide variety of materials, like the highly effective perovskites and adaptable organic semiconductors for photovoltaic applications, the responsive and flexible piezoelectric polymers for mechanical energy conversion, the multipurpose carbon-based nanostructures (Zawrah et al. 2023; Zhang and Olin 2020; Shao et al. 2020), and the creative bio-inspired systems that learn from the energy mechanisms found in nature (Figs. 6.2–6.6).

Table 6.1 Comparative summary table of sustainable energy harvesting materials

Material type	Energy source	Efficiency (typical)	Sustainability benefits	Challenges	References
Organic Semiconductors	Solar	~18% PCE	Flexible, biodegradable, low-cost	Stability, long-term degradation	Roy et al. (2024)
Halide Perovskites	Solar	>25% PCE	Scalable, high-efficiency	Lead toxicity, moisture sensitivity	Roy et al. (2024) *(includes perovskite-based discussion)*
Piezoelectric Polymers	Mechanical	10–60 μW/cm^2	Lightweight, biocompatible	Lower piezoelectric constants	Nor et al. (2021), Xu et al. (2024)
Carbon Nanostructures	Mechanical, Thermal, EM	Variable	Abundant, recyclable, hybridizable	Processing cost, dispersion issues	Ramanujam et al. (2022)
Bioinspired Systems	Biochemical, Mechanical	~μW to mW scale	Biodegradable, bio-compatible	Low energy density, stability	Clementi et al. (2022) *(covers bio-inspired and hybrid piezo systems)*

The fundamentals of energy harvesting will be covered for each material class, along with new developments in material synthesis, structural design, and device integration that are expanding the realm of energy conversion (Figs. 6.7 and 6.8). This review critically examines the practical factors necessary for the broad adoption of these technologies, going beyond the material science itself. We will look at important topics like their long-term stability under different operating conditions, their scalability for mass production, and their compatibility with existing technological infrastructures. The paper will also examine the emerging interdisciplinary trends that are driving advancement in this area, such as the revolutionary roles of additive manufacturing in the creation of devices, green chemistry in sustainable synthesis processes, and artificial intelligence in materials discovery. Future advancements that we look forward to include the development of multifunctional hybrid materials that can harvest from various sources, the production of self-powered and self-healing systems that extend the lifespan of devices (Cao et al. 2023; Cheng et al. 2023; Oh et al. 2024) (Fig. 6.11), and the smooth integration of energy harvesters into commonplace applications like wearable and implanted electronics (Figs. 6.9 and 6.10). This chapter's ultimate goal is to give researchers, engineers, and policymakers a complete grasp of the latest one as well as an outlook on how sustainable materials will develop in energy harvesting in the future. In the quest for a decentralised, intelligent, and ecologically conscious energy future, we hope to support the group's endeavour to fully utilise sustainable materials by shedding light on the enormous potential of these innovations (Fig. 6.1).

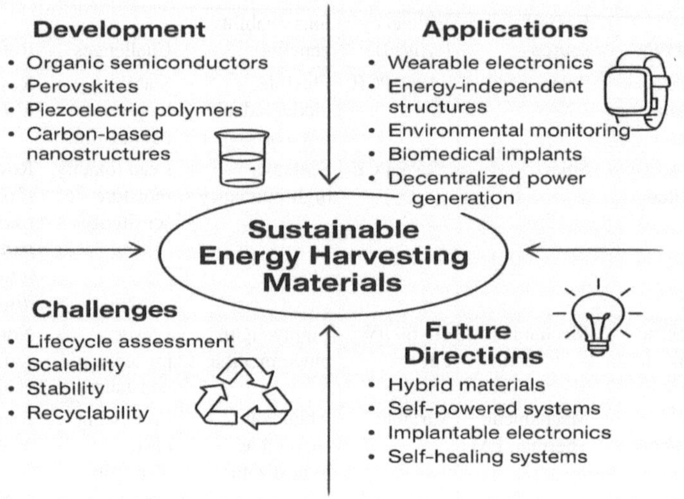

Fig. 6.1 An overview of the setting of sustainable energy harvesting materials. The figure divides developments into four main areas: (1) Development—showcasing new materials like carbon--based nanostructures, organic semiconductors, perovskites, and piezoelectric polymers; (2) Applications—such as energy-independent structures, wearable electronics, biomedical implants, environmental monitoring, and decentralised power generation; (3) Challenges—focusing on scalability, stability, recyclability, and lifecycle assessment; and (4) Future Directions—thinking about hybrid materials, self-powered systems, implantable electronics, and self-healing technologies

6.2 Novelty of the Study

Compared to traditional reviews, this chapter places sustainable energy harvesting (SEH) in a larger socio-technical and environmental framework by highlighting the combination of materials science with artificial intelligence (AI), additive manufacturing, and green chemistry. This chapter is unique in that it uses a multi-disciplinary framework. By integrating sustainable materials science with cutting-edge technological enablers like artificial intelligence (AI), additive manufacturing, and green chemistry, this study is novel in that it positions energy harvesting as both a technological advancement and an ecological and societal solution. The current study stresses sustainability throughout the whole lifetime, from raw material extraction to end-of-life recyclability or biodegradation, whereas previous assessments have mostly concentrated on efficiency increases or device layouts. For instance, the investigation of lead-free perovskites and biodegradable organic photovoltaics responds to the urgent demand for environmentally safe substitutes, while the use of bioinspired and biocompatible materials demonstrates potential future uses in implantable electronics and healthcare. Additionally, the study maps the melding of advancements in a variety of fields, including organic semiconductors, perovskites, piezoelectric polymers, carbon nanostructures, and bioelectrochemical systems, into hybrid multifunctional devices that integrate mechanical, thermal, biochemical, and solar energy harvesting. This chapter presents a forward-looking roadmap (2025–2035) that predicts not only improved performance but also energy equality through decentralised microgrids and self-powered medical equipment, including AI-driven materials discovery pipelines and 3D-printable nanocomposites into the conversation. Thus, the idea is in expressing a revolutionary vision—sustainable energy harvesting as the foundation of next-generation energy systems that are effective, environmentally benign, scalable, and socially impactful—rather than listing current advancements.

6.3 Basics of Energy Harvesting

Energy harvesting produces power from ambient energy that is naturally present. The significant structures consist of:

- Photovoltaics (PV): Using semiconductors to convert light into electricity.
- Piezoelectric Effect: An electric charge is produced by mechanical strain.
- Thermoelectric Effect: Electrical potential is generated by thermal gradients.
- Triboelectric Effect: Charge separation is produced by friction between materials.
- Bioelectrochemical Systems: Power is produced by biological processes in microorganisms or biomolecules (Fig. 6.2).

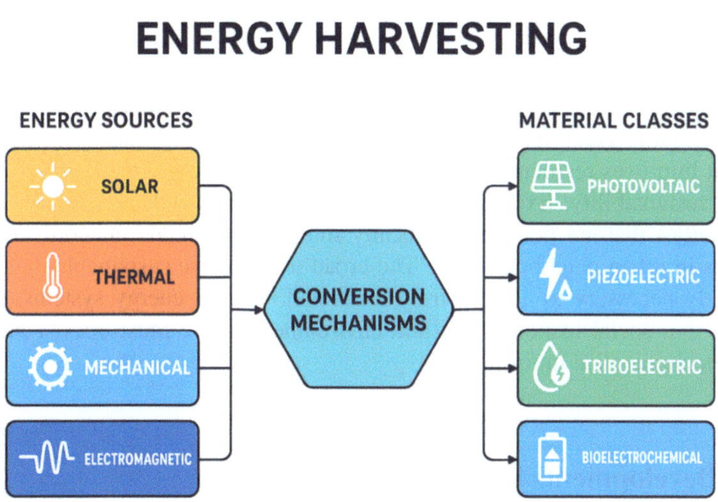

Fig. 6.2 Energy generation framework: Sustainable material classes enable the conversion of natural energy sources (solar, thermal, mechanical, and electromagnetic) through particular processes (photovoltaic, piezoelectric, triboelectric, and bioelectrochemical)

6.4 The Need for Sustainable Materials

Because conventional technologies like solar panels, batteries, and wind turbines sometimes rely on materials that are costly to create, difficult to dispose of, or ecologically hazardous, sustainable materials are essential in energy harvesting. Heavy metals like cadmium, lead, or lithium, for instance, are found in many traditional batteries. If they are disposed of incorrectly, they may pollute land and water, endangering both human health and animals. The urgent need for greener alternatives is demonstrated by the fact that populations close to major battery dumping facilities frequently experience hazardous exposure in real life. Biodegradable polymers, organic solar cells, and recyclable metals are examples of eco-friendly products that provide a solution by lowering production energy and waste. For example, scientists have created solar panels with perovskite materials, which are less expensive and simpler to recycle than silicon panels (Gallegos et al. 2024). Portable solar-powered chargers constructed from recycled plastics demonstrate the influence of sustainable materials in our daily lives by reducing environmental impact while simultaneously producing clean energy. In a similar vein, wind turbines built from recyclable, modular parts simplify maintenance and minimise waste going to landfills at the end of their useful lives (Khalid et al. 2023). Innovations like flexible solar sheets and small-scale piezoelectric devices employ inexpensive, locally accessible, and non-toxic materials, even in developing nations, improving energy access equity while preserving the environment (Xu et al. 2024). We can design energy systems that are not only technologically sophisticated but also socially and environmentally responsible by focussing on biodegradable, reasonably priced, and very efficient materials. By ensuring that energy harvesting does not jeopardise ecosystems, human health, or the availability of future resources, this approach fosters a circular economy in which materials are recycled, repurposed, and safely reintroduced into the environment. Adopting green materials is not just a theoretical approach; it is a practical one that improves resilience, lessens ecological footprints, and supports long-term energy sustainability. Examples from the real world include microgrids in rural communities that are powered by locally sourced sustainable materials and solar rooftops in urban neighbourhoods. The broad use of these sustainable materials is ultimately necessary to achieve truly clean and resilient energy systems, demonstrating that social responsibility and environmental care must coexist alongside energy technological innovation.

6.5 Developments in Sustainable Energy Harvesting Materials

Considerable progress has been achieved in creating a wide variety of sustainable materials, each with special qualities appropriate for various energy harvesting systems. Because of their great performance, affordability, and environmental friendliness, these materials may be used in a variety of energy harvesting systems.

6.5.1 *Important Elements of These Changes*

A Wide Variety of Materials: Numerous materials, including organic, inorganic, and hybrid ones, have been created by researchers specifically for various energy collecting methods. Examples include thermoelectric materials for heat energy, piezoelectric materials for mechanical energy, and perovskites for solar cells.

Reliable Qualities: The environmental effect of many new materials is decreased because they are recyclable, biodegradable, or derived from plentiful natural resources. They frequently use little energy during production and have few harmful consequences.

6.6 Properties for Their Implementation

- Materials with certain characteristics are needed for each energy harvesting system:
- Materials with strong light absorption and charge transfer are required for solar cells.
- Materials that distort under mechanical stress are necessary for piezoelectric generators.
- Materials that can sustain a temperature gradient and effectively conduct electricity are necessary for thermoelectric generators.
- Increased Durability and Efficiency: Modern materials are more feasible for practical use due to their increased stability, longer operating lifespans, and better energy conversion efficiencies.
- Integration with IoT and wearables: Particularly in isolated or off-grid areas, flexible and lightweight materials have been created for incorporation into wearable electronics, sensors, and self-powered devices.

6.7 Photovoltaic Organic Semiconductors

With a power conversion efficiency (PCE) of more than 18%, non-fullerene acceptors (NFAs) and donor-acceptor copolymers are the driving forces behind recent developments in organic photovoltaics (OPVs). They are perfect for wearable and distributed electronics because of their versatility, low processing temperature, and potential biodegradability. Because of their mechanical flexibility, solution processability, and tuneable electronic characteristics, organic semiconductors have attracted a lot of interest and are therefore perfect for flexible and transparent solar cells. They are very appealing for distributed energy generation due to their low fabrication costs and large-area production potential. At the end of their lifecycle, many organic materials' inherent biodegradability also helps to reduce their effect on the surrounding.

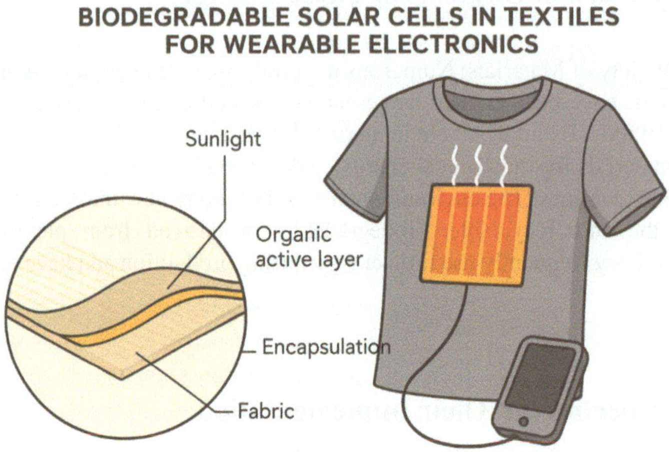

Fig. 6.3 Illustration of the case study: Textiles incorporating biodegradable solar cells for wearable electronics. A T-shirt with an organic photovoltaic layer inserted in it, shielded by biodegradable encapsulation, as seen in the graphic. It uses sunshine to power tiny gadgets

An Example of a Case Study
In order to charge low-power Internet of Things devices, a recent prototype showed solar textiles that could generate tens of milliwatts per square centimetre. Following disposal, the cloth broke down in a matter of months in a composting environment, demonstrating its suitability for use in circular economy applications (Fig. 6.3).

6.8 Perovskites: A Revolution in Photovoltaics

Because of their remarkable energy conversion capacities, perovskite solar cells have become a disruptive technology that is quickly catching up to and even surpassing traditional silicon-based solar cells. The creation of stable and lead-free perovskite formulations, enhancements to device architecture, and scalable manufacturing methods are some of the major developments. Even though lead toxicity is still an issue for some compositions, these issues are being addressed by continuous research into less hazardous substitutes and effective encapsulation techniques. Perovskite precursors' affordability and abundance further strengthen their reputation for sustainability. Because of their high PCEs (>25%) and ease of processing, halide perovskites have revolutionised photovoltaic science. However, lead-free substitutes and improved encapsulation actively address issues like lead toxicity and degradation (Fig. 6.4).

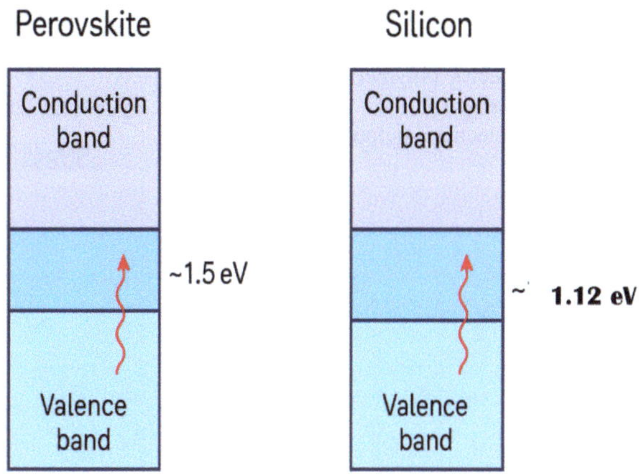

Fig. 6.4 Shows the band structures of silicon (~1.12 eV) and perovskite (~1.5 eV) semiconductors in comparison, showing how electrons are excited across the bandgap when illuminated

6.9 Piezoelectric Polymers for Harvesting Mechanical Energy

Piezoelectric polymers are great options for turning mechanical motions and vibrations into electrical energy. Examples of these include polyvinylidene fluoride (PVDF) and its co-polymers. They are appropriate for wearable electronic devices, self-powered sensors, and biomedical implants because of their versatility, low weight, biocompatibility. Recent developments include improving their crystalline structure, adding nanomaterials to improve performance, such as graphene or carbon nanotubes, and creating novel device designs that optimise energy conversion efficiency and strain transfer. These polymers are sustainable because they are non-toxic and relatively simple to process. Flexible energy harvesters can benefit from the piezoelectric properties of polymers like PVDF and its co-polymers. Dielectric constants and strain transfer are being improved by the inclusion of nanomaterials (like graphene and carbon nanotubes) (Fig. 6.5).

6.10 Carbon-Based Nanostructures

Mechanical, thermal, and electromagnetic energy can all be harvested by carbon nanostructures (like graphene and CNTs). They are crucial to hybrid and multi-modal systems due to their scalability, tunable conductivity, and mechanical strength. The electrical, mechanical, and thermal properties of carbon-based nanostructures—like graphene, CNTs, and carbon quantum dots—are exceptional, making them extremely adaptable for a range of energy harvesting applications.

- **Thermoelectric applications**: Due to their high power factors, graphene and carbon nanotubes (CNTs) can be designed to effectively transform waste heat into electrical power.

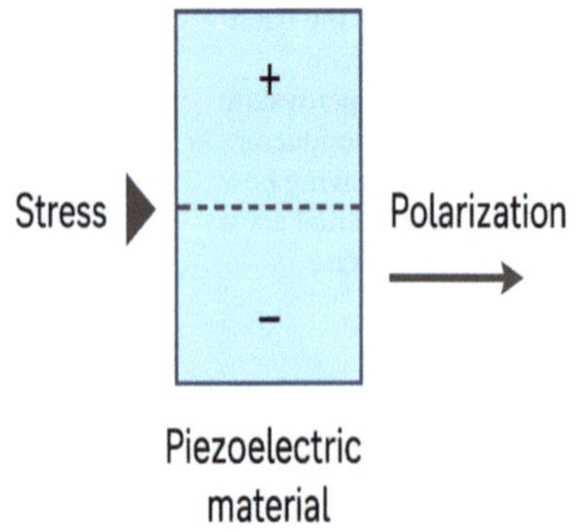

Fig. 6.5 Piezoelectric effect diagram: mechanical stress causes piezoelectric polymers to become polarised, producing an electrical output

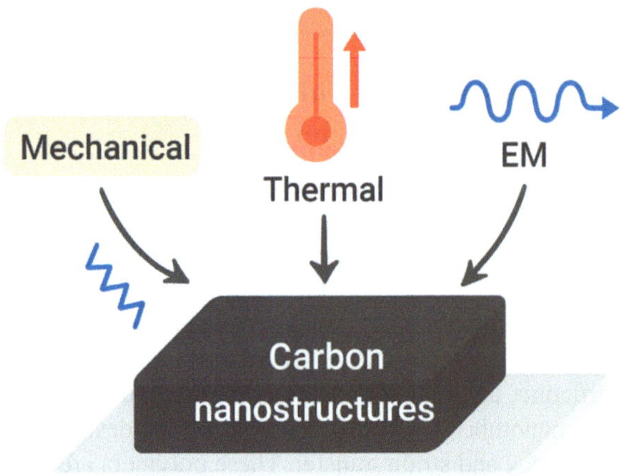

Fig. 6.6 Carbon nanostructures and hybrid energy collection: These multipurpose systems allow for the simultaneous capture of electromagnetic (EM), temperature gradients, and mechanical vibrations

- **Piezoelectric and triboelectric nano-generators**: The performance of piezoelectric and triboelectric nanogenerators, which capture energy from friction and mechanical deformation, is improved by the distinct mechanical characteristics of these nanostructures, particularly in composite materials.
- **Electromagnetic shielding and harvesting**: They can absorb and transform electromagnetic radiation due to their high electrical conductivity. Their sustainability is aided by the abundance of carbon and the capacity to create these materials using scalable and frequently environmentally friendly techniques (such as chemical vapour deposition) (Fig. 6.6).

6.11 Bioinspired Systems: The Energy Blueprint of Nature

Bioinspired energy harvesting materials, which take their cues from biological systems, are an exciting and naturally sustainable field. Among the examples are:

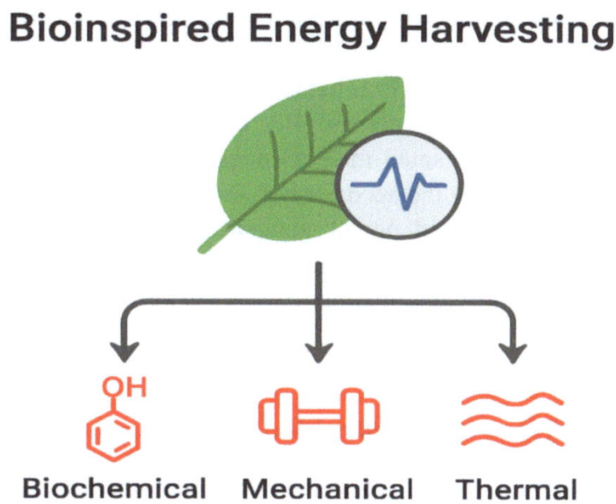

Fig. 6.7 Bioinspired energy harvesting: devices that use mechanical, thermal, and biological processes to generate sustainable power are inspired by natural systems (Wu et al. 2021; Sohail et al. 2024; Jiang et al. 2020)

- ***Plant-derived cellulose***: Because of its superior mechanical qualities and biodegradability, cellulose can be utilised to make flexible solar cell substrates or as a dielectric in triboelectric nanogenerators.
- ***Protein-based structures***: Studies investigate how proteins and peptides can be used to simulate natural energy conversion processes by using their piezoelectric or thermoelectric qualities.
- ***Bioelectrochemical fuel cells*** are a highly sustainable, decentralised source of power that uses microorganisms to directly convert biomass into electricity. With their unmatched biodegradability and frequent use of easily accessible, renewable feedstocks, these systems reduce their negative effects on the environment (Fig. 6.7 and Table 6.1).

6.12 Uses and Possible Impacts

Numerous revolutionary applications in a range of industries are being made possible by the developments in sustainable energy harvesting materials:

- **Wearable electronics and IoT devices**: By incorporating lightweight and flexible energy harvesters that transform body movement or ambient light into electricity, wearable health monitors, smart textiles, and self-powered sensors can be made possible without the need for frequent battery replacement.
- **Energy-independent structures**: Energy-independent structures, i.e., smart buildings and infrastructures can be achieved through the integration of transparent solar cells into windows, thermoelectric generators that use waste heat from HVAC systems, and piezoelectric sensors embedded in roads that collect energy from traffic.

- **Environmental monitoring and remote sensing**: autonomous wireless sensor networks set up in hazardous or remote areas for infrastructure inspection, agriculture, and environmental monitoring.
- **Biomedical implants**: Biocompatible and biodegradable energy harvesters can power implanted medical devices such as nerve stimulators and pacemakers, eliminating the need for invasive surgeries to change batteries.
- **Decentralised power generation**: Off-grid power options for disaster relief, rural communities, and portable electronics can be obtained through flexible solar films and modular energy harvesting units.

6.12.1 Other Case Studies

The implementation of sustainable energy harvesting (SEH) technology is the basis of a future in which electricity systems are decentralised, environmentally benign, and seamlessly incorporated into everyday life. It is not just a scientific endeavour. Wearable electronics, biomedical gadgets, remote sensing, smart infrastructure, and decentralised off-grid power systems are just a few of the applications. Sustainable harvesting materials like organic semiconductors, perovskites, piezoelectric polymers, carbon nanostructures, and bioinspired systems are well-suited to meet the specific needs of each of these domains in terms of energy density, flexibility, biocompatibility, and stability.

6.12.2 Electronic Gadgets That Can Be Worn

One of the most exciting areas of energy harvesting applications is wearable electronics. Lightweight, adaptable, and sustainable energy solutions are becoming more and more in demand as augmented reality (AR) systems, fitness trackers, smart fabrics, and health monitoring devices proliferate (Matos-Peralta et al. 2023; Pan et al. 2024; Dong et al. 2024). Conventional wearable batteries are large, need to be charged frequently, and eventually add to the trash generated by electronics. By capturing ambient energy like sunshine, body heat, and motion, SEH materials overcome these difficulties (Xiong et al. 2018; Ke et al. 2023; Ali et al. 2024; Liu et al. 2018).

6.12.3 Integration of Solar Power and Intelligent Textiles

Smart textiles turn garments into portable solar panels by using perovskite-coated or organic photovoltaic (OPV) fibres. These have the capacity to provide enough energy to run Bluetooth connection modules, e-ink screens, and low-energy Internet of Things sensors. The case study on biodegradable solar textiles (Xiong et al. 2018; Ke et al. 2023) shows that it is possible to incorporate ultra-flexible OPVs into textiles that are comfortable and breathable while producing power. For instance, a garment with organic solar cells integrated into it may concurrently monitor body temperature using built-in sensors and recharge a smartphone during the day.

6.12.4 Harvesting Motion and Vibration Energy

Body motions can be converted into electrical power by weaving piezoelectric polymers, such as PVDF nanofibers, into fabrics. Vibrations caused by walking, stretching, or even a heartbeat can produce power at micro-watt to milliwatt levels. For medical monitoring, when continuous operation without frequent battery replacements is essential, this method is very beneficial.

Case Study: Triboelectric nanogenerators (TENGs) placed in shoes that produce enough power to charge tiny devices while walking were created by a Georgia Tech team.

6.12.5 Thermal Energy Capturing of Human Body

Carbon nanostructure-based thermoelectric devices use the temperature differential between ambient air and human skin to produce power. It can support ultra-low-power sensors despite having a limited power density.

For biomedical devices that need to run continuously, like ECG patches, this collecting method is extremely important.

Obstacles and Prospects

1. Ensuring that electronics withstand mechanical stress and laundry cycles is known as washability and durability.
2. Biocompatibility: Preventing encapsulants or coatings from irritating the skin.
3. Creating effective micro-supercapacitors to store gathered energy is known as energy management.

6.12.6 Advanced Buildings

Energy-autonomous buildings, which both create and consume less energy, are becoming more and more common in modern construction. Self-sustaining infrastructures are made possible by energy harvesting materials, which are essential to this transition.

6.12.7 Transparent Photovoltaic (PV) Windows

It is possible to design semi-transparent photovoltaic coverings for windows using halide perovskites and organic semiconductors. These harness solar power without obstructing vision. PV windows significantly increase the usable area by taking advantage of vertical surfaces, unlike rooftop solar panels (Nor et al. 2021).

Case Study: To drastically lower the net energy load of high-rise buildings, researchers at UCLA created transparent perovskite windows with an efficiency of >10%.

6.12.8 Floors and Roads Using Piezoelectric Technology

Vehicles or footfall cause continuous vibrations on roads, pavements and building floors. These structures may be made to convert mechanical stress into electrical energy by embedding piezoelectric polymers or ceramics (Clementi et al. 2022). For instance, a piezoelectric flooring system in Tokyo's subway stations showed that LED screens and tiny sensors could be powered only by foot circulation.

6.12.9 HVAC Systems Using Thermoelectric Power

Temperature differences between interior and outdoor spaces can be converted into useable power using thermoelectric materials that are included in building insulation. On the other hand, they may also be utilised for solid-state cooling, which lessens the need for traditional air conditioners (Clementi et al. 2022).

6.12.10 Connectivity to Smart Grids

Energy collected from these dispersed sources may be used to power IoT devices (such as occupancy sensors and smart thermostats) and enhance energy management by feeding into smart grids at the building level.

Obstacles and Prospects

1. Scaling Costs: Perovskites are deposited across a large region for windows.
2. Durability: Piezoelectric road systems' mechanical wear.
3. Efficiency versus Beauty: Juggling window transparency and photovoltaic performance.

6.12.11 Monitoring Wirelessly

Agriculture, environmental monitoring, and disaster relief all use remote sensing equipment. SEH is crucial since these sensors are frequently positioned in areas where battery replacement or recharging is not feasible (Roy et al. 2024).

6.12.12 Applications in Agriculture

- Sensors for soil moisture are powered by tiny solar modules that are encased in biodegradable polymers.
- Wireless communication systems combined with energy-harvesting traps are used for pest detection.
- Drones equipped with solar-powered batteries for prolonged operation are used for crop monitoring.

6.13 Observation of the Environment

- Ocean Buoys: Using OPVs to capture sunlight and piezoelectric devices to capture wave motion.
- Forest Fire Sensors: Biofuel cell-powered, deployable, biodegradable nodes that track humidity and temperature in far-flung locations.

6.13.1 Relief from Disasters

Power infrastructure restoration is frequently delayed following natural catastrophes. Nanogenerators or deployable solar textiles offer rapid, environmentally friendly power sources for medical devices, illumination, and communication.

Case Study: Portable OPV sheets were used to supply emergency illumination and sensor power during the Fukushima accident in 2011.

Obstacles and Prospects

1. Durability: Equipment must survive extreme weather conditions, such as UV, salinity, and strong winds.
2. Energy Density: Hybrid harvesting systems are necessary when there is little sun irradiation or sporadic motion.
3. Eco-Impact: In delicate habitats, biodegradable designs reduce ecological impact.

6.13.2 Implants for Biomedicine

Life-saving technology includes implanted medical equipment, including insulin pumps, cochlear implants, pacemakers, and nerve stimulators. However, relying on batteries requires replacing them with dangerous operations. Self-powered implants that capture energy from the human body are SEH's game-changing answer. But the majority of these devices use traditional batteries, which ultimately run out and require surgery to replace, which is expensive, risky, and can cause issues. Self-powered implants, made possible by self-energy harvesting (SEH) technologies, provide a game-changing answer by directly transforming bodily energy—such as blood flow, body heat, or kinetic movements—into electrical energy. For instance, thermoelectric devices may use body heat to power insulin pumps, while piezoelectric materials incorporated into pacemakers can harness the heart's rhythmic action to create electricity. These developments lengthen the lifespan and dependability of implants, lessen the need for recurrent intrusive procedures, and lessen reliance on traditional batteries. In addition to improving patient safety, self-powered implants also increase accessibility and reduce healthcare expenses, particularly in areas where regular surgical procedures are impractical. Biodegradable implants that safely decompose after their useful lives, hybrid devices that combine many energy sources, and intelligent sensors that continually power themselves while monitoring physiological variables are some of the future prospects in SEH for biomedicine. In animal models, real-world studies have already shown that piezoelectric nanogenerators can power small-scale biomedical devices, pointing to a time when implants can really become sustainable, autonomous, and completely integrated with human physiology (Lee et al. 2024; Shi et al. 2016; Yang et al. 2021; Zhao et al. 2020; Preatoni et al. 2022).

6.13.3 Pacemakers' Piezoelectric Harvesting Machines

The heartbeat's mechanical energy may be transformed into electrical energy using piezoelectric polymers. This energy is enough to keep the pacemaker running, which significantly lowers the need for surgery (Jiang et al. 2020; Wu et al. 2021).

Case Study: In a live animal model, Jiang et al. (2020) showed how an implanted piezoelectric nanogenerator could effectively power a pacemaker.

6.13.4 Implants That Use Thermoelectrics

Body heat serves as a consistent energy source. Microwatts of electricity can be produced using thermoelectric modules made of biocompatible materials, which makes them perfect for gradually but steadily powering implants.

6.13.5 *Biofuel Cells*

Body fluids containing glucose and oxygen are converted into energy by enzyme-based biofuel cells. These are naturally biocompatible and provide implants with sustained power. For instance, a tiny wireless transmitter was effectively powered by a glucose biofuel cell implanted in a rat.

Obstacles and Prospects

1. Biocompatibility: Preventing toxicity and immunological reaction.
2. Miniaturisation: Making sure gadgets are comfortable to fit within the body.
3. Creating materials that endure for decades inside the body is known as longevity.

6.13.6 *Dispersed Power*

Centralised power systems are frequently absent or unstable in rural or disaster-affected areas, depriving people of steady electricity. By offering modular, off-grid power sources that are scalable, sustainable, and low-maintenance, decentralised energy systems present a workable answer. By placing power generation closer to the point of use and lowering transmission losses, these systems—such as solar mini-grids and independent solar household systems—can be customised to match the unique energy requirements of isolated communities. For example, by lowering reliance on imported fossil fuels and increasing resilience to natural catastrophes, decentralised renewable energy solutions have been crucial in boosting energy security and advancing sustainable development in areas like sub-Saharan Africa(climate change writers.com). Decentralised systems also provide local communities with more authority by encouraging self-governance and generating revenue via local energy production and management (EIS). Because of these advantages, decentralised energy systems are essential to attaining universal energy access and advancing sustainable development objectives (Figs. 6.8 and 6.9).

6.14 Solar Fabrics Off the Grid

In off-grid settings like rural homes, refugee camps, and disaster recovery zones, deployable organic photovoltaic (OPV) and perovskite-based solar textiles are becoming game-changing options for delivering instantaneous and sustainable energy. These cutting-edge materials enable quick deployment and integration into a variety of applications by fusing the energy-generating potential of solar cells with the flexibility and light weight of textiles. Because of their lightweight, flexible, and perhaps biodegradable nature, organic solar textiles—including OPVs—are perfect for incorporation into textiles and fabrics. Numerous uses are made possible by

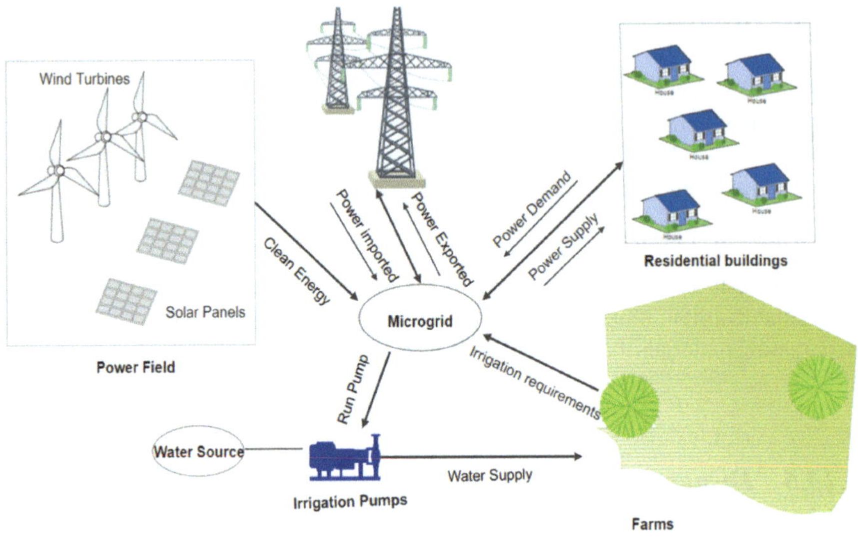

Fig. 6.8 A solar PV/WT hybrid energy system proposed by a study for a rural community (Mustafa et al. 2022)

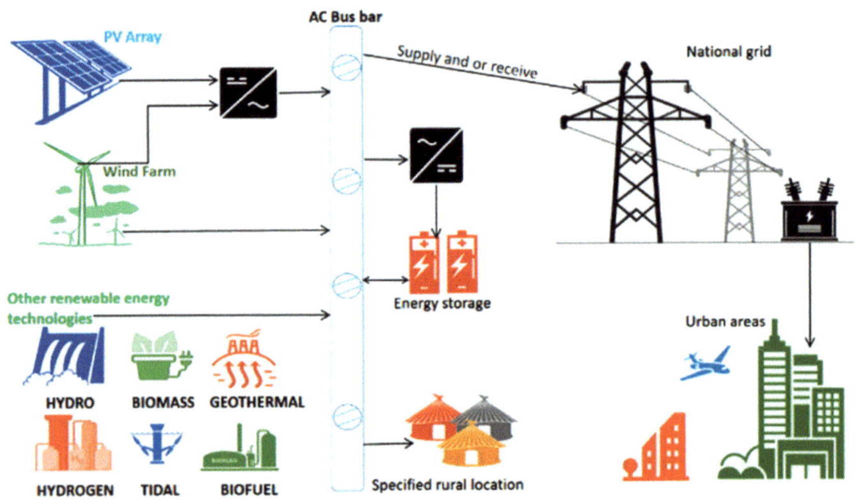

Fig. 6.9 Mini-grid technology for renewable energy proposed in sub-Saharan Africa (Babayomi et al. 2023)

their ability to be printed or weaved into fabrics, including portable chargers, clothes, and tents that run on solar power. To provide electricity while on the road or in isolated areas, for example, scientists have created incredibly tiny and light solar cells that can be easily affixed to any surface, even textiles (MIT News).

Solar fabrics made of perovskites provide yet another exciting option for off-grid electricity. Perovskite solar cells are appropriate for large-scale deployment in off-grid locations because of their high efficiency and ability to be produced using inexpensive, scalable techniques. Their capacity to provide more economical and efficient energy might completely reshape off-grid power in the future, lowering the cost and increasing the accessibility of clean electricity for rural populations throughout the globe. These technologies are already being used in the real world. For instance, a 2-MW solar photovoltaic plant has been installed in Jordan's Azraq refugee camp, giving 20,000 Syrian refugees residing in around 5000 shelters energy (IKEA Foundation). This project demonstrates the viability and significance of using renewable energy sources into humanitarian contexts.

Additionally, developments in solar textile technologies are still ongoing. Leading the way in the development of solar-powered textiles are businesses such as Pvilion, which combine fabric with photovoltaic cells to produce electrically powered goods. Their products include building facades, military tents, solar canopies, and wearables like clothes and bags.

6.14.1 Microgrids That Are Hybrid

Self-sufficient microgrids are made possible by combining thermoelectric cookstoves, solar textiles, and piezoelectric harvesters (roads, pathways). Energy equality is enhanced by these methods, particularly in low-income nations.

Case Study: By reducing the need for kerosene lights, portable solar lanterns and phone chargers produced from OPVs have improved access to education and health outcomes in sub-Saharan Africa.

Obstacles and Prospects

1. Cost: Ultra-low-cost, large-area manufacturing is required.
2. Maintenance: Making sure that gadgets require little care, even in harsh environments.
3. Connecting dispersed harvesters to operational microgrids is known as integration.

6.14.2 Comparative Analysis and Synthesis

In every application, SEH materials show the capacity to:

1. Use biodegradable harvesters in place of batteries to cut down on electrical waste.
2. Allow sensors and gadgets to run continuously without needing to be recharged.
3. Reduce your reliance on fossil fuels and help achieve climate objectives.

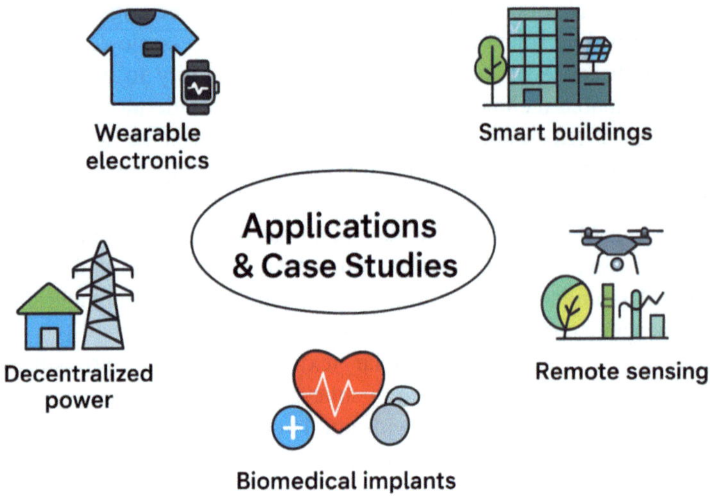

Fig. 6.10 Shows case studies and applications of sustainable energy harvesting, with a focus on five areas: decentralised power, wearable electronics, smart buildings, remote sensing, and biomedical implants

However, improvements in lifetime analysis, material stability, and economic scalability are necessary for broad implementation. These issues should soon be resolved by the multidisciplinary fusion of sophisticated manufacturing, green chemistry, and artificial intelligence (Fig. 6.10).

6.14.3 Challenges and Aspects of Sustainability

Although the development of sustainable materials for capturing energy is encouraging, a comprehensive approach to sustainability calls for tackling a number of important issues and problems:

- Environmental Footprint and Lifecycle Assessment
  To measure the effect of the environment of these resources, by the extraction of raw materials to producing, operation, and recycling or disposal at the end of their useful lives, a thorough lifecycle assessment (LCA) is essential. The synthesis of some advanced materials may still require energy-intensive procedures or the use of hazardous solvents, even though materials such as organic semiconductors and bioinspired systems have inherent biodegradability. Reducing the overall ecological footprint throughout the lifecycle is the aim.
- Affordability and Scalability
  Sustainable energy harvesting technologies need to be affordable and scalable to mass production in order to be widely adopted. Even though many new materials are less expensive than their conventional counterparts, it is still very difficult to

scale up their synthesis and device fabrication without sacrificing functionality or adding additional environmental costs. Commercial viability depends critically on cost-effective manufacturing techniques.

- Durability and Long-Term Stability
  Energy harvesting devices must be stable and long-lasting in real-world settings in order to be used in practice. Over time, variables such as exposure to fluctuating temperatures, humidity, UV rays, and mechanical stress can impair the performance of materials. Creating strong encapsulation techniques and materials that are inherently stable is crucial to extending operational life and lowering replacement frequency, which improves sustainability.
- Waste Management and Recyclability
  One important component of sustainability is the possibility of recyclability at the end of a device's life. Minimizing waste and advancing a circular economy can be achieved by focusing on circular designing products and materials that are simple to disassemble and recover valuable parts from. Knowing how biodegradable materials break down and making sure they don't release toxic by-products into the environment are equally crucial.
- Multidisciplinary Patterns and Prospects
  The convergence of various scientific and technological disciplines is critical to the advancement of renewable energy capture.
- The Exploration of Materials and Artificial Intelligence
  Materials design and discovery are being revolutionised by Artificial Intelligence (AI), especially ML and data analytics. Development of high-performance, sustainable energy harvesting materials with desired properties can be greatly accelerated by AI algorithms, which can quickly screen enormous material databases, predict new material properties, and optimise synthesis pathways.
- Complex Structures with Additive Manufacturing
  Unprecedented potential for producing intricate, personalised, and effective energy harvesting devices is provided by additive manufacturing (3D printing). It makes it possible to precisely position various materials, fabricate complex structures for improved energy conversion, and quickly prototype new designs, all of which result in more effective material use and less waste.
- Sustainable Synthesis through Green Chemistry
  The development of environmentally friendly synthesis pathways for sustainable energy harvesting materials depends heavily on the concepts of green chemistry. This entails using renewable feedstocks, optimising atom economy, reducing the use and production of hazardous materials, and creating safer chemicals and procedures.

6.15 Future Advancements: An Overview of the Horizon

A number of exciting developments are indicated by the trajectory of sustainable energy harvesting:

- **Multifunctional Hybrid Materials**
  The efficiency and adaptability of energy harvesting systems will be greatly increased by the development of multipurpose hybrid materials that can concurrently capture energy from several sources (like solar and thermal). Devices will be able to function well in a variety of settings and conditions thanks to these materials. The application of 3D-printed nanocomposites that integrate triboelectric and piezoelectric effects is investigated by Razack and Sadasivuni (2025). These hybrid materials are perfect for wearables and biomedical implants because they are very customisable. According to the authors, these systems are lightweight, scalable, and safe for the environment, particularly when combined with biodegradable polymers. Hybrid self-powered systems that integrate thermoelectric, photovoltaic, and triboelectric harvesters into flexible substrates are described by Tang et al. (2023). These systems are particularly well-suited for smart textiles and promise functionality in a variety of settings.
- **Self-Healing and Self-Powered Systems**
  The creation of self-powered and self-healing systems will result from the combination of energy harvesting devices with advanced features. Self-powered systems take out the necessity of external energy resource, making them fully autonomous, while self-healing materials can fix damage on their own, increasing device lifespan and decreasing waste. With polymer layers that exhibit self-healing properties following damage, Jiang et al. (2020) presented implantable triboelectric nanogenerators (iTENGs) that can power cardiac and nerve stimulators on their own. In therapeutic and long-term health monitoring systems, these technologies are becoming indispensable. For wearable and industrial IoT systems, Gołąbek and Strankowski (2024) talk about smart sensors that are integrated with self-sufficient energy harvesting and repair mechanisms.
- **Combining Wearable and Implantable Electronics**
  The seamless integration of sustainable energy harvesting materials into wearable and implantable electronics will be fuelled by their ongoing miniaturisation, flexibility, and biocompatibility. This will enable continuous operation without frequent battery replacements, revolutionising personal monitoring, healthcare, and human-computer interfaces. Self-powered skin electronics are thin, flexible devices that use biomechanical motion to monitor health (Wu et al. 2021). In a small patch-like form, these systems combine sensors, harvesters, and storage elements. In order to eliminate the need for invasive battery replacements, Sohail et al. (2024) describe energy solutions for implantable medical devices (IMDs), such as pacemakers and neural stimulators, which depend on body heat and movement for continuous operation.

6.15.1 Roadmap for Materials for Sustainable Energy Harvesting (2025–2035)

See Fig. 6.11.

Timeline Design

- Milestone 1
  2025–2028: Immediate
- Free of lead perovskites
- Wearables with biodegradable OPVs
- IoT sensor piezoelectric nanofibers

solar panel | T-shirt | Internet of Things sign

- Milestone 2
  2028–2031: Intermediate
- Multipurpose hybrid harvesters that combine piezo, tribo, and thermal
- PV windows that are transparent in smart buildings
- Biomedical implants powered by biofuel cells

DNA strand | glass-paned building | hybrid energy

- Milestone 3
  2031–2035: Extended
- Completely autonomous biomedical implants
- AI-enhanced SEH products
- Dispersed hybrid microgrids for emergency and rural assistance

decentralised grid | □ AI chip | pacemaker

Fig. 6.11 Sustainable energy harvesting materials roadmap for 2025–2035, illustrating the shift from near-term, biodegradable, and lead-free devices to mid-term, multifunctional hybrid systems, and long-term, biomedical, and decentralised energy uptake

6.15.2 *Sustainable Energy Harvesting: Ethical and Social Aspects (SEH)*

Technologies for sustainable energy collection have significant ethical and societal ramifications in addition to being a technological fix. Efficiency and affordability have historically been the primary criteria used to evaluate energy technology, but the advent of sustainable materials forces us to consider the wider implications for both people and the environment.

1. Reduction of E-Waste:
   The ability of SEH to lessen electronic waste, or "e-waste," is one of its biggest ethical benefits. Traditional batteries, such as the lithium-ion coin cells found in pacemakers, wearable technology, and Internet of Things gadgets, need to be changed on a regular basis and contribute significantly to hazardous waste streams. The UN Global E-waste Monitor 2020 reports that fewer than 20% of the 53.6 million tonnes of e-waste produced each year are formally recycled (Forti et al. 2020). SEH can directly lessen this hazardous load by substituting or

augmenting throwaway batteries with biodegradable harvesters (such as organic solar cells or biofuel cells). Given that e-waste disproportionately affects low-income communities, where a large part of recycling and dumping occurs, like in Ghana and India, this is consistent with environmental justice principles (Grant and Oteng-Ababio 2019).

2. Access to Energy Equity:
One of the most urgent issues facing the world today is energy inequality. There are still 350 million people in South Asia and more than 600 million in sub-Saharan Africa without consistent access to electricity (Citaristi 2022). Rural areas are left behind by costly and slow-growing centralised electricity infrastructures. Localised, low-maintenance electricity is provided via decentralised SEH solutions, such as solar textiles, piezoelectric microgrids, or biofuel-powered agricultural sensors. This can enhance healthcare (vaccine refrigeration), education (school illumination), and economic fairness (small business support). In an ethical sense, SEH transforms from a technology into an empowerment tool that helps close the global energy divide and empowers communities to produce and control their own electricity (Sohail et al. 2024).
3. Issues with Autonomy and Privacy:
IoT devices and self-powered sensors provide a more subtle but significant ethical dilemma. Real-time health, mobility, environmental, or infrastructure monitoring is made possible by SEH, which allows sensors to operate continuously without human intervention. Smart cities, healthcare, and agriculture all benefit from this, however privacy concerns are raised: The data belongs to whom? Who has access to it and how is it kept? For instance, a self-powered biometric sensor in wearable apparel may continuously check temperature and heart rate; this is excellent for health, but it could be invasive if governments or insurance firms take advantage of it. The dangers of data colonialism and consent violations in digital health technology have been highlighted by academics (Crawford and Joler 2018). To guarantee that the advantages of autonomy do not result in exploitation or monitoring, SEH ethically requires concurrent investment in digital rights, permission frameworks, and data protection.
4. Responsibility Across Generations:
The fact that SEH materials, especially biodegradable and recyclable systems, represent intergenerational responsibility is an ethical aspect that is frequently disregarded. Researchers are creating technology that will not be inherited as trash issues by future generations by decreasing dependency on hazardous, non-recyclable materials (such as lithium-ion batteries or lead-based perovskites). This places SEH in a context of sustainability ethics, which states that innovation in the present must not jeopardise the environment in the future (Roy et al. 2024; Jiang et al. 2020).
5. Innovative Inclusivity:
Lastly, inclusive innovation should be given top priority in the creation of SEH materials. This entails making sure that advantages are available, reasonably priced, and flexible enough to be applied in a variety of situations rather than being exclusive to affluent countries. Research labs, local communities, and politicians may work together to make SEH a worldwide public benefit rather than a "luxury technology" (Ali et al. 2024).

6.16 Conclusion

An important step towards a future with clean and resilient energy is the development of sustainable materials for energy harvesting. Piezoelectric polymers, carbon-based nanostructures, organic semiconductors, perovskites, and bioinspired systems are just a few examples of the materials science breakthroughs that are constantly expanding the limits of energy conversion efficiency, affordability, and environmental friendliness. Scalability, long-term stability, and comprehensive lifecycle management are still issues, but there are encouraging avenues for overcoming them thanks to interdisciplinary trends that combine materials science with artificial intelligence, additive manufacturing, and green chemistry. The upcoming advancements in self-powered and self-healing systems, multipurpose hybrid materials, and the widespread use of energy harvesters in daily life mark a significant paradigm shift away from conventional, resource-intensive systems and towards decentralised, intelligent, and genuinely sustainable energy solutions. In the quest for a future driven by clean, renewable, and ecologically conscious energy, we can fully utilise sustainable materials by adopting these innovations. The ethical and social aspects of SEH strengthen the discussion in the welfare of people and the earth, extending it beyond labs and businesses. Decentralised microgrids promote energy fairness, e-waste reduction guarantees environmental justice, and cautious control of self-powered sensors protects privacy. When taken as a whole, these features demonstrate that SEH's real promise is not only cleaner power but also a more equitable and sustainable future.

References

Ali I, Islam MR, Yin J, Eichhorn SJ, Chen J, Karim N, Afroj S (2024) Advances in smart photovoltaic textiles. ACS Nano 18:3871–3915

Babayomi OO, Olubayo B, Denwigwe IH, Somefun TE, Adedoja OS, Somefun CT et al (2023) A review of renewable off-grid mini-grids in sub-Saharan Africa. Front Energy Res 10:1089025

Cao X, Xiong Y, Sun J, Xie X, Sun Q, Wang ZL (2023) Multidiscipline applications of triboelectric nanogenerators for the intelligent era of internet of things. Nano-Micro Lett 15:14

Cheng T, Shao J, Wang ZL (2023) Triboelectric nanogenerators. Nat Rev Methods Primers 3:39

Clementi G, Cottone F, Di Michele A, Gammaitoni L, Mattarelli M, Perna G et al (2022) Review on innovative piezoelectric materials for mechanical energy harvesting. Energies 15:6227

Cordaro A, Müller R, Tabernig SW, Tucher N, Schygulla P, Höhn O et al (2023) Nanopatterned back-reflector with engineered near-field/far-field light scattering for enhanced light trapping in silicon-based multijunction solar cells. ACS Photonics 10:4061–4070

Citaristi I (2022) International energy agency—iea. In The Europa directory of international organizations 2022 (pp. 701–702). Routledge

Crawford K, Joler V (2018) Anatomy of an AI system: The Amazon Echo as an anatomical map of human labor, data and planetary resources. AI Now Institute

Das KS (2020) Microbial fuel cells: a path to green, renewable energy. In: Practices and perspectives in sustainable bioenergy: a systems thinking approach. Springer, New Delhi, pp 195–206

Dong H, Yu F, Bi Z, Zhang C, Liu X, Geng Y, Ohnishi S, Li H (2024) Life cycle environmental and economic assessment of Tetra Pak recycling technologies. Resour Conserv Recycl 202:107355

Forti V, Baldé CP, Kuehr R, Bel G (2020) The global e-waste monitor 2020. United Nations University, Bonn/Geneva/Rotterdam
Gallegos MV, Gil-Escrig L, Zanoni KP, Bolink HJ, Damonte LC (2024) Recycling and reusing ITO substrates from perovskite solar cells: a sustainable perspective. Solar Energy Mater Solar Cells 277:113117
Gołąbek J, Strankowski M (2024) A review of recent advances in human-motion energy harvesting nanogenerators, self-powering smart sensors and self-charging electronics. Sensors 24:1069
Grant R, Oteng-Ababio M (2019) Electronic-waste circuitry and value creation in Accra, Ghana. In: Value chains in sub-Saharan Africa: challenges of integration into the global economy. Springer, Cham, pp 115–131
Greenman J, Thorn R, Willey N, Ieropoulos I (2024) Energy harvesting from plants using hybrid microbial fuel cells; potential applications and future exploitation. Front Bioeng Biotechnol 12:1276176
Jiang D, Shi B, Ouyang H, Fan Y, Wang ZL, Li Z (2020) Emerging implantable energy harvesters and self-powered implantable medical electronics. ACS Nano 14:6436–6448
Ke H, Gao M, Li S, Qi Q, Zhao W, Li X et al (2023) Advances and future prospects of wearable textile- and fiber-based solar cells. Solar RRL 7:2300109
Khalid MY, Arif ZU, Hossain M, Umer R (2023) Recycling of wind turbine blades through modern recycling technologies: a road to zero waste. Renew Energy Focus 44:373–389
Lee DM, Kim J, Hyun I, Kim SW (2024) Self-powered medical implants using triboelectric technology. Acc Mater Res 5:533–543
Liu P, Gao Z, Xu L, Shi X, Fu X, Li K et al (2018) Polymer solar cell textiles with interlaced cathode and anode fibers. J Mater Chem A 6:19947–19953
Matos-Peralta Y, Reali M, Camus A, Niyonkuru D, Hoseinsizadeh M, Gao Z, Santato C (2023) Nature as the source of materials for more sustainable organic electronics. Mater Matters 1:18–22
Mustafa J, Almehmadi FA, Alqaed S, Sharifpur M (2022) Building a sustainable energy community: design and integrate variable renewable energy systems for rural communities. Sustainability 14:13792. https://doi.org/10.3390/su142113792
Nor NM, Hamzah HH, Razak KA (2021) Recent advancement in sustainable energy harvesting using piezoelectric materials. In: Sustainable materials for next generation energy devices. Elsevier, pp 221–248
Oh E, Zuniga MMG, Suhr J (2024) Trends and perspective for eco-friendly composites for next-generation automobiles. Compos Res 37:115–125
Pan X, Wu Y, Wang Y, Zhou G, Cai H (2024) Mechanical energy harvesting based on the piezoelectric materials: Recent advances and future perspectives. Chem Eng J 154249
Preatoni E, Bergamini E, Fantozzi S, Giraud LI, Orejel Bustos AS, Vannozzi G, Camomilla V (2022) The use of wearable sensors for preventing, assessing, and informing recovery from sport-related musculoskeletal injuries: a systematic scoping review. SensS, 22(9):3225
Ramanujam BTS, Haridass R, Muralidharan P, Nanjundan AK, Dubal D, Annamalai PK (2022) Electroactive polymers and their carbon nanocomposites for energy harvesting. In: Nanogenerators. CRC Press, pp 361–396
Razack RK, Sadasivuni KK (2025) Advancing nanogenerators: the role of 3D-printed nanocomposites in energy harvesting. Polymers 17:1367
Roy R, Ashmika R, Chakraborty A (2024) Polymeric energy harvesting materials: a comprehensive overview. In: Emerging materials, technologies, and solutions for energy harvesting, pp 33–71
Shao J, Willatzen M, Wang ZL (2020) Theoretical modeling of triboelectric nanogenerators (TENGs). J Appl Phys 128:11
Shi Q, Wang T, Lee C (2016) MEMS based broadband piezoelectric ultrasonic energy harvester (PUEH) for enabling self-powered implantable biomedical devices. Sci Rep 6:24946
Sohail A, Ali A, Shaukat H, Bhatti FM, Ali S, Kouritem SA et al (2024) Integrating self-powered medical devices with advanced energy harvesting: a review. Energy Strategy Rev 52:101328
Tang W, Sun Q, Wang ZL (2023) Self-powered sensing in wearable electronics—a paradigm shift technology. Chem Rev 123:12105–12134

Wu M, Yao K, Li D, Huang X, Liu Y, Wang L et al (2021) Self-powered skin electronics for energy harvesting and healthcare monitoring. Mater Today Energy 21:100786
Xu J, Buin A, Ip AH, Li W, Voznyy O, Comin R et al (2015) Perovskite–fullerene hybrid materials suppress hysteresis in planar diodes. Nat Commun 6:7081
Xu M, Wen Y, Shi Z, Xiong C, Zhu F, Yang Q (2024) Piezoelectric biopolymers: advancements in energy harvesting and biomedical applications. Polymers 16:3314
Xiong J, Cui P, Chen X, Wang J, Parida K, Lin MF, Lee PS (2018) Skin-touch-actuated textile-based triboelectric nanogenerator with black phosphorus for durable biomechanical energy harvesting. Nat Commun 9(1):4280
Yang SY, Sencadas V, You SS, Jia NZX, Srinivasan SS, Huang HW et al (2021) Powering implantable and ingestible electronics. Adv Funct Mater 31:2009289
Yeatman EM (2009) Energy harvesting: small scale energy production from ambient sources. In: Active and passive smart structures and integrated systems 2009, vol 7288, pp 17–23
Yildiz F (2009) Potential ambient energy-harvesting sources and techniques. J Technol Stud 35:1
Zawrah MF, Taha MA, Youness RA (2023) Advanced ceramics: stages of development. In: Advanced ceramics. Springer Nature, Cham, pp 1–46
Zhang R, Olin H (2020) Material choices for triboelectric nanogenerators: a critical review. EcoMat 2:e12062
Zhao J, Ghannam R, Htet KO, Liu Y, Law MK, Roy VA et al (2020) Self-powered implantable medical devices: photovoltaic energy harvesting review. Adv Healthc Mater 9:2000779

Chapter 7
Emerging Polymers for Green Energy: Roadmap for Sustainable Development

Kwanruthai Butsriruk and Mingjing Qu

Abstract The pristine evolution of polymer-like celluloid was discovered in the nineteenth century, during a period of resource scarcity, and thus, polymer became a critical innovation for modern green energy technologies. This chapter describes the assisted polymers to enhance sustainable power systems. To date, polymers have been applied in various manufacturing sectors, such as wind turbine blades with lightweight polymer materials, solar panels with encorage parts of polymers, increased battery potential through electrolytes, and preparing membranes in fuel cells to facilitate proton exchange. Despite the polymers having abundant advantages from the aforementation, however, the sustainability is still addressed. Mostly producing polymers are produced from fossil fuels and have a notorious impact on he filed of recycling. For this reason, the principles of design for circularity have been focused on bio-based feedstocks, recyclable thermoplastics (e.g., vitrimers), and depolymerization strategies, to raise performance and response to the environment. Case studies highlight interdisciplinary breakthroughs, either recyclable thermoplastic wind blades or organic photovoltaics. This chapter concludes by outlining grand challenges: closing the recycling loop, overcoming performance-sustainability trade-offs, and ensuring long-term stability in harsh environments. A paradigm shift toward "sustainable polymers" is urged, where materials innovation aligns with circular economy principles to support a truly green energy transition.

Keywords Polymer · Recyclable · Green energy · Microplastic pollution

List of abbreviations includes technical terms and acronyms found in the chapter. Common chemical abbreviations for polymers are included to aid reading of literature, even if not all were explicitly used in the chapter's text.

K. Butsriruk (✉)
College of Engineering, Xi'an International University, Shaanxi, China

M. Qu
School of Business, Xi'an International University, Shaanxi, China

S. K. Verma et al. (eds.), *Sustainable and Natural Polymers*, Clean Energy Production Technologies, https://doi.org/10.1007/978-981-92-0092-4_7

Abbreviations

EV	Electric Vehicle
GHG	Greenhouse Gas (usually in the context of emissions like CO_2 equivalent)
LCA	Life Cycle Assessment (or Analysis)
Nafion	A brand name for a perfluoro sulfonic acid polymer (used as a proton exchange membrane in fuel cells)
OPV	Organic Photovoltaics (solar cells using organic materials)
P3HT, PTB7, etc.	Shorthand names for specific conjugated polymers used in OPV (not spelled out here, but commonly known in the field)
PA	Polyamide (e.g., Nylon)
PC	Polycarbonate (engineering plastic, often for transparent parts)
PCBM, ITIC, Y6	Specific organic molecules used as acceptors in OPV
PE	Polyethylene
PET	Polyethylene Terephthalate (polyester used in bottles, fibers)
PHA	Polyhydroxyalkanoate (biodegradable polyesters made by microbes)
PLA	Polylactic Acid (a biodegradable polyester from renewable resources)
PP	Polypropylene
PTAA	Poly [bis (4-phenyl) (2,4,6-trimethylphenyl)] amine
PVC	Polyvinyl Chloride (a common plastic, used in pipes, cables, etc.)

7.1 Opening Vignette: "From Celluloid to Solar Cells"

Late nineteenth century, a billiards parlor in New York: A crack echoes as an ivory billiard ball break. John Wesley Hyatt, a young inventor, recalls a $10,000 prize offered for a substitute to ivory—the diminishing tusks of elephants can no longer satisfy demand (Morris 2023). In his workshop, Hyatt treats cotton-derived cellulose with camphor and creates a revolutionary material: celluloid, the world's first synthetic plastic (Morris 2023). Advertised as the *"savior of the elephant and the tortoise,"* celluloid can mimic luxury materials like ivory and horn, liberating manufacturing from nature's limits (Reilly 1991). Little could Hyatt have known that this pliable "plastic" would spark a materials revolution, one that would shape industries and touch every facet of modern life (Morris 2023).

7.1.1 Early Twenty-First Century, a Sunny Rooftop

A technician unrolls a sheet of flexible solar cells that gleam blue-black in the sunlight. These film-like solar panels are made with polymer materials—literally plastic solar cells. They are lightweight, bendable, and can be installed like wallpaper (Maoz et al. 2025). As the sun's rays hit the panel, electrons flow, generating clean electricity. It's a scene far removed from a 1800s billiard hall, yet deeply connected. The same class of materials that once replaced ivory in billiard balls now helps replace fossil fuels in energy production. Polymers have journeyed from celluloid to solar cells, evolving from curiosities to key enablers of green technology (Rasmussen 2021; Yelshibay et al. 2024).

This opening vignette highlights two moments, a century apart, bound by innovation in polymers. The first synthetic plastic solved an environmental crisis of its day—saving natural resources by providing a human-made alternative (Abrahms-Kavunenko 2023). Today, advanced polymers tackle a new crisis: enabling sustainable energy and reducing reliance on fossil fuels. The story of plastics is thus a story of ingenuity and adaptation. In the chapters that follow, we will trace how polymers were born, how they rose to ubiquity, and how they are now being re-imagined to power a greener future (Patel and Deshmukh 2012; Davidson et al. 2024). From Hyatt's celluloid to cutting-edge solar cells, plastic pathways to power continue to unfold.

7.2 The Birth of Polymers

What are polymers? The word "polymer" comes from Greek meaning "many parts," an apt description for molecules made of long chains of repeating sub-units. Polymers abound in nature—cellulose in plant cell walls, silk from silkworms, and natural rubber from trees. In fact, "plastic" originally meant something pliable or easily shaped, but today it refers to this category of long-chain materials (Morris 2023). In the late nineteenth century, chemists learned to create synthetic polymers by linking small molecules (monomers) into long chains, often using carbon from plentiful petroleum feedstocks (Morris 2023). These synthetic polymer chains can be much longer than natural ones, and it is their immense length and internal structure that give plastics their remarkable strength, light weight, and flexibility (Young and Lovell 2011). In other words, the very "plasticity" (moldability) of plastics comes from their polymer nature.

7.2.1 The First Synthetic Plastic: An Environmental Innovation

By the 1860s, demand for natural ivory (for billiard balls, piano keys, combs, etc.) was devastating elephant populations (Morris 2023). In 1869, John W. Hyatt, spurred by a contest to find an ivory substitute, treated cellulose from cotton with camphor and produced celluloid. Celluloid could be molded into shapes and dyed to imitate expensive natural materials. This invention was revolutionary—for the first time, human manufacturing was not constrained by what could be harvested or mined from nature (Combs 2018). As contemporary advertisements proclaimed, celluloid would "protect the natural world from the destructive forces of human need," saving elephants and tortoises by replacing their ivory and shell with plastic (Balser et al. 2000).

7.2.2 In Hindsight, the Irony Is Clear

The quest to protect nature launched the Age of Plastics. Early plastics like celluloid and its cousin Parkesine (demonstrated by Alexander Parkes in 1862 from cellulose nitrate) showed that synthetic materials could reduce pressure on scarce natural resources (Rasmussen 2021). Inexpensive celluloid also democratized consumer goods—material wealth became a bit more accessible when combs and brush handles no longer required rare ivory (Freinkel 2011a).

7.2.3 Bakelite and the Polymer Boom

The next milestone came in 1907. Leo Baekeland, a Belgian-American chemist, invented Bakelite, the first fully synthetic plastic not derived from existing natural polymers (Morris 2023). Baekeland had been searching for a replacement for shellac, a natural resin used as an electrical insulator, to support the rapidly electrifying United States (Crespy et al. 2008). His creation—Bakelite, a phenol-formaldehyde resin—was electrically insulating, heat-resistant, and could be mass-produced in molds. Marketed as "the material of a thousand uses," Bakelite found its way into telephones, radios, electrical sockets, jewelry, and more. It ushered in the era of thermosetting polymers (plastics that harden into an infusible solid) and demonstrated the enormous commercial potential of synthetics. Bakelite's success catalyzed major chemical companies to invest in polymer research. By the 1920s and 1930s, a deluge of new plastics arrived: PVC (polyvinyl chloride) in 1926, made pliable by additives for use in pipes and cable insulation; polystyrene (commercialized 1930); nylon (polyamide or PA) in 1935, a synthetic silk for stockings and parachutes; Teflon (PTFE) in 1938, a non-stick, heat-resistant marvel initially used in wartime nuclear projects. Each new polymer often emerged to meet a specific need—nylon for scarce silk, PVC for a durable rubber alternative, Teflon for extreme chemical resistance—but soon found myriad applications by virtue of plastics' versatility.

7.2.4 *World War II: Plastics Prove Their Mettle*

During World War II, the plastics industry underwent explosive growth. Traditional materials like metal, rubber, and glass were strategically rationed or in short supply, so synthetic replacements became crucial. Nylon, invented in 1935 by Wallace Carothers at DuPont, was repurposed from women's hosiery to military parachutes and ropes. Plexiglas (acrylic) provided shatter-resistant aircraft canopies in place of glass. Synthetic rubber (Buna-S, Buna-N) was produced at scale to compensate for cut-off natural rubber supplies. A Time magazine article in 1942 noted that due to the war effort, "plastics have been turned to new uses and the adaptability of plastics demonstrated all over again". Indeed, between 1940 and 1945, U.S. plastic production tripled (Nicholson and Leighton 1942). By war's end, the polymer genie was out of the bottle: manufacturers and consumers alike had grown accustomed to plastics in everyday goods.

7.2.5 *Post-War Plastic Prosperity*

In the prosperity of the 1950s, plastics conquered consumer markets. As author Susan Freinkel quipped, "product after product, market after market, plastics challenged traditional materials and won" (Freinkel 2011b). Plastics replaced steel in cars, paper and glass in packaging, and wood in furniture. Their appeal was obvious: they were cheap, light, could be fashioned into any color or shape, and often lasted longer with less maintenance. By the 1960s, polymer chemistry delivered polypropylene (PP) (1954), polycarbonate (PC) (1953), and other "miracle materials" that further expanded applications. Plastics were seen as symbols of modernity and human ingenuity—a literal material embodiment of the mid-twentieth century's optimism about technology.

7.2.6 *Early Warnings and the Seeds of Sustainability*

Even at the height of plastic's golden age, not everyone was optimistic. In 1963, the sci-fi writer Isaac Asimov presciently warned that "the rapid accumulation of plastics" in the environment might become a serious problem (though few listened at the time). By the late 1960s and 1970s, public perception of plastics began to sour as

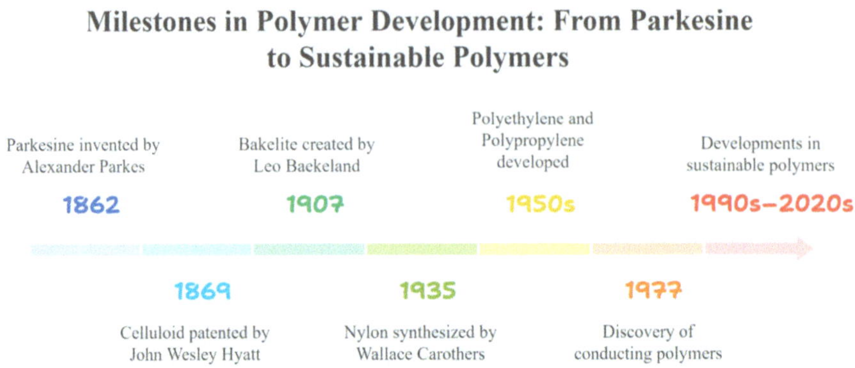

Fig. 7.1 A timeline of polymer development. (Developed by the authors)

awareness of pollution grew. Discarded plastic items were found polluting oceans, and iconic incidents like the 1969 burning Cuyahoga River highlighted society's chemical waste. Plastics, once praised as saviors of nature, were now criticized as symbols of throwaway culture—cheap, "plastic" substitutes that never biodegrade (Lakhiar et al. 2024). The stage was set for a more nuanced understanding: polymers had given humanity unprecedented material abundance and convenience, but at a cost to environmental health that would need reckoning.

To facilitate understanding of the development of polymers, Fig. 7.1 presents a timeline of each period for polymer development.

7.3 Why Polymers Matter for a Greener Energy Future

Why focus on polymers in the context of clean energy? The answer lies in the unique properties of plastics and the diverse roles they play in modern technology. Polymers have become critical enablers of renewable energy and energy efficiency, often in ways invisible to the end-user. As the world pushes for decarbonization, these materials are helping to make green technologies lighter, cheaper, more durable, and more effective (Ramesh et al. 2024). A few key reasons illustrate why polymers matter for a greener future:

7.3.1 Lightweighting and Energy Savings

Replacing metal and glass with plastics in cars, airplanes, and other transport can dramatically reduce weight, improving fuel efficiency. For instance, a 10% reduction in vehicle weight yields about a 6–8% improvement in fuel economy. In electric vehicles (EVs), lightweight polymer composites offset the heavy battery mass, extending driving range (Joost 2012; Cuenca et al. 2000). This principle has been applied in designs like the BMW i3's carbon-fiber-reinforced polymer body, which significantly cuts weight to improve performance per battery charge. In buildings, polymer foams (like polystyrene or polyurethane insulation) reduce heating and cooling needs by providing excellent thermal insulation—savings that translate into lower energy consumption for climate control. Indeed, plastics being "lightweight and good for insulation" means they "help save fossil fuels used in heating and in transportation" by reducing energy losses (Andrady and Neal 2009).

7.3.2 *Versatility in Renewable Energy Devices*

Many renewable energy systems fundamentally rely on polymer components. Wind turbines use polymer-based composites (resins reinforced with glass or carbon fibers) for their enormous blades. These composites offer an ideal combination of strength and low weight, enabling blades to stretch 75+ m long to capture more wind without collapsing under their own weight (Firoozi et al. 2024). Solar panels incorporate polymers too; the backing sheets and encapsulant layers that protect delicate silicon cells from weather are often durable plastics (like TEDLAR® PVF film or EVA resin). Emerging organic solar cells—sometimes dubbed "plastic solar cells"—actually use semiconducting polymers as the active layer, converting light to electricity. Fuel cells and batteries also depend on polymers: the electrolyte membrane in a proton-exchange membrane fuel cell (such as the Nafion® membrane in hydrogen fuel cells) is a sulfonated fluoropolymer crucial for conducting protons while separating gases. In advanced batteries, polymer separators (microporous films of polyolefin) keep electrodes apart and enhance safety, and new solid polymer electrolytes are being researched to replace flammable liquid electrolytes in next-gen lithium batteries. Even electric vehicle charging stations employ engineering plastics for components like wire insulation and housing, thanks to their electrical insulating properties and outdoor durability (Du et al. 2019).

7.3.3 *Enhancing Efficiency and Functionality*

Polymers often bring specialized functionality that improves the performance of green technologies. For example, anti-reflective or self-cleaning polymer coatings on solar panels can boost light capture. In wind turbines, certain polymer composites provide damping that reduces vibration and fatigue on blades (Murray et al. 2021).

7.3.4 Manufacturing and Cost Advantages

In the field of economics and production, polymers have been frequently produced with low-cost supply and have been fabricated with large-scale production, these are promising to respond to clean energy technology. Typically, plastics act as one of the polymer fields capable of tunable complex shapes via extrusion or printing approaches, which facilitate organizing plastic patterns more than metal and ceramic matrices. For example, the roll-to-roll printing approach offers remarkably cheap prices for manufacturing polymer-based solar films, which is similar printing of newspapers (Gnida et al. 2022). Similarly, polymer components enable modular, mass-produced wind turbine parts and battery packs. Cost reduction is a key reason solar and wind have become competitive in recent years, and polymers have quietly aided those cost declines by enabling simpler fabrication and integration. The versatility of polymers also speeds innovation: engineers can prototype new designs quickly using 3D-printed plastics or resin molds, accelerating the development of better energy devices (Mohanty et al. 2022).

7.3.5 Polymers as Sustainable Materials (Potentially)

Paradoxically, while conventional plastics are made from fossil fuels, polymers can also be part of the solution to move beyond those same fuels. Bio-based polymers derived from renewable biomass (like polylactic acid from corn, or bio-polyethylene from sugarcane ethanol) can replace petro-plastics in many applications, reducing the carbon footprint of materials. Moreover, polymers can be designed to be intrinsically sustainable—for example, made to be recyclable or biodegradable by design (topics we will explore in Sects. 7.4 and 7.6). In an energy context, imagine biodegradable polymer solar panels that can be safely composted at end-of-life instead of landfilled, or wind turbine blades built from bio-resins that sequester CO_2 during the growth of their plant feedstock. These ideas are driving current research so that the polymer-enabled clean energy revolution does not repeat the environmental mistakes of the first plastics revolution.

In summary, polymers matter for green energy because they provide the material foundation on which cleaner technologies are built. These methods can provide lightweight materials, which save energy, are strong to rough use, providing dependability, and lower supplies, leading to put them into action. As one industry source puts it, companies providing energy, whether renewable or traditional, "depend on high-performance plastics for a number of critical applications" in generation, storage, and efficiency. In fact, polymers now play a critical role in producing renewable energy through wind turbines and solar cells. The challenge ahead is to ensure that these same polymers are produced, used, and disposed of in sustainable ways, so that materials for green energy are themselves green. The next sections delve into

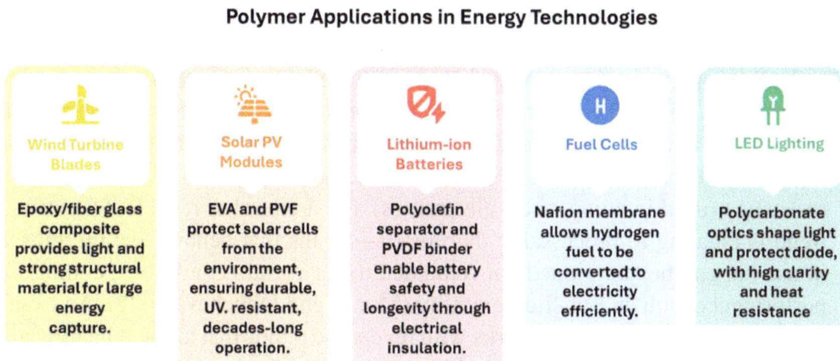

Fig. 7.2 Polymer applications in energy technologies. (Developed by the authors)

how polymer science and technology are rising to that challenge, without sacrificing the benefits that plastics have brought to energy applications.

Figure 7.2 depicts a schematic illustration of how polymers contribute to a clean energy system. For instance, an infographic showing a wind turbine (with labels on the composite blade and polymer parts in the nacelle), a solar panel (with polymer encapsulant and backing film highlighted), an electric car (with polymer battery components and lightweight body indicated), and a building (with insulation and plastic windows). The figure underscores the pervasive presence of polymers in modern energy and efficiency technologies.

7.4 Fundamental Design Principles

How do scientists and engineers design polymers fit for both high performance and sustainability? This section outlines some fundamental design principles guiding modern polymer development, especially as they apply to green energy and environmental responsibility. We will touch on molecular structure, property tuning, and new design philosophies like "circularity by design."

7.4.1 Molecular Architecture and Properties

At the most basic level, the properties of a polymer depend on how its molecules are put together. Key variables include the length of the polymer chains (molecular weight), the geometry of the chain (linear, branched, or crosslinked network), and the chemistry of the repeating units. By controlling these, chemists tailor materials for specific applications:

- *Thermoplastics* vs. *Thermosets:* Thermoplastics are polymers made of independent linear or branched chains that soften and melt upon heating (allowing reshaping and recycling). Polyethylene (PE), PVC, and nylon are examples. Thermosets, like Bakelite or epoxy resin, consist of chains that are chemically crosslinked into a network; once set, they will not melt but instead decompose with heat. Thermosets offer superior high-temperature stability and chemical resistance (used in composites, electronics), but are hard to recycle. One design trend is creating re-processable thermosets—resins that behave like thermosets in use but can be "un-zipped" or remolded under certain conditions, combining performance with end-of-life flexibility (Morici and Dintcheva 2022). An example is vitrimers, polymer networks with reversible bonds that allow reheating and reshaping. This is promising for things like wind turbine blades: traditionally thermoset, but future versions might use vitrimer chemistry to allow recycling.
- *Copolymers and blends:* Actually, advanced polymers do not only repeat one type to be produced, but are composed of two or various monomers polymerized together to form copolymers or polymer blends, corresponding to different properties that can be tuned. The ionomer m embrane Nafion in fuel cells is exemplified because its main structure is similar to Teflon, with both stable mechanical and chemical properties. Nevertheless, the sulfonic groups in Nafion do obstruct compatible traits of one structure in a copolymer. The block polymers on a photovoltaic device consist of electron-rich and electron-deficient regions, enabling absorption of photons and the separation of charges. Designed with a structure of two block polymers capable of combining multiple functions in one material through the precise arrangement of monomers or mixing polymers. These are challenges of maintaining polymer morphology at the nano scale to control each component phase without separation, which is a major theme in polymer physics. Advances in this area have led to all-polymer solar cells where both donor and acceptor are polymers, offering better stability than when mixing polymers with small molecules (Gnida et al. 2022).
- *Additives and composites:* Raw polymers often do not meet all requirements, so additives (plasticizers, stabilizers, fillers) are included. For example, adding certain plasticizers to PVC makes it flexible for cable insulation, while without plasticizers, it's rigid for pipes. In designing polymer insulation foams, a blowing agent (which creates bubbles) is chosen to yield a foam structure that traps gas and insulates heat.

7.4.2 Fiber Reinforcement Is a Critical Design Strategy

Embedding glass, carbon, or natural fibers in a polymer matrix yields a composite with vastly higher strength and stiffness. This is how relatively weak polymers create wind turbine blades and hydrogen storage tanks that rival steel in performance.

The fiber-polymer interface must be engineered (often fibers are surface-treated to bond well with the resin).

7.4.3 From a Design Standpoint, a Composite Is a Multi-Scale Material

At the molecular level, the polymer may need toughness, at the micro-level fiber distribution matters, and at the macro-level, the part shape is optimized. Achieving the right balance is an art and science—one being continually refined for applications like longer turbine blades or lighter vehicle frames.

7.4.4 Green Chemistry and Process Design

Fundamental design isn't only about the polymer's formula; it's also about how it's made. Green chemistry principles encourage using safer reagents, renewable feedstocks, and energy-efficient processes. For example, traditional polymerization of polycarbonate (PC) involved toxic phosgene gas; modern processes avoid that in favor of transesterification routes. Biopolymers like PLA are now produced via fermentation and catalysis starting from corn sugar—a very different pathway than petro-plastics.

7.4.5 The Design of Catalysts Is Crucial

To save energy and decrease plastic trash waste, the advanced catalysts are thus introduced to polymerizations at a lower temperature, being focused on by scientific researchers as a pivotal principle for sustainable polymers (Verebélyi et al. 2023). Moreover, catalysts allow the synthesis of polymers with significantly precise structures like block, length, tacticity, etc., to improve properties and increase recyclability, including can also aid in producing polymers without mixing structures that are difficult to separate.

7.4.6 Another Aspect Is Process Intensification

Reducing the steps and the solvent in the polymer production in continuous flow reactions, like smaller polymer plants in the future, modular, and renewable electricity, can drastically cut the carbon footprint of polymer manufacturing.

7.4.7 *Life-Cycle Thinking and Circularity*

The assessment of the life cycle and considerations of life destination counted from the begin design step are perhaps the new principle transformation. Before the performance and cost of polmyers were primary prospective while impact environment of polymers without aspect. Now, many researchers realized about life cycle of the design engineering for new polymers to ensure from the outset that materials can corresponding to demand of a circular economy. This means asking many questions like: Where do the monomers come from? (Ideally from renewable or waste resources rather than new fossil resources.), How much energy and water does production use? What happens after the polymer's end-of-life?, Can it be recycled mechanically? Chemically broken down and repurposed? Will it safely biodegrade in the environment or a compost facility? By answering these early, a polymer can be designed with, say, chemical linkages that permit depolymerization back to monomers (facilitating true recycling), or with functional groups that allow enzymatic degradation if it escapes to nature. An example is designing polyesters or polyurethanes that incorporate bonds cleavable by mild chemical or biological conditions, enabling recycling. The "sustainable polymers" framework defines such materials as those from renewable/recycled feedstocks and which can be responsibly managed at end-of-life (recycled or biodegraded) with reduced environmental impact across their life cycle. Meeting all these criteria simultaneously is challenging, but it is the goal. A visual representation often used is a life-cycle loop (Fig. 7.3) where biomass grows (absorbing CO_2), is converted to polymer, which is used and eventually either recycled into new polymer or biodegrades back to CO_2 and water,

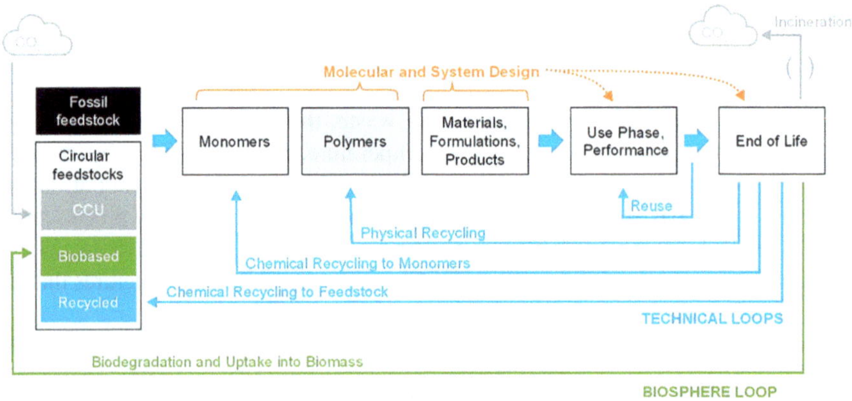

Fig. 7.3 The demonstration of polymer life-cycle design that depicts loop of converted plant biomass to monomers, and applied to a plastic product via polymerization, which is used in an application, like part of a solar panel or wind turbine, then after use it either goes to recycling to reform to be polymer or biodegradation to return to carbon dioxide (CO_2) and water, which plants necessary use again. This "circular polymer economy" diagram should highlight design inputs at each stage: e.g., catalyst design in polymerization, stabilizers added for use-phase durability, and functional groups. (Copyright from Angewandte Chemie in 2022 (von Vacano et al. 2022))

completing the cycle. The frontier of the polymer field in science refers to designing polymers in practice to follow ideal thinking.

7.4.8 Balancing Performance with Sustainability

The trade performance and eco-friendliness has been required for designing polymers. While bio-based or recyclable polymers have historically not matched the strength, heat resistance, or other benchmarks set by incumbent plastics. For example, early bio-polyesters, such as the qualities of PLA, have brittle and low heat distortion temperatures, leading to limited use. Thus, identifying application-appropriate targets is the principle that not every sustainable polymer must do everything instead of PET or PE does each can be optimized for certain uses. Newer sustainable polymers warrant rational design to close the performance gap. Since bio-based polyamides and polyesters with properties on par with engineering plastics are now reported by many researchers through careful control of monomer structure and polymer architecture. The concept of "drop-in" sustainable polymers has emerged with materials that can directly replace traditional plastic in existing processes and products, for instance, bio-polyethylene that is chemically identical to fossil polyethylene. These require little compromise in performance or infrastructure. Another strategy is hybrid materials, e.g., blending a biodegradable polymer with a small fraction of conventional polymer or fiber reinforcement to boost performance while retaining partial sustainability (though end-of-life handling then becomes more complex). Ultimately, advanced polymer design strives to create materials that check all the boxes: economical, high-performance, and environmentally benign. Achieving that trifecta is no small feat and demands creativity in molecular design, which is why multidisciplinary research—drawing on polymer chemistry, materials engineering, and environmental science—is so important (a topic highlighted in Sects. 7.7 and 7.9).

In summary, designing polymers for the future involves a holistic approach. It starts at the monomer level (choosing building blocks that are sustainable and creating the desired polymer architecture), extends through processing (green manufacturing techniques), and ends with foresight into recycling or degradation. Principles like life-cycle thinking and rational design for sustainability have joined traditional polymer design principles of optimizing mechanical or thermal properties. For a general scientific audience, the key takeaway is that polymers are "made-to-order" materials—by altering their composition and structure, we can give them almost any property we want. Now we are learning to "program" not just properties in service, but also what happens before and after service (the birth and rebirth of the polymer). The design polymers philosophy route would enable applied green renewable energy technologies, with no impact on the planet's surroundings and well-being.

7.5 Current Application Landscape

To date, polymers are still favorably used around facets of the clean energy landscape. In this section, the applications of polymers in several fields in current use, such as solar panels, wind blades, and vehicle or phone batteries, and so on. To clarify and appreciate each domain, these applications have been integrated to identify the polymer materials used, work function, and capitalization.

7.5.1 Solar Energy (Photovoltaics)

Modern solar panels or photovoltaic modules are designed with advanced techniques. The combination of materials and polymers is similar sandwich, with polymers acting as bread and solar cells as the middle bread. For example, a standard silicon (Si) PV module uses ethylene-vinyl acetate (EVA) polymer as an encapsulant layer that comprehensively covers the fragile silicon cells to cushion and bond, including to reinforce a stronger glass front sheet. Another side of a polymer backsheet often uses a fluoropolymer like PVF Tedlar® or a multi-layer PET laminate to serve as electrical insulation and weather protection (Hong and Xu 2025).

These polymer layers are critical for module reliability: they must resist UV light, moisture ingress, and temperature cycling for over 25 years outdoors. Improved encapsulant polymers have steadily increased panel lifetimes by reducing cell cracking and corrosion. In newer thin-film solar panels (like CdTe or perovskite solar cells), flexible polymer substrates can replace glass entirely, enabling lightweight, rollable solar panels. The burgeoning field of organic photovoltaics (OPV) relies on conjugated polymer semiconductors (like P3HT or PTB7) as the active layer that absorbs sunlight. While not yet as efficient as silicon, OPVs have reached about 18% efficiency in lab devices (Gnida et al. 2022), and they offer unique benefits: they can be printed from solution onto plastic films, allowing colorful, flexible solar sheets that can stick on windows or curved surfaces. Polymers also serve as antireflective coatings on glass (nano-textured PMMA layers that reduce reflection) and even as solar concentrators (plastic sheets doped with luminescent dyes that guide light to solar cells at the edges). In the emerging perovskite solar cells, a thin polymer (such as PEDOT: PSS or PTAA) often acts as a hole transport layer, improving performance and stability (Butsriruk et al. 2023; Butsriruk 2021). The big picture: without polymers, modern solar panels would be heavier, less durable, and more expensive. By using plastics, we get lightweight modules that can be deployed on rooftops without heavy support and that last decades through heat and cold.

7.5.2 Wind Energy

Wind turbines are giants of composite engineering. The blades of utility-scale wind turbines (now often 70+ m long) are typically made from glass fiber-reinforced epoxy or polyester resins, sometimes with carbon fiber reinforcements in critical sections. These fiber-reinforced polymer (FRP) composites offer an unparalleled combination of lightweight, high strength, and fatigue resistance. Each blade is basically a hollow shell made by layering fiberglass cloth and resin in a mold, which then cures into a stiff structure. The use of polymers allows engineers to create the precise aerodynamic shapes needed for efficiency, with internal spars and ribs also made of composites. Polymer composites have enabled blade lengths to grow steadily, and longer blades capture more energy from the wind, significantly lowering the cost per kWh of wind power. Inside the turbine nacelle, polymers abound as well: nylon or POM (acetal) gears in small auxiliary drives, polyurethane potting compounds protecting sensors, and polymeric electrical insulation for the generator. Even the tower often has a polymer protective coating (a specialized paint) to withstand salt spray or desert sand. One interesting application is polymer bushings and bearings in wind turbines, which can self-lubricate and damp vibrations, reducing maintenance. Ensinger (a plastics manufacturer) notes that wind turbine bearings use polymers with friction-reducing properties to improve reliability (Peng et al. 2023). The current trend is exploring thermoplastic blades (see Case Study 7.1) to allow recycling, using resins like PET or PA reinforced with fibers. Additionally, offshore wind turbines benefit from polymers in their cables (polyurethane sheathing) and floats. In a brief, from root design to complete construction, polymers are intrinsic to wind turbine architecture and function.

7.5.3 Energy Storage: Batteries and Supercapacitors

The structure of a lithium-ion battery cell is used in a phone or an electric car, and is normally composed of several polymer components, of which the most critical is the separator, a polyolefin film thickness of 10–20 μm is typically made of polypropylene (PP) or polyethylene (PE), placed between the battery's anode and cathode. The separator domain is to electrically extract the electrodes while allowing ions to pass through the microscopic pores. These separators are a product of advanced polymer processing, which have exceedingly uniform and often have a shutdown feature with pores that melt closed at around 130 °C to stop current in an overheating scenario. Another polymer in Li-ion batteries is the binder, like polyvinylidene fluoride (PVDF) or carboxymethyl cellulose, that acts as glue to bind the active material powders onto the metal foil substrate. Though only a few percent of the battery by weight, binders have a high impact on performance and longevity. Thus, newer silicon-anode batteries use specialized polymer binders that can

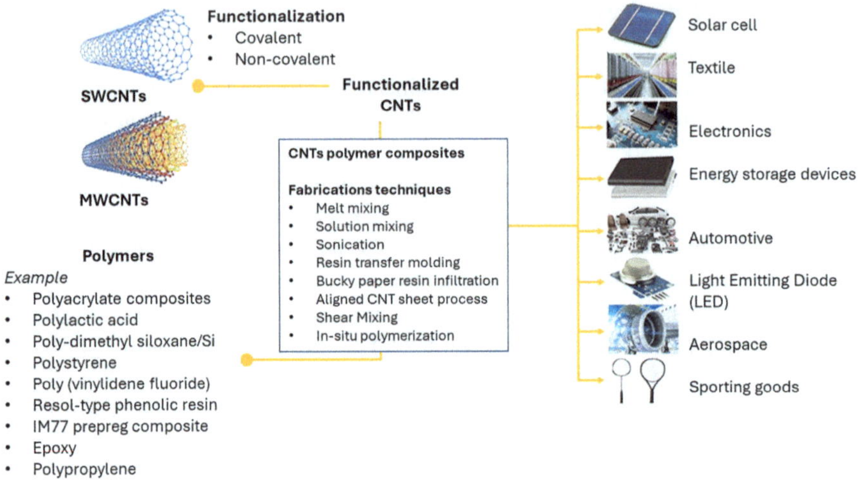

Fig. 7.4 The presentation of several products from polymer materials, like a part of the solar cell, textile, electronics, energy storage devices, vehicle, diode, aerospace, and sporting goods, was reproduced from Polymers in 2021 (Mohd Nurazzi et al. 2021)

accommodate expansion. Battery packs and cells are also housed in endure polymer casings of between PC/ABS blends or polypropylene that are lightweight and resistant to impact when compared to metal cans (Shen et al. 2023).

The broad takeaway from this application survey is that polymers are everywhere in sustainable energy systems, functioning as structural materials, electrical components, thermal managers, and protective layers. They often serve unsung supporting roles that nevertheless are mission-critical: a wind turbine blade's resin might not be as flashy as a new semiconductor material, but without it, the turbine couldn't spin and produce power. In many cases, polymers have expanded the possibilities of energy technology, making devices more adaptable (think flexible solar cells), more durable (weather-resistant coatings), and more affordable (mass-produced plastic parts). However, with this ubiquity comes responsibility: the reliance on polymers means we must ensure they are produced and disposed of sustainably. The next section evaluates how current polymer use in energy stacks up on a sustainability scorecard, and what improvements are underway.

Figure 7.4 illustrates the conceptual diagram for polymers in clean energy applications, such as a sun for solar panel, a wind turbine, a battery, a fuel cell, and a house (efficiency).

7.6 Sustainability Scorecard

Polymers have clearly become indispensable in clean energy technologies, but how do they fare when it comes to sustainability and environmental impact? This section examines the sustainability scorecard for polymers in a greener energy context, identifying successes, shortcomings, and emerging solutions. We evaluate polymers across their life cycle: resource extraction, production energy, use-phase benefits, and end-of-life outcomes. The goal is to see where we stand and how to improve, moving toward truly sustainable polymer use in the energy sector.

Raw Materials and Production Currently, the vast majority of polymers (over 90%) are made from fossil fuel feedstocks—oil or natural gas. This corresponds to polymer production to greenhouse gas emissions, and non-renewable resource use. In 2019, about 4–8% of global oil and gas was used as feedstock and energy for making plastics (Andrady and Neal 2009). That means that not only reduce fossil fuel burning for energy, but it also addresses fossil fuel transformation into materials. Traditionally, the production of plastics is energy-intensive and emits significant CO_2.

The example of polymers manufacturing with 1 kg of polyethylene (PE) might emit on the order of 2–3 kg CO_2 when accounting for everything, neither oil extraction, cracking to ethylene, and polymerization. Whereas some plastics, like polycarbonates or nylons, have even higher footprints since more complex synthesis. On the other hand, the production of plastics typically uses less energy and emits less CO_2 than producing metals like aluminum or steel per unit mass. For this reason, replaced heavier materials with plastic to gain material efficiency. But because people on the planet have used huge volumes of plastic, global production exceeded 460 million tons in 2021, enormous impact on the total environment (Dokl et al. 2024). Life-cycle studies of polymers in 2023 show that the manufacturing of plastics contributed ~3.3% of global greenhouse gas emissions (Jones et al. 2023).

Some improvements are underway. Bio-based polymers are entering the market—for instance, bio-polyethylene made from sugarcane ethanol (as done in Brazil) or polyethylene furanoate (PEF) made from plant-derived furans (Suarez et al. 2023). These can have a lower cradle-to-gate carbon footprint if renewable energy powers their production. However, bio-based doesn't automatically mean low emissions: fertilizer use, land changes, and processing energy can erode the benefits. Still, a 2024 study comparing life-cycle metrics found that switching to bio-based or recycled feedstocks can significantly cut fossil energy use and greenhouse gas (GHG) emissions for many plastics (Dokl et al. 2024). Another angle is green manufacturing: using renewable energy in polymer plants, improving catalysts to require less process energy, and recycling heat. Some polymer companies now build "bio-refineries" integrated with bioproduct and energy production to maximize efficiency (Kumar et al. 2024). On our scorecard, we'd give current polymer production a middling grade on sustainability—heavy reliance on fossil inputs and significant emissions, but with a clear path to improvement through renewable feedstocks and process optimization.

7.6.1 Use-Phase Impacts (Positive and Negative)

Polymers often improve sustainability during the use phase of products, especially in energy applications. For example, lightweight plastics in cars and airplanes save many times the energy (and emissions) over the vehicle's life than was used to create the plastics. A study found that substituting plastic parts for metal in cars and trucks markedly reduces fuel consumption, outweighing the environmental cost of making the plastic (Andrady and Neal 2009). Similarly, plastic insulation saves huge amounts of heating fuel, which can save energy in 1 year and reduce many times the energy required to produce it. In this way, plastics enable such energy savings and renewables deployment that aim to arguably prevent more emissions than they emit. Some analysts have suggested that modern society's use of plastics in the form of packaging, insulation, and car parts is related to energy savings, reduction of food waste, and other efficiencies that still have societal benefits. However, the drawback still occurs persistent and pervasive waste problem around the world. Durable plastics can be used in single-time or short-life applications, creating waste mountains that nature cannot digest. Each year, around 11 million tons of plastic waste are dropped into the oceans, harming marine life and then breaking down into microplastics that have been found from the deepest trenches to Arctic snow. The toxic additives in these microplastics have affected ecosystems and the health of humans, which remains an active research concern. From a sustainability side, the contamination of the environment is a harmful signal to use polymers nowadays (Rohrer 2007). Actually, a fair scorecard has to truly weigh both direct benefits and externalities, like energy saved by a plastic component and plastic debris polluting water and soil, respectively.

7.6.2 End-of-Life or the Achilles' Heel of Plastics

Unfortunately, the statistics of discarded plastics in 2015 showed that only about 9% of all plastic waste had been recycled even once, 12% incinerated, and nearly 79% collected in landfills or the natural environment (Geyer et al. 2017). In the field of energy technologies, a similar challenge with solar panels is that old wind turbine blades are currently mostly cut up and landfilled. While solar panels contain polymers along with glass and metals, they are beginning to see recycling initiatives. Also, decommissioned lithium batteries require careful handling and currently focus on recovering metals, with the plastic components often burned for energy or wasted. The problem is that most polymers were never designed to be recycled, and many are part of complex products. Traditional mechanical recycling—where plastic is melted and remolded—is hampered by contamination, polymer mixing (e.g., multi-layer films), and property degradation (each re-melt can break polymer chains, lowering quality). Thus, recycled plastic often ends up in lower-grade uses (downcycling), and eventually still becomes waste. The scorecard here is poor: despite decades of awareness, globally, we still recycle only a small fraction of plastics. In 2019, Europe led with about 30% recycling rates, while the U.S. was around 8% (Geyer et al. 2017).

From a design perspective, previously described in Sect. 7.4, the best route solution is to design to follow inherently loop polymers. Among exciting developments are polymers that can depolymerize on command: e.g., poly (diketoenamine) (PDK) plastics that can be fully reverted to monomers through the use of acid or advanced biopolymers that rapidly compost under some conditions, like enzyme-triggered breakdown. Additionally, integration of bio-sourced materials is studied to make polymers that can be metabolized by microbes after being spent, becoming part of natural biogeochemical cycles rather than persistent trash. Polyhydroxyalkanoate (PHA) is a polymer produced by certain bacteria, and already fulfills this, being both bio-sourced and truly biodegradable in marine environments; however, this polymer type's high cost and limited performance confine it to niche uses so far (Zhou et al. 2023).

7.6.3 Case in Point

Consider that tens of thousands of aging wind turbine blades will retire in the coming decades, and these cannot be recycled; they can only be melted down. This is why companies are racing to invent recyclable blade materials, like the thermoplastic resin in the Case Study of Sect. 7.1, and methods to deal with existing ones by grinding them for use as filler in cement, or pyrolyzing them to recover fibers. On the scorecard, wind blades would currently score low due to most ending up in landfills; however, the next generation of designs is poised to improve that with recyclable polymers (Li et al. 2023). Furthermore, solar panels actually contain relatively small amounts of polymer; this of mass mostly glass and aluminum, but still pose a recycling challenge due to the encapsulant and backsheet polymers that must be separated to recover high-value silicon and metals (Roche et al. 2024). The solvent or thermal methods to delaminate PV modules are still of interest research is ongoing on; meanwhile, certain companies simply shred panels and incinerate the plastic portion to recover metals/glass. Also, a more sustainable polymer encapsulant could dissolve or depolymerize after being spent, which could greatly enhance solar panel recyclability.

The effectiveness of a race development of polymer recycling: the performance of polymers works together with response to utilized demand and innovation to ensure do not also undermine the sustainability production of the clean energy technologies. Encouragingly, as detailed in the next section, many interdisciplinary efforts, neither materials science nor engineering, are converging to address these challenges. The "eco-conscious strategies" to enhance polymer sustainability

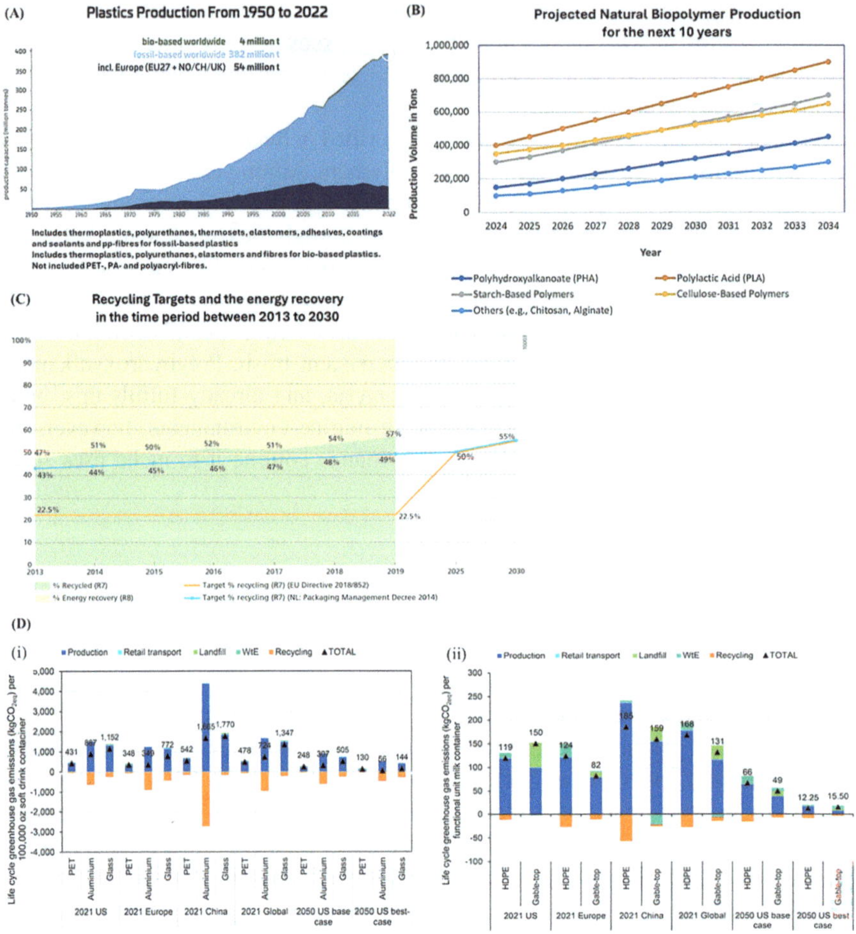

Fig. 7.5 (**a**) shows the volume of polymer production between 1950 and 2022, (**b**) shows the prediction of various polymer productions within a decade starting from 2024, both (**a**) and (**b**) are Copyright from Polymers in 2024 (Beena Unni and Muringayil Joseph 2024), (**c**) The percentages for each year (gathered and forecasted from 2013 to 2030 based on Afvalfonds Verpakkingen's annual monitoring) show plastic recycling targets (orange line) and energy recovery (blue line), copyright (Calisto Friant et al. 2022), and (**d**) Reproduced from Environmental Science & Technology in 2024 (Meng et al. 2024), comparison of GHG emissions (i) between 100 k oz of aluminum cans and PET bottles, and (ii) between Gable-top cartons and HDPE bottles. Both datasets are collected and predicted in Western Europe and China from 2020 to 2050

include everything from greener synthesis and recycling technologies to using life-cycle assessments as standard practice. By viewing polymers through the lens of a circular economy—keeping them in use as long as possible and regenerating new materials from old—we can dramatically improve their sustainability scorecard (Fig. 7.5a–d). In doing so, we ensure that polymers truly live up to the "green" promise of the renewable energy systems they are part of, rather than detracting from it.

7.7 Interdisciplinary Case Studies

Innovations at the nexus of polymers and energy often arise from interdisciplinary collaboration—chemists, engineers, physicists, environmental scientists (and more) working together. In this section, we present two case studies as "boxed" sidebars, illustrating how such collaborations are tackling real-world challenges. Each case highlights a specific breakthrough or project where polymer science and energy technology intersect, demonstrating the value of a broad, integrated approach.

Case Study 7.1 Reinventing the Wind Turbine Blade—A Recyclable Composite Revolution

1. **Challenge:** Wind turbine blades, typically made from thermoset fiberglass composites, are extremely difficult to recycle (Fig. 7.6a). As wind power booms, tens of thousands of blades (each 40–60 m long or more) will retire in the coming decades (Cheng et al. 2025). The prospect of mountains of decommissioned blades ending up in landfills or being incinerated has raised serious environmental concerns. Traditionally, once the epoxy resin in a blade is cured, the material cannot be remelted or reprocessed—a fundamental materials challenge. The question arose: can we make a wind blade out of polymers that would allow end-of-life recycling, without sacrificing performance?

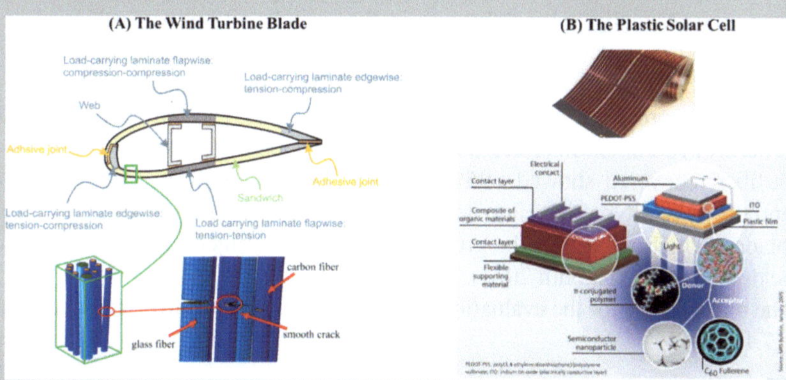

Fig. 7.6 (**a**) The diagram of each part of the wind blade of the turbine, after gravitational forces and torque loads are considered, where the glass fiber is a material in the blades, is reproduced from Materials in 2017 (Mishnaevsky et al. 2017). (**b**) The photo of a plastic solar device specimen (above) and the scheme of a plastic solar cell structure (below) were represented in 2024 (Zboncak 2024)

Case Study 7.1 (continued)

2. **Interdisciplinary Solution:** Around 2018, a partnership formed between materials scientists at Arkema Inc. (a chemical company) and engineers at General Electric's LM Wind Power and the U.S. National Renewable Energy Lab (NREL). Their goal: develop a thermoplastic resin system for blades. Arkema had a novel liquid thermoplastic resin called Elium®, which behaves like a thermoset during manufacturing (infusing into fibers and then hardening), but crucially, it can be unzipped/dissolved with the right process (Hoodin 2024). Chemists and polymer scientists at Arkema worked on optimizing this acrylic-based resin so it could cure fast in blade molds and bond well to glass fibers. Meanwhile, mechanical and aerospace engineers at LM Wind Power designed full-size blade molds and manufacturing processes to accommodate this new resin (which has a different viscosity and cure profile than epoxy). NREL researchers reported contributions to testing the manufactured blades to obtain structural performance and longevity (Hoodin 2024).

In 2023, the prototype 77-m wind turbine blade using the Elium thermoplastic composite was produced (Hoodin 2024). The produced wind blade met all mechanical requirements, like stiffness, strength, fatigue life, demonstrating that the thermoplastic composite was on par with conventional epoxy composite in structural performance. Actually, a breakthrough in the recycling tests was achieved at the end-of-life of the product, which can be made by cutting the wind blade and immersing it in a solvent at elevated temperature to dissolve the thermoplastic resin. Then, fiberglass reinforcement could be separated out cleanly, and the resin precipitated back out from solution. In effect, they proved the blade to be 100% recyclable—the glass fiber recovered showed >95% of its original strength, and the resin could potentially be repolymerized into new products. This was a stark contrast to thermoset blades, which at best can be ground into filler for cement, a form of very low-value downcycling (Hoodin 2024).

The underscores for the evaluation of material and design can be calculated from Eq. (7.1)

$$P = \frac{1}{2}\rho A V^3 C_p \eta \tag{7.1}$$

where P presents as the power generated in units of watts, ρ is referred to as the density of air in units of kilograms per m^3, A is the area of the blade sweeping in units of m^2, V is the speed of wind in units of meter per second, C_p shows the value of the coefficient of the turbine performance, which depends on the materials and design qualities, and η is referred to as the potential of the turbine system (Firoozi et al. 2024).

Case Study 7.2 The Quest for Plastic Solar Cells—When Chemists and Physicists Team Up

A solar panel, sometimes called plastic solar cells, uses carbon-based polymers and molecules to convert sunlight into electricity (Fig. 7.6b) (Lin et al. 2024). To achieve this, chemists got to work designing new π-conjugated polymers with tailored energy levels and band gaps to absorb more sunlight and match the fullerene acceptor (Dutta 2022). One breakthrough was the polymer P3HT (poly(3-hexylthiophene)), which, when combined with fullerene PCBM in 2005, yielded ~5% efficiency—a record at the time (Uddin et al. 2024).

7.7.1 Progress Through Collaboration

By the 2010s, a whole ecosystem of researchers formed—teams often included computational chemists (predicting properties of new molecules), synthetic chemists (actually making novel donor polymers or non-fullerene acceptors), spectroscopists and microscopists (probing the morphology and charge dynamics in blends), and engineers (integrating OPV modules into real products like solar windows or chargers). This convergence paid off with rapid efficiency gains: new polymer donors like PTB7, and later PM6 and D18 (to name a few), paired with new non-fullerene acceptors (e.g., ITIC and Y6 families developed around 2015–2019 by Chinese chemists), pushed single-junction OPV efficiencies over 18% in labs (Gnida et al. 2022). For context, that's on par with the early silicon solar panels, and it happened due to clever molecule design (chemistry) plus deep understanding of device physics—for instance, the field learned to engineer the energy levels to minimize voltage losses and to use additive processing to optimize morphology (Gnida et al. 2022).

7.7.2 Status and Outlook

As of the mid-2020s, polymer-based solar cells have achieved more than 15% efficiency with promising stability for several years under outdoor conditions in small areas, and further companies are once again looking to commercialize OPV for niches like indoor solar cells through sensors from room light or as tinted window films that generate power (Zhang et al. 2023). The journey isn't over, but the progress so far showcases interdisciplinary synergy. Had it been only chemists working on their own or physicists on theirs, OPV would not be where it is. The melding of fundamental science and practical engineering that synthetic routes enable new materials, combined with device optimization, to elevate plastic solar cells from a curiosity to a viable technology.

7.8 Grand Challenges and Frontier Directions

The green energy routes of polymer products in the future have several grand challenges and frontier research directions. This last substantive section discusses big hurdles and emerging ideas to set the stage for producing polymers that can truly usher in a sustainable energy paradigm. Many of these challenges are inherently interdisciplinary, tying together threads of chemistry and engineering.

7.8.1 *Grand Challenge 1: Making Polymers Sustainable by Design (Closing the Loop)*

As detailed in the sustainability scorecard, one of the huge challenges is achieving a circular lifecycle for polymers that not only serve their function in energy devices but also re-enter the materials stream after use without downgrading or causing pollution. The frontier direction for built-in recyclability or degradability of polymers, for example, new polymer categories can be depolymerized and then back as monomers at end-of-life with a specific trigger, like heat, a catalyst, or light. Researchers have explored dynamic covalent chemistries, either vitrimers or reversible bonds, as well as chemically recyclable polymers, such as polycarbonate (PC) ethers or certain polyesters that can degrade to monomer under certain conditions. Another encouraging aspect is that biologically assimilable polymer materials and when they leak into the environment, can be broken down by microbes into harmless biomass or nutrients. Advanced research on enzyme-polymer interactions has tried to use engineered enzymes, like PETase, that can efficiently digest common plastics, including tweaking polymer structures to be more enzyme-friendly without losing performance (Kumar et al. 2024).

7.8.2 *Grand Challenge 2: Performance Vs. Sustainability Trade-Offs*

Nowadays, many of the sustainable polymers (biobased, biodegradable, etc.) mostly underperform relative to the best petro plastics (Dhanraj et al. 2022). For instance, bioplastics, like Polylactic acid (PLA) or starch blends, might be attribution of brittle or have low thermal resistance, and mechanically recycled bioplastics are often free from impurities or inferior properties (Nizamuddin et al. 2024). To address this gap, designing sustainable polymers via engineering could match or exceed the performance of traditional ones. Synergistic blends and composites (e.g., reinforcing a biodegradable polymer with cellulose fibers or nano-clays to give it strength) or new polymer architectures altogether might be used (Kumari et al. 2023). The example of this work mentions bio-derived polyamides (PA) and polyesters, which mimic the molecular structure of high-performance petroleum polymers (Nizamuddin et al. 2024). By using monomers from renewable sources (like castor oil-derived sebacic acid or sugar-derived furanics) and polymerizing them into materials analogous to nylon or PET, researchers have made biopolymers with impressive thermal and mechanical traits. Catalyst development plays a role too: advanced catalysts allow polymerizing monomers that were previously hard to use, unlocking new materials. Also, nano-engineering could help: incorporating functional nanoparticles or additives that compensate for weaknesses of sustainable polymers (e.g., adding UV stabilizers to biodegradable plastics so they survive longer in use but still degrade later, or using toughening agents to improve impact resistance of brittle biopolymers). The frontier concept of "sustainable composites"—combining natural fibers (like hemp, flax, bamboo) with bio-resins—also holds promise for replacing fiberglass in certain applications (Andrew and Dhakal 2022). Ultimately, the goal is no-compromise materials, where choosing a sustainable option doesn't mean sacrificing durability or efficiency of the device.

7.8.3 Grand Challenge 3: Polymer Stability in Demanding Applications

Green energy applications can be harsh: think of solar panels under constant UV irradiation, batteries with reactive chemistry. Polymers in these settings must endure without degrading—or conversely, we must accept and manage their degradation. For instance, one reason organic solar cells haven't fully broken through is concerns about longevity—polymers can photo-oxidize, pinholes can form in encapsulants, etc. (Chen et al. 2025). Likewise, in high-temperature applications (like concentrated solar power plants or geothermal systems), many polymers fail because of heat. The challenge is to extend polymer operating windows via better stabilizers, coatings, or fundamentally more robust polymers (Chen et al. 2025). Frontier research includes developing self-healing polymers that can recover from damage (imagine a wind turbine blade coating that automatically heals after a lightning strike or sand erosion) and new antioxidant and UV absorber chemistries that can protect polymers for decades. Another approach is hybrids—e.g., polymer-derived ceramics for ultra-high temp, or polymer-metal composites that can withstand radiation in nuclear applications (Ionescu et al. 2021; Wu et al. 2023). There's also interest in real-time monitoring of polymer components (like sensors embedded in a composite blade to detect the onset of fatigue or environmental stress cracking) so that maintenance can be done proactively, avoiding catastrophic failure and extending life. The frontier is not just making polymers last longer, but making smarter polymer systems that can signal their health or adapt to conditions. For example, a recent study demonstrated a battery electrolyte that changes color if it starts degrading (a built-in indicator for replacement) (Sattar et al. 2024).

7.8.4 Frontier Technology Spotlight

A few exciting polymer innovations on the horizon deserve mention:

7.8.4.1 Polymer Electrolyte Fuel Cells 2.0

Research into new polymer membranes that operate at higher temperature (>120 °C) or lower humidity than Nafion could drastically improve fuel cell vehicles (higher efficiency, simpler cooling). Materials like protic ionic liquid-infused polymers or acid-base blend membranes are on the frontier (Kumar and Alam 2024; Hwang and Sung 2023). These require chemists to synthesize stable structures and engineers to test them in real cells.

7.8.4.2 Artificial Muscles and Energy Harvesters

Emerging fields like soft robotics and energy harvesting often use piezoelectric or dielectric elastomer polymers (Wang et al. 2022). These can act as generators (converting mechanical motion to electrical energy)—imagine roads embedded with polymer mats that harvest energy from car vibrations, or wave energy converters using giant rubber "muscles" that flex with ocean waves (Guo et al. 2023). The frontier challenge is durability and scaling for these novel energy devices.

In confronting these challenges, one recurring theme is the need for interdisciplinary collaboration and systems thinking. No longer can a polymer scientist operate in isolation, focusing only on monomer chemistry, for example. They must collaborate with environmental scientists to consider end-of-life, with mechanical engineers to ensure the material meets application needs, and with business experts to ensure the solution can be commercialized. The "grand challenges" are grand precisely because they span multiple domains. However, this is also what makes the frontier exciting, as solutions will likely come from unexpected synergies with several fields. A catalyst discovered in the scope of labs might co-work with an engineering breakthrough in reactor design to enable a new recycling process that knowledge of biologists in the field of enzymes might integrate with a polymer in the field of chemists to yield a compostable plastic for solar panel encapsulation.

The frontier clearly points toward a beyond paradigm of polymer development sustainability, as well as equal to performance and cost. If the twentieth century was about maximizing performance and volume of plastics, the twenty-first century (especially in the context of green energy) will be about refining those achievements to align with environmental stewardship. Many call this the dawn of the "Age of Sustainable Polymers" (Mohanty et al. 2022). It is a pivotal moment, and polymers—the very materials that helped create some of our environmental problems—are poised to be key contributors to solving them when guided by innovative, interdisciplinary science.

7.9 Conclusions and Outlook: Toward a Polymeric Energy Paradigm

From the birth of celluloid to the frontiers of recyclable composites and plastic solar cells underscores a powerful narrative: polymers have continuously reinvented themselves to meet society's needs. In the early days, they mimicked scarce natural materials, offering humanity new possibilities (Beena Unni and Muringayil Joseph 2024). In the modern era, they became the workhorse materials, enabling technologies from electrification to renewable energy. Now, as we face the urgent need for sustainability, polymers are poised to transform once again, this time into eco-conscious materials fitting a circular economy (Wiśniewska et al. 2023).

Looking ahead to the "polymeric energy paradigm," we can envision a future where sustainable polymers are a linchpin of the global energy system:

- Wind turbine blades and solar panels could be made from fully recyclable or bio-based polymers, so renewable energy infrastructure leaves no legacy of waste (Korniejenko et al. 2021). Imagine wind farms where old blades are routinely collected and remade into new ones—a circular supply chain (Leon Jr. 2023).
- Energy storage devices might use solid polymer electrolytes that are safer and derived from renewable feedstocks, reducing dependency on scarce minerals and eliminating toxic solvents (Thrisha and Saratha 2024).
- Lightweight polymer composites in vehicles will enable electrification and efficiency such that transportation emissions plummet, with those polymers themselves being recovered and reused at vehicle end-of-life (Yan and Xu 2025).
- Even in heavy industries, new high-temperature polymers or polymer composites could replace metals in certain applications, saving energy in production and enabling novel designs (for example, polymer parts in wind turbines or hydroelectric turbines that self-heal or adjust shape) (Melentiev et al. 2022).

To reach this paradigm, a few parting calls to action:

- **Invest in R&D and Scale-up for Sustainable Polymers:** Many promising materials and processes exist at lab scale. These need pilot plants, demonstration projects, and ultimately full commercialization. Governments and industries should continue and increase support for these, much like how investment in silicon solar technology eventually drove it to mass viability. The gap between lab innovation and market adoption must be bridged.
- **Encourage Cross-Sector Collaboration:** The energy sector and polymer/materials sector should work hand in hand. For example, wind energy companies can form consortia with chemical companies and recyclers (as seen in the ZEBRA blade project) to ensure, from the design phase that products are recyclable and recycling is planned. Similar coalitions in solar and battery industries could proactively tackle polymer sustainability in those products. Conferences, joint task forces, and knowledge-sharing platforms between traditionally separate industries will help.

- **Educate and Involve the Public:** The perception of plastics has understandably soured among the public due to pollution issues. Communicating the positive role polymers play in enabling green tech is important, as is engaging the public in solutions (like participating in recycling programs, supporting products made with sustainable polymers, etc.). Informed consumers can drive demand for greener materials and hold companies accountable. Education at all levels (schools, museums, media) about materials sustainability can cultivate a generation that understands the nuances—that it's not "plastics = bad", but rather "how we make and manage plastics" that matters.

In conclusion, the evolution of polymers from celluloid to solar cells is more than a catchy phrase—it encapsulates how human ingenuity with these materials has shaped our world for the better, and occasionally for the worse. Now we stand at a crossroads where that same ingenuity must be applied to reshape polymers themselves for a sustainable future. The story is one of adaptation: plastics saved elephants by replacing ivory; now they might help save the planet by enabling and accelerating the clean energy transition, provided we reimagine plastics in a way that aligns with ecological boundaries. As one review article put it, "The future success of this industry will rely on the development of sustainable polymers – materials derived from renewable feedstocks that are safe in both production and use and that can be recycled or disposed of in ways that are environmentally innocuous." Achieving that is no small task, but the rapid advances and collaborative efforts documented in this chapter give reason for optimism.

The Plastic Pathways to Power we have traced lead to a future where polymers and energy technology co-evolve symbiotically: clean energy enabling greener plastics (e.g., renewable-powered recycling plants), and greener plastics further enabling clean energy (e.g., lighter materials improving efficiency). By prioritizing eco-conscious strategies now, in research, industry, we can ensure that polymers continue to be allies in innovation rather than sources of pollution. The call to action is clear: it will take concerted, cross-disciplinary effort from scientists and engineers to drive this next materials revolution. If we succeed, future generations might look back and coin a new term for our era—not the Plastic Age as a detraction, but the Sustainable Polymer Age, wherein humanity learned to harmonize its materials with the planet's health while unlocking new wonders of technology.

Let us collaborate, innovate, and lead the charge toward that polymeric energy paradigm. The elephants and tortoises were early beneficiaries of plastic innovation; now it's time to ensure all of Earth's inhabitants benefit from the re-invention of plastics for a greener, more empowered world.

References

Abrahms-Kavunenko S (2023) Toward an anthropology of plastics. J Mater Cult 28(1):3–23

Andrady AL, Neal MA (2009) Applications and societal benefits of plastics. Philos Trans R Soc Lond B Biol Sci 364(1526):1977–1984

Andrew JJ, Dhakal HN (2022) Sustainable biobased composites for advanced applications: recent trends and future opportunities—a critical review. Compos Part C: Open Access 7:100220
Balser K et al (2000) Cellulose esters. Ullmann's Encyclopedia of Industrial Chemistry
Beena Unni A, Muringayil Joseph T (2024) Enhancing polymer sustainability: eco-conscious strategies. Polymers 16(13):1769
Butsriruk K (2021) Design and fabrication of perovskite solar cells for tandem structures การออกแบบและประดิษฐ์เซลล์สุริยะเพอรอฟสไกต์สำหรับโครงสร้างแบบแทนเดม. Chulalongkorn University
Butsriruk K et al (2023) Surface treatment of PTAA hole transport layer for inverted perovskite solar cells. J Phys: Conf Ser 2431(1):012045
Calisto Friant M et al (2022) Transition to a sustainable circular plastics economy in the netherlands: discourse and policy analysis. Sustainability 14(1):190
Chen G-L et al (2025) Boosting the long-term stability of all-polymer solar cells by using natural cellulose as an interlayer. J Mater Chem C 13(13):6581–6587
Cheng X et al (2025) A review of thermoplastic composites on wind turbine blades. Compos Part B: Eng 299:112411
Combs MT (2018) Camphor, a plastic history: China, Taiwan, and Celluloid, 1868–1937. University of California, Irvine
Crespy D, Bozonnet M, Meier M (2008) 100 Years of Bakelite, the material of a 1000 uses. Angew Chem Int Ed 47(18):3322–3328
Cuenca R, Gaines L, Vyas A (2000) Evaluation of electric vehicle production and operating costs. Argonne National Lab, IL (US)
Davidson EC, Tian H, Priestley RD (2024) Polymers for the clean energy transition. JACS Au 4:4544–4545
Dhanraj ND, Hatha AAM, Jisha MS (2022) Biodegradation of petroleum based and bio-based plastics: approaches to increase the rate of biodegradation. Arch Microbiol 204(5):258
Dokl M et al (2024) Global projections of plastic use, end-of-life fate and potential changes in consumption, reduction, recycling and replacement with bioplastics to 2050. Sustain Prod Consump 51:498–518
Du B et al (2019) High thermal conductivity insulation and sheathing materials for electric vehicle cable application. IEEE Trans Dielectr Electr Insul 26(4):1363–1370
Dutta K (2022) 13—π-Conjugated polymers for application in proton exchange membrane fuel cells. In: Kumar V et al (eds) Conjugated polymers for next-generation applications. Woodhead Publishing, pp 447–503
Firoozi AA, Firoozi AA, Hejazi F (2024) Innovations in wind turbine blade engineering: exploring materials, sustainability, and market dynamics. Sustainability 16(19):8564
Freinkel S (2011a) A brief history of plastic's conquest of the world. Sci Am 29
Freinkel S (2011b) Plastic: a toxic love story. Text Publishing
Geyer R, Jambeck JR, Law KL (2017) Production, use, and fate of all plastics ever made. Sci Adv 3(7):e1700782
Gnida P et al (2022) Polymers in High-efficiency solar cells: the latest reports. Polymers 14(10):1946
Guo Y et al (2023) Dielectric elastomer artificial muscle materials advancement and soft robotic applications. SmartMat 4(4):e1203
Hong G, Xu H (2025) Equivalent stiffness model for glass–glass photovoltaic modules in cable-suspended photovoltaic systems. Energies 18(11):2854
Hoodin K (2024) ZEBRA project developed closed-loop recycling process for wind turbine Blades. Compos Manuf
Hwang W, Sung Y-E (2023) Recent developments of polymer electrolyte membrane fuel cell design. J Electrochem Sci Technol 14(2):120–130
Ionescu E et al (2021) Polymer-derived ultra-high temperature ceramics (UHTCs) and related materials. In: Baino F, Tomalino M, Tulyaganov D (eds) Ceramics, glass and glass-ceramics: from early manufacturing steps towards modern frontiers. Springer, Cham, pp 281–323

Jones MW et al (2023) National contributions to climate change due to historical emissions of carbon dioxide, methane, and nitrous oxide since 1850. Sci Data 10(1):155
Joost WJ (2012) Reducing vehicle weight and improving U.S. energy efficiency using integrated computational materials engineering. JOM 64(9):1032–1038
Korniejenko K et al (2021) Tackling the circular economy challenges—composites recycling: used tyres, wind turbine blades, and solar panels. J Compos Sci 5(9):243
Kumar V, Alam MN (2024) Polymeric materials in energy conversion and storage. Polymers 16. https://doi.org/10.3390/polym16223132
Kumar A et al (2024) Prospects of sustainable polymers. Sci Rep 14(1):9430
Kumari S et al (2023) Petroleum-based plastics versus bio-based plastics: a review. Nat Environ Pollut Technol 22:1111–1124
Lakhiar IA et al (2024) Plastic pollution in agriculture as a threat to food security, the ecosystem, and the environment: an overview. Agronomy 14(3):548
Leon M Jr (2023) Recycling of wind turbine blades: recent developments. Curr Opin Green Sustain Chem 39:100746
Li Q et al (2023) Insights on the role of particulate filler on the polymer-modified cement-based coating and its application. J Build Eng 76:107395
Lin J et al (2024) Substrate modifications for stability improvements of flexible perovskite solar cells. Energy Technol 12(5):2300958
Maoz M et al (2025) Recent advances in flexible solar cells; materials, fabrication, and commercialization. Sustainability 17(5):1820
Melentiev R et al (2022) Metallization of polymers and composites: state-of-the-art approaches. Mater Des 221:110958
Meng F, Brandão M, Cullen JM (2024) Replacing plastics with alternatives is worse for greenhouse gas emissions in most cases. Environ Sci Technol 58(6):2716–2727
Mishnaevsky L et al (2017) Materials for wind turbine blades: an overview. Materials 10(11):1285
Mohanty AK et al (2022) Sustainable polymers. Nat Rev Methods Primers 2(1):46
Mohd Nurazzi N et al (2021) Fabrication, functionalization, and application of carbon nanotube-reinforced polymer composite: an overview. Polymers 13(7):1047
Morici E, Dintcheva NT (2022) Recycling of thermoset materials and thermoset-based composites: challenge and opportunity. Polymers 14(19):4153
Morris CA (2023) History and future of plastics. https://www.sciencehistory.org/education/classroom-activities/role-playing-games/case-of-plastics/history-and-future-of-plastics/#:~:text=treating%20cellulose%20derived%20from%20cotton,tortoiseshell,%20horn,%20linen,%20and%20ivory. Accessed 15 Jun 2025
Murray RE et al (2021) Structural validation of a thermoplastic composite wind turbine blade with comparison to a thermoset composite blade. Renew Energy 164:1100–1107
Nicholson JL, Leighton GR (1942) Plastics come of age. Harper's Mag 306
Nizamuddin S et al (2024) Bio-based plastics, biodegradable plastics, and compostable plastics: biodegradation mechanism, biodegradability standards and environmental stratagem. Int Biodeterior Biodegradation 195:105887
Patel D, Deshmukh SP (2012) Polymer in sustainable energy. J Miner Mater Charact Eng 11(7):661–666
Peng H et al (2023) Review of wind power bearing wear analysis and intelligent lubrication method research. Coatings 14(1):30
Ramesh M et al (2024) Powering the future: a comprehensive review of polymer composite energy storage applications. Eng Proc 61(1):24
Rasmussen SC (2021) From parkesine to celluloid: the birth of organic plastics. Angew Chem 133(15):8090–8094
Reilly JA (1991) Celluloid objects: their chemistry and preservation. J Am Inst Conserv 30(2):145–162
Roche GH et al (2024) Impurity tolerance of polymer solar cells: the crucial role of palladium catalyst and its ligands. Solar Energy Mater Solar Cells 269:112760

Rohrer J (2007) Personal development as the meaning of life. Plastic bags and plastic bottles—CO_2 emissions during their lifetime
Sattar S et al (2024) In operando optical analysis of electrolyte colour change and its correlation to capacity fade in Li-ion cells at elevated temperatures. Sensors 24(23):7686
Shen H et al (2023) Application and development of silicon anode binders for lithium-ion batteries. Materials (Basel) 16(12)
Suarez A et al (2023) Is sugarcane-based polyethylene a good alternative to fight climate change? J Clean Prod 395:136432
Thrisha K, Saratha R (2024) Natural polymer-based electrolytes for energy storage devices—an overview. Ionics 30(3):1245–1266
Uddin MN et al (2024) Investigation of photovoltaic properties of poly(3-hexylthiophene): polynaphthalene-bithiophene heterostructure devices. Opt Mater 157:116385
Verebélyi K et al (2023) Highly efficient cationic polymerization of β-pinene, a bio-based, renewable olefin, with $TiCl_4$ catalyst from cryogenic to energy-saving room temperature conditions. Int J Mol Sci 24(6):5170
von Vacano B et al (2022) Sustainable design of structural and functional polymers for a circular economy. Angew Chem 135
Wang Y et al (2022) Dielectric elastomer actuators for artificial muscles: a comprehensive review of soft robot explorations. Resour Chem Mater 1(3):308–324
Wiśniewska P, Saeb MR, Bencherif SA (2023) Biomaterials recycling: a promising pathway to sustainability. Front Biomater Sci 2
Wu C et al (2023) Ultrafast high-temperature sintering of polymer-derived ceramic nanocomposites for high-temperature thin-film sensors. Chem Eng J 463:142518
Yan L, Xu H (2025) Lightweight composite materials in automotive engineering: state-of-the-art and future trends. Alexandria Eng J 118:1–10
Yelshibay A et al (2024) Conducting polymers in solar cells: insights, innovations, and challenges. Organics 5(4):640–669
Young RJ, Lovell PA (2011) Introduction to polymers. CRC Press
Zboncak K (2024) Block diagram of infrared plastic solar cell an emerging tec
Zhang L et al (2023) Recent progress of all polymer solar cells with efficiency over 15%. Solar RRL 7(12):2300219
Zhou W et al (2023) Polyhydroxyalkanoates (PHAs) synthesis and degradation by microbes and applications towards a circular economy. J Environ Manag 341:118033

Chapter 8
Innovative Polymer Materials for Solar Energy and Industrial Applications

Mansi Sharma and Ashutosh Kumar

Abstract The global transition toward the sustainable energy is crucial to manage the associated issues of rising energy demands and climate change. The present chapter outlines the role of advanced and innovative solutions for sustainability, focusing on the pivotal contribution from solar energy. There are various stages of advancement of renewable energy materials and the evolving technological innovations in photovoltaics (PV). Thus the crucial analysis for next-generation materials including perovskite solar cells (PSCs), dye-sensitized solar cells (DSSCs), quantum dot solar cells (QDSCs), and organic photovoltaics (OPVs) has been discussed and highlighted in the present work to address the effective integration of solar technologies. The emerging trends among these technologies with the remarkable progress in power conversion efficiencies and interface engineering aiming at minimal device losses have been presented and discussed. The chapter further explores the practical integration of solar energy technology into industrial processes and highlights the potential solutions with the combined use of solar PV and solar thermal energy for advanced green buildings. Moreover, it provides a comprehensive roadmap for policymakers to design effective regulations for the distribution and adoption of advanced solar technologies considering the assessment of lifespan and environmental impacts. The study directly contributes to the achievement of Sustainable Development Goals for affordable and clean energy (SDG 7) and industry, innovation, and infrastructure (SDG 9).

Keywords Organic photovoltaics · Perovskite solar cells · Quantum dot solar cells · Sustainable materials · Energy policy frameworks · Net zero energy

---

M. Sharma (✉)
School of Engineering and Technology, Manav Rachna International Institute of Research and Studies, Faridabad, India
e-mail: mansisharma.set@mriu.edu.in

A. Kumar
Department of Physics, CSJM University, Kanpur, Christ Church College, Kanpur, Uttar Pradesh, India

S. K. Verma et al. (eds.), *Sustainable and Natural Polymers*, Clean Energy Production Technologies, https://doi.org/10.1007/978-981-92-0092-4_8

8.1 Introduction

8.1.1 Sustainable Material System for Energy

Sustainability has emerged as a critical solution for resource management to deal with evolved challenges of growing population and energy imbalance (Hariram et al. 2023; Glavič 2021). Globally, these challenges are association with several social, economic, and environmental factors such as scarcity of resources, construction, and infrastructure and most significantly to the climate change. The rising development targeting energy and material demands has resulted in a drastic reduction in environmental footprint (Tetteh et al. 2021; Shamoon et al. 2022; Khan et al. 2021). Thus, it is highly required to work on mitigation of depleting resources and underscore the ways to decarbonize natural resources. Since natural resources depend on water and biodiversity, countries are now following global sustainable development goals by adopting simple and effective strategies to protect the environment and ensure resources for the future (Khan et al. 2021; Dhingra 2023; Mishra 2024). These strategies majorly deal with (1) the advanced alternatives of fossil fuels, (2) the progressive technology for energy production, and (3) energy saving designs. The practical implementation of these strategies with the concept of renewable energy system demands the integration of sustainable design inspired by energy saving measure for large scale energy system.

8.1.2 Objectives and Scope of the Study

In green energy technologies, the renewable energy materials play a significant role in mitigating harmful emissions to ensure minimal environmental impact. The major sources of renewable systems are solar, wind, thermal, hydro, etc., where emerging areas of solar energy technologies have significantly reshaped the industrial expansion of renewable system. These technologies have evolved through more innovative designs with enhanced efficiency and eco-friendly constructions to meet the target of energy demand (Dhingra 2023; Olabi et al. 2023; Østergaard et al. 2021; Jamil et al. 2016).

The present chapter aims to highlight the dynamic interplay of material innovations and the industrial advancement in solar energy systems. It focuses on the commercialization of energy technologies and explores clean energy pathways via integration of advanced energy materials. The chapter briefly outlines the application of solar energy materials and technology for the management and conservation of natural resources for a sustainable future.

8.1.3 Overview of Solar Energy Technologies

The energy from the sun directly in the form of solar energy is a keystone of the global energy transition toward renewable energy. The technologies of capturing and converting one form of energy to another are diversified by following unique mechanisms and applications; among these, the primary emerging solar-based systems are: solar photovoltaic (PV) and solar thermal energy technologies (Østergaard et al. 2021; Khare et al. 2023; Sakthivadivel et al. 2021). The present section provides an overview of these technologies and examines the emerging trend of concentrated energy systems in solar utilities.

8.1.4 Photovoltaic Systems: Principles and Applications

Photovoltaic effect is the principle underlying photovoltaic technology and explains the contribution of photon in generation of photocurrent. The incident photons of given frequency knocks out the valence electron from atoms of a semiconducting material above the threshold and generate electricity. Photovoltaic systems are considered as more reliable, clean, and a direct method of generating energy from natural sources like the sun (Al-Ezzi and Ansari 2022; Dambhare et al. 2021). In the past two decades, a remarkable growth has been observed in the PV generation which indicates their significant contribution in global market to meet the targeted energy demand by local authorities of nations. These PV devices are easy to design and offer the advantage of low maintenance, which makes them a standalone system in external environment for energy conversion. The typical output from a PV device varies from micro- to megawatt, extending its application in communication, power source, aerospace transportation, solar homes, etc. (Kumar 2024). The PV systems consist of various components including electrical connections, mechanical structures.

8.1.5 Concentrated Solar Power (CSP) Systems

The mechanism of capturing solar energy and its utilization for generating heat for household and industrial processes is recognized as solar thermal technology (Tasmin et al. 2022). The heat generated via these technologies is being utilized to drive the turbines for generation of electricity. In recent years the solar thermal systems have majorly contributed in the design and infrastructural improvements in the conventional architecture and building construction, enabling future net zero energy independence. These architectures are the blend of solar photovoltaic systems and solar thermal system where the residual solar energy which has not been utilized for energy conversion is sent to integrated solar thermal systems of building for

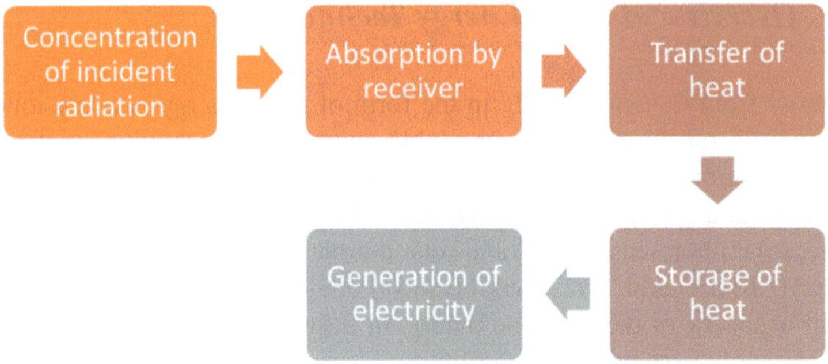

Fig. 8.1 Flow diagram for mechanism of concentrated solar power system

effective heat generation. Figure 8.1 presents the basic flow for a concentrated solar power system. The basic design of such systems exist in two temperature ranges—low temperature system and high temperature systems—depending on the energy utility applications. The integration of both photovoltaic system with the solar thermal system is the result of concentrated solar power (CSP) technology (Weinstein et al. 2015). In 1980, the first CSP plant was developed at utility scale (Müller-Steinhagen and Trieb 2004; Sawin 2013; Pitz-Paal 2020). Between 1984 and 1995 the technology was at its peak commercialization. Figure 8.1 illustrates the basic five steps of conversion in CSP system.

In CSP system, the technology utilizes the combination of lenses or mirrors for concentrating the incident direct solar radiation to generate useful energy in the form of heat and electricity. There are various commercially adopted researched configurations of CSP including parabolic trough, linear Fresnel, central receiver tower, etc. Each configuration owns a particular advantage in the energy market. Out of these, the designing of solar furnace is being widely used as a tool in research laboratory building.

8.1.6 Emerging Trends in Solar Energy Utilization

In recent years, renewable energy technology has depicted the enhanced efficiencies and better affordability of the solar power systems. While exploring the key trends in solar energy utilization, a major advancement is the enhanced power conversion efficiency of solar systems exceeding 29% (Al-Ashouri et al. 2020). These technological innovations have significantly contributed to mitigating the economic restrictions on the expansion of solar installation and made them available for diverse applications extending from rooftop to large-scale solar energy farms. Moreover, the concept of bifacial design for capturing the sunlight efficiently from both sides has opened the scope of expanding solar market even in reflective environments like water surfaces and sand. The output of such design is nearly 30% more compared to the basic design. Additionally, tandem designs and advanced perovskite solar cells are attracting strong interest because they achieve efficiencies above 29% (Liu et al. 2020). However, the intermitted availability which limits the solar energy capture throughout the day remains a potential challenge (Wallerand 2018). The large spatial and temporal variations along with the diluted solar energy adds a barrier to complete utility compared to generation (Yogi Goswami 1998). To address this intermittency, the integrated solar energy storage systems have emerged as a significant technological advancement in the energy storage technologies. These systems are designed to ensure the balanced consumption inline with the needs and have effectively evolved as demand–response technologies. Furthermore, in recent years the advanced energy networking with the adoption of digital and automated systems has also evolved in smart grids for managing the generation and distribution of energy. In smart grid technology, artificial intelligence (AI) plays a crucial role by providing real-time tracking for power dispatch and predictive loads and manages the coordination between the hybrid power systems such as wind and solar (Chen et al. 2021). These innovative solutions with smart energy management and dynamic load balancing have resulted in improved grid response to variable demands and thus help in reducing energy waste.

8.2 Sustainable Materials in Solar Energy

Light-absorbing materials are the prime requirement of solar photovoltaics, where the material is able to depict specific light absorption. The bandgap of the absorbing material should meet the requirement of photon absorption under specific incidence where the incident photon energy must be greater than or equal to the energy band gap (Chen et al. 2021). When photons interact with semiconducting material in the solar cell, the excitation of electrons leads to generation of electron-hole pairs. The generation is further limited by certain loss factors, and under standard illumination conditions this limit is defined as Shockley–Queisser limit, which imposes theoretical restrictions on the maximum power conversion efficiency of photovoltaic cells (Zanatta 2022). Thus, the optimization of cell design under limited power conversion range is a pivotal challenge in research based on conventional materials like silicon. Therefore, this limitation has motivated photovoltaic design with the alternative low-cost materials and widens the scope for organic and polymer-based solar cells (Ali et al. 2025). Moreover, the single-junction cell design has been further advanced to maximize power conversion efficiencies with the development of tandem and heterojunction solar cells.

8.2.1 Organic Photovoltaics (OPVs)

Organic photovoltaics is one of the emerging cost-effective technology for capturing solar energy (Wu et al. 2021). Organic solar cells are non-toxic, lightweight, and flexible, making them promising for green energy production. In the past decades, the technology has attracted significant scientific interest as an eco-friendly renewable energy system with comparatively easier processing primarily based on solution processing methods compared with other photovoltaic systems (Wu et al. 2021; Zhu et al. 2025). However, the device is limited by poor spectral absorption, low carrier mobility, and short diffusion length. Moreover, the device also faces challenges in terms of long-term stability as a result of structural disorder. In view of these shortcomings, the device was often entitled with the low power conversion efficiencies and substantial energy losses. Thus, the fabrication of highly efficient, large–area, and stable OPV modules remains as a key challenge in the commercialization of technology.

Despite the discussed challenges, the technology still possesses significant advantages over other PV technologies in view of tunable morphology and design technology (Öz et al. 2025). In recent years, the development of PV was conceptualized with the tailored bandgap conditions and the understanding of matching of energy level of prime donor polymers with fullerene-based acceptors. Later on, with the development of non-fullerene acceptors, the research has diverted the focus toward acceptor materials from donors with the idea of manipulating the photon-to-electron conversion. These acceptors are found to be more effective in light absorption making them available for flexible and high-transport networks. Additionally, the device offers high FF with large open circuit voltage and high short circuit current (Fig. 8.2).

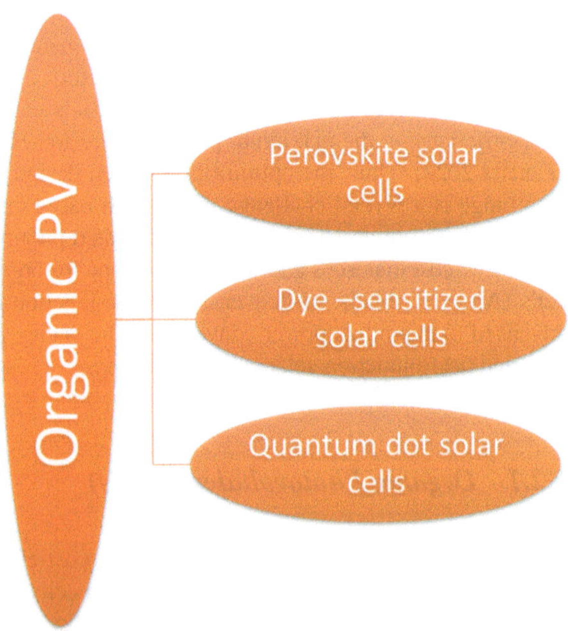

Fig. 8.2 Classification of organic photovoltaic cells

8.2.2 *Perovskite Solar Cells: Advancements and Challenges*

Perovskite solar cell (PSC) technology has gained significant attraction worldwide in view of its promising cost-effective thin-film technology for high-efficiency photovoltaics. The incorporation of metal halide perovskite material has benefited the designing of higher light absorption along with long carrier diffusion length (Wang et al. 2019; Jiang et al. 2019; Jung et al. 2019; Wu et al. 2019a). The general formula for perovskite material is ABX_3, where A is a cation of large radius, B is a cation of small radius, and X is an anion. (MA+), FA+, Cs+, and Ca+ and Pb2+, Sn2+, and Ti4+ are some of the examples of large radius and small radius cations, respectively, used in PSCs (Borriello et al. 2008; Li and Zhang 2020). O_2, F-, Cl-, etc., are incorporated as anions in the structure. Since 2006, the development of PSCs has encountered a rapid progress in power conversion efficiencies starting from 2.2%. Approaching the efficiencies of crystalline silicon solar cells, PSCs have reached the efficiency of 25.5% (Green et al. 2012). Moreover, the PSC sub-modules have also been recently tested for practical use with the remarkable efficiencies of about 18% (Green et al. 2012). The advancement in technology with the perovskite-based tandem structures has depicted significant progress in PSCs with further improvements in carrier diffusion length and offers a remarkable reduction in defect density, thus the power conversion efficiency limits have increased beyond the single-junction PSCs (Al-Ashouri et al. 2020; Gu et al. 2020; Xiao et al. 2020).

Over the years, the technology has progressed in terms of better long-term stability under incident light and heat. The enhanced PCE has found to be highly correlated with the method of synthesis as well as the properties of materials used as absorber layer, electron transfer layer, structural stability, and interfaces. The synthesis of perovskite solar cells has been explored using various techniques including spin coating (Lee et al. 2012), chemical vapor deposition (Chen et al. 2014), dip coating (Burschka et al. 2013), spray pyrolysis (Barrows et al. 2014), thermal evaporation (Malinkiewicz et al. 2014), and atomic layer deposition (Sutherland et al. 2014). The flexibility for the choice of fabrication processing makes the technology more popular compared to any other.

8.2.3 *Dye-Sensitized Solar Cells*

In photovoltaic technologies, the dye-sensitized solar cells are a well-known inexpensive category of sustainable solar design in view of their simple processing. The primary component of dye-sensitized solar cell (DSSC) is the monolayer sensitizer which is absorbed on a semiconductor oxide. Titanium dioxide is one of the commonly used semiconductor in DSSC which is a popular paint base. In DSSC, the absorption of light takes place in dye and the separation of charge carriers is the result of photo-induced energy transfer from dye to semiconducting solid (Wang et al. 2006; Adedokun et al. 2016). In recent year, the design of DSSCs has been evolved effectively with the inclusion of advanced materials for the dyes and the optimizations for the size of nanoparticles. The extraction of dye sensitizers from natural resources is one of the crucial aspect in the managing the cost of the device. Additionally, fabrication at ambient temperature and environmental insensitivity of material adds to the advantages of technology making the device more feasible for large-scale and low-energy production.

Furthermore, the advantages are well reflected in the integration of DSSCs with various storage devices such as batteries and supercapacitors. The combination, aided by the use of non-toxic material, addresses the challenge of solar intermittence and, thus, improves the device's practical applicability (Saadat et al. 2022; Areerob et al. 2022). However, recent research has been aimed to address the challenge of long-term stability with the utilization of emerging co-sensitizers and nano-structuring of photoanode to enhance light harvesting and charge transport to reach the commercialization of the device with the highest power conversion efficiency.

8.2.4 Quantum Dot Solar Cells

In solar cells, the power conversion efficiency is considered as one of the crucial parameter for performing the optimization of design and structural properties of different layers. In the realm of managing the lost energy in solar cells the concept of carrier multiplication has been provided as a promising method for reclaiming the lost energy in the form of heat (Lee 2024). In carrier multiplication, the single photon is able to generate multiple electron hole pairs if the energy of the photon is greater than or equal to twice the energy bandgap of semiconducting absorber. The process results in the generation of more number of charge carriers available for conduction.

In bulk semiconductors, the optimization for carrier multiplication is mainly dominated by the fast carrier relaxation which leads to recombination without contributing to photocurrent. The theory of quantum dots emerges as an interesting substitute for bulk in order to unveil carrier multiplication (CM) effect. As a result of quantum confinement effects, semiconducting nanocrystals, which are advanced materials, boost CM efficiency owing to improved Coulomb interactions. These cells represent a remarkable advancement in third-generation solar cells with a potential for high conversion efficiencies with tunable bandgaps. Quantum dots (QD) are nanostructured materials having a diameter typically in the range of 2–10 nm. Due to their small size, the structures exhibit a dominating quantum confinement effect, which results in unique optoelectronic properties with the precise control over the bandgap by size. Thus, by optimizing the size of semiconducting nanocrystals, their absorption can be tuned to match the solar spectrum. Another advantage of QD solar cells is scalable manufacturing with low processing costs (Kramer and Sargent 2011). For QD photovoltaics the research has evolved with various advanced materials like CdSe, CdTe, PbS, and PbSe, which are widely studied among all QD materials owing to the high efficiencies of more than 15% and good stability. However, there are critical concerns associated with the toxicity of the material. The research efforts result in exploring the field with perovskite QDs, copper indium sulfide/selenide, III-V QDs, etc., depicting promising efficiencies, but still lagging behind due to stability issues under light (Pietryga et al. 2016). Furthermore, the carrier mobility has been observed to be relatively low in QD structure due to carrier hopping transport mechanism between individual dots. The technology majorly lags behind silicon-based photovoltaic cell; however, the integration of QD absorber can offer unique applications and serve as a suitable alternative for solar wearables, solar windows, portable devices, etc.

8.2.5 Progress in Organic Photovoltaics (OPVs)

The emerging trend in the material choice and the design for organic photovoltaics is provided in Table 8.1.

8.3 Innovations in Industrial Development

8.3.1 Integration of Solar Energy in Industrial Processes

The industrial sector is known as one of the major user of energy. Thus, the dominant portion of the utilities is dependent on the conventional resources like fossil fuels. In the past decades, solar energy in industrial processes have evolved as one of the emerging field focusing the mitigation of industrial carbon footprints as a result of fossil fuel consumption, which is crucial in the context of global climate goals (Utami and Dewi 2024). These processes draw a promising pathway to achieve sustainable and cost-effective industrial operations. Solar energy integration includes the harnessing of solar thermal energy for heat and electricity required to supplement the conventional power sources. In industrial processes with the requirement of medium to low temperature such as drying, washing, space heating, etc., the solar thermal integration is highly adopted with the technologies like flat plate collectors for solar radiation and concentrated solar power systems (Mekhilef et al. 2011; Martínez-Rodríguez et al. 2021).

The industrial thermal applications are primarily categorized on the basis of the required temperature range. Thus, specific technologies are used for meeting the required level of consumption in variable range of temperature. For low-temperature uses (typically 100 °C) like water heating, food processing, and washing, flat-plate collectors are used. Similarly, to achieve medium temperature between 100 and 200 °C, the evacuated tube collectors are considered as more efficient for serving the application in chemical processes. Furthermore, the technology has also

Table 8.1 Key highlights of the materials and technique for OPVs

	Advancement	Material	Key results	Reference
Perovskite solar cells	Interface engineering for improved carrier extraction	$TiO_2/CH_3NH_3PbI_3$ with intermolecular exchange process (IEP)	20% PCE	Yang et al. (2015)
	Mixed-cation (Cs/FA/MA) strategy to enhance stability and bandgap tuning	$FA_{0.85}MA_{0.10}Cs_{0.05}Pb(I_{0.85}Br_{0.15})_3$	21.1% PCE, enhanced thermal stability	Saliba et al. (2016)
	2D/3D mixed perovskite structures for moisture resistance	$(BA)_2(MA)_3Pb_4I_{13}$ (2D/3D hybrid)	Improved humidity and illumination stability	Tsai et al. (2016)
	Formamidinium-based $FAPbI_3$ stabilization via additives	$FAPbI_3$ co-doped with Cs^+ and MDA^{2+}	23.7% PCE (certified)	Min et al. (2019)
	Introduction of quasi-2D perovskites for stability and high VOC	Multidimensional hybrid $GA(MA)_nPb_nI_{3n+1}$ ($n = 5$)	22.26% PCE, 1200-h stability	Zhang and Park (2022)
	Tandem perovskite/silicon solar cells	Perovskite ($FAPbI_3$)/Si tandem architecture	29.1% PCE with V_{oc} approaching 1.92 V	Al-Ashouri et al. (2020)
	Ambient printing of perovskite films	Utilizing 1-butyl-3-methylpyridine thiocyanate (BPySCN) as functional additive	23.70% PCE (0.09 cm^2) and 23.01% PCE (117.0 cm^2	Liu et al. (2024)
Dye-sensitized and quantum dot solar cells	Development of zinc porphyrin-dye sensitizer with cobalt-based electrolyte	Porphyrin dye with Co(II/III) redox couple	12.3% PCE	Yella et al. (2011)
	Molecular engineering of porphyrin dye	Zn-porphyrin-based dyes	13% PCE	Mathew et al. (2014)
	Incorporation of graphene–TiO_2 nanocomposites for better charge transport with optimized ratio of graphene in TiO_2	Graphene–TiO_2 hybrid photoanode	Enhanced efficiency up to 22%	Wang and Cho (2017)
	Cobalt-based redox electrolytes to improve voltage and stability	Naphthalene-based donor–acceptor organic dyes	Enhanced long-term stability under light	Wu et al. (2019b)
	Synthesis of metal-free organic dyes for eco-friendly DSSCs	Organic chromophores-based dyes	12.34% PCE	Lin et al. (2020)
	Quasi-solid dye-sensitized solar cells	Electrospun polyvinylidene fluoride (PVDF) membrane	Improved flexibility and stability	Cheng et al. (2019)
	Introduction of nickel-based redox shuttles	Ni(II/III) complexes	20.4% PCE, under fluorescent illumination	Kaur et al. (2024)

been well explored with the integration of concentrated solar thermal systems to meet the high temperature demands of over 1000 °C. In concentrated system, the design adopts the use of mirrors to concentrate the sunlight onto a receiver, which thus transfers the heat to the required unit. The receivers are basically having parabolic troughs and linear Fresnel reflectors for covering temperatures up to 500 °C. For achieving very high temperature of over 500 °C for industrial processes like annealing, calcination, and melting of metals, the solar tower with a field of mirrors is being utilized.

Other than utilizing the sunlight for heating, there is simultaneous utilization of photons for electricity with the help of photovoltaic systems. On an industrial scale, there are various models adopted for the integration of solar PV. The well know is the on-site rooftop and ground-mount PV model where the solar system is being installed on the adjacent land and the roofs of the factory and commercial units to promote the direct consumption with lower transmission losses. Other than the direct consumption plan, there is solar power purchase agreements which allow the consumer to purchase the electricity often available on their industrial site at lower rate from a third party who takes the responsibility of owning, operating, and taking care of the maintenance of the solar system. Furthermore there are off-site solar farms which are expanding on a large scale to facilitate the industrial power demands wheeled through the grid.

8.3.2 Case Studies: Industrial Applications of Solar Technologies

The concrete example of the integration of solar PV and solar thermal technologies into a single facility is the green building architecture with net zero consumption. Similarly, the integration of both the technologies is also seen in a dairy farm. In recent years, the sustainable building design has attracted most of the interest of global market. This sustainable architecture aiming for net zero energy consumption is the base of green building projects. The initiative has been a part of Asia-Link program where the potential benefit of green technologies are highlighted to make people aware globally for the application and integration of the technology for energy conservation (Maka and Alabid 2022). The green architect employs the passive design for the building which balances out production and consumption. The integration of solar energy in passive solar design demonstrates the application of generation of electricity and regulation of thermal utilities for heating demands (Chel and Kaushik 2018; Bock 2019). However, the design implementation requires specific sites for construction after analyzing the local climate conditions based on which the output evaluation can be made (Bock 2019; Maurer et al. 2017).

8.3.3 Environmental and Economic Impacts

Life cycle assessment (LCA) studies are crucial for understanding the upstream and downstream processes associated with manufacturing, installation, and end-of-life management for emerging PV technologies. The study includes the assessment of energy paybacks, toxicity level, carbon footprints, and the resource consumption aiming at overall sustainability of PV technology. The distinctive LCA starts from the extraction and generation of material for solar systems for PV and thermal. The typical operational lifetime of a solar silicon-based panel is 25–30 years. On an average there is 0.5–1% annual degradation due to environmental impacts due to climate and pollution. In LCA, the energy payback time (EPBT) is one of the crucial factors for contributing to the assessment. In polymer-based solar cells or for organic photovoltaics, the EPBT is estimated in typically in the order of months (3–12 months). The range is highly dependent on the factors like efficiency, method of synthesis, and the available geographical location. Lower EPBT is mainly attributed to the lower manufacturing cost and makes them best-performing PV technology among all. Moreover, in recent years the flexible and transparent design technology in organic PV has been found to depict very short EPBTs in the range of 51 days to 1 year, which is an attractive solution for adoption in sustainable future (Bock 2019).

Economically, solar integration has impacted the global market significantly. The inexpensive technology cost and the rising prices of conventional materials for energy has proposed the levelized cost of energy and cost competitiveness in the market. Furthermore, in view of volatile prices of fossil fuels, solar energy materials have evolved with the benefit of net zero fuel cost and long-term price stabilities, which allows the energy security and maintains the price stability in stakeholders (Anctil et al. 2020).

8.4 Policy and Regulatory Frameworks

Aiming at the expansion of renewable and sustainable energy resources for mitigating the environmental impacts and sharing the loads of rising energy demand, there is need to adopt specific agendas and strategies for energy planning and distribution. The contribution of these strategies is to encourage the worldwide sustainable practices for utilization of energy. The responsibility of design and planning has been a global agenda where nations have participated with the proposal of various regulatory frameworks to reward the renewable initiatives in various commercial and household projects. The frameworks have aimed to consider the revolution toward sustainability where the climate impact can be minimized (IRENA 2023). In 2015, in view of the concerns for green energy society, 17 Sustainable Development Goals (SDGs) were adopted by the United Nations focusing on the evolution of sustainable systems for ensuring the necessary energy production for society (Swain and

Karimu 2020). Simultaneously, the focus of these is to preserve the resources for future generations (Nam-Chol and Kim 2019).

In this planning, the energy policy frameworks play an important role for balancing out the present needs and the future demand. The implementation of these frameworks in government as well as in public sectors ensures the optimal utilization of resources and also supports the coordination between various institutions for achieving affordable and secure energy. These frameworks represent a systematic guide to production, distribution and consumption of energy.

National energy policies incorporating subsidies, public policies, international treaties, and legislation aiming at the reduction in greenhouse emissions represent the example of energy policy framework. These policies also promote the transitions to renewable energy via presenting the socio-economic benefits with the incentivization in the energy sector. The energy incentives target the private as well as public sectors with the proposal for profit making and cost-effective energy retrofit solutions. These incentives are mainly designed to balance out the global cost savings, and depending on the local climate conditions the profit is tailored to meet the optimum energy-saving demand. The common incentives include the grants, energy subsidies, and tax deductions considering the requirements and benefits of stakeholders. Thus the financial benefits are sufficient to propose a profit-making energy market which facilitates the broader adoption of renewable resources.

8.5 Future and Concluding Remarks

The present study underscores a standard transition driven by technological innovations in sustainable materials for solar energy. The evolution and advancement in third-generation PV technology have been demonstrated. The observed high efficiency beyond 29% in tandem perovskite configuration and the advancement in organic PV from non-fullerene acceptors to cutting-edge quantum dot solar cell technology are significant advancements extending the boundaries of performance and stability. Moreover the non-toxic nature of these PV materials has also been highlighted as one of the significant contribution. Another highlight is the contribution of direct integration of solar energy in industrial process. Simultaneous to the on-site PV models to ensure the effective distribution and enhanced energy security, the solar systems are also responsible for managing the heat demand of the industry or the given infrastructure. The heat share in solar energy systems is managed with specific receiving and transfer units based on the concentrator mirror arrangement.

These emerging solutions with advanced materials have significantly contributed to the better lifespans as revealed in the life assessment cycle. Additionally, the innovation in non-toxic materials for organic PV with the benefits of extremely short EPBT present the technology as an attractive competitor in the energy market. To support the overall adoption in view of the rising consumption of energy, the policy frameworks are instrumental in creating stable markets with attractive incentives and tax benefits for stakeholders. However, the field is still has a lot of scope for exploring the scalability, long-term stability, and efficiency. By overcoming the existing challenges of toxicity and low efficiency, the organic solar technologies have the potential to evolve as a complementary power source and contribute further to the global economy.

References

Adedokun O, Titilope K, Awodugba AO (2016) Review on natural dye-sensitized solar cells (DSSCs). Int J Eng Technol IJET 2(2):34–41

Al-Ashouri A, Köhnen E, Li B, Magomedov A, Hempel H, Caprioglio P, Márquez JA, Morales Vilches AB, Kasparavicius E, Smith JA, Phung N (2020) Monolithic perovskite/silicon tandem solar cell with>29% efficiency by enhanced hole extraction. Science 370(6522):1300–1309

Al-Ezzi AS, Ansari MNM (2022) Photovoltaic solar cells: a review. Appl Syst Innov 5(4):67

Ali AO, Elgohr AT, El-Mahdy MH, Zohir HM, Emam AZ, Mostafa MG, Al-Razgan M, Kasem HM, Elhadidy MS (2025) Advancements in photovoltaic technology: a comprehensive review of recent advances and future prospects. Energy Convers Manag X:100952

Anctil A, Lee E, Lunt RR (2020) Net energy and cost benefit of transparent organic solar cells in building-integrated applications. Appl Energy 261:114429

Areerob Y, Hamontree C, Sricharoen P, Limchoowong N, Nijpanich S, Nachaithong T, Oh WC, Pattarith K (2022) Synthesis of novel $MoWO_4$ with ZnO nanoflowers on multi-walled carbon nanotubes for counter electrode application in dye-sensitized solar cells. Sci Rep 12(1):12490

Barrows AT, Pearson AJ, Kwak CK, Dunbar AD, Buckley AR, Lidzey DG (2014) Efficient planar heterojunction mixed-halide perovskite solar cells deposited via spray-deposition. Energy Environ Sci 7(9):2944–2950

Bock M (2019) A building integrated solar thermal collector with active steel skins. Energy Build 201:134–147

Borriello I, Cantele G, Ninno D (2008) Ab initio investigation of hybrid organic-inorganic perovskites based on tin halides. Phys Rev B Condens Matter Mater Phys 77(23):235214

Burschka J, Pellet N, Moon SJ, Humphry-Baker R, Gao P, Nazeeruddin MK, Grätzel M (2013) Sequential deposition as a route to high-performance perovskite-sensitized solar cells. Nature 499(7458):316–319

Chel A, Kaushik G (2018) Renewable energy technologies for sustainable development of energy efficient building. Alex Eng J 57(2):655–669

Chen Q, Zhou H, Hong Z, Luo S, Duan HS, Wang HH, Liu Y, Li G, Yang Y (2014) Planar heterojunction perovskite solar cells via vapor-assisted solution process. J Am Chem Soc 136(2):622–625

Chen C, Zheng S, Song H (2021) Photon management to reduce energy loss in perovskite solar cells. Chem Soc Rev 50(12):7250–7329

Cheng F, Ou Y, Liu G, Zhao L, Dong B, Wang S, Wen S (2019) Novel quasi-solid-state electrolytes based on electrospun poly (vinylidene fluoride) fiber membranes for highly efficient and stable dye-sensitized solar cells. Nanomaterials 9(5):783

Dambhare MV, Butey B, Moharil SV (2021) Solar photovoltaic technology: a review of different types of solar cells and its future trends. J Phys Conf Ser 1913(1):012053. IOP Publishing

Dhingra R (2023) Explorations into sustainable development with forest and biodiversity conservation. Selfypage Developers Pvt. Ltd

Glavič P (2021) Evolution and current challenges of sustainable consumption and production. Sustainability 13(16):9379

Green MA, Emery K, Hishikawa Y, Warta W, Dunlop ED (2012) Solar cell efficiency tables (version 39). Prog Photovolt Res Appl 20(1):12–20

Gu S, Lin R, Han Q, Gao Y, Tan H, Zhu J (2020) Tin and mixed lead–tin halide perovskite solar cells: progress and their application in tandem solar cells. Adv Mater 32(27):1907392

Hariram NP, Mekha KB, Suganthan V, Sudhakar K (2023) Sustainalism: an integrated socio-economic-environmental model to address sustainable development and sustainability. Sustainability 15(13):10682

IRENA (2023) Renewable power generation costs in 2022. International Renewable Energy Agency

Jamil M, Ahmad F, Jeon YJ (2016) Renewable energy technologies adopted by the UAE: prospects and challenges—a comprehensive overview. Renew Sustain Energy Rev 55:1181–1194

Jiang Q, Zhao Y, Zhang X, Yang X, Chen Y, Chu Z, Ye Q, Li X, Yin Z, You J (2019) Surface passivation of perovskite film for efficient solar cells. Nat Photonics 13(7):460–466

Jung EH, Jeon NJ, Park EY, Moon CS, Shin TJ, Yang TY, Noh JH, Seo J (2019) Efficient, stable and scalable perovskite solar cells using poly (3-hexylthiophene). Nature 567(7749):511–515

Kaur R, Dalpati N, Delcamp JH, Farnum BH (2024) Nickel-based two-electron redox shuttle for dye-sensitized solar cells in low light applications. ACS Appl Energy Mater 7(9):3645–3655

Khan N, Jhariya MK, Raj A, Banerjee A, Meena RS (2021) Eco-designing for sustainability. In: Ecological intensification of natural resources for sustainable agriculture. Springer, Singapore, pp 565–595

Khare V, Chaturvedi P, Mishra M (2023) Solar energy system concept change from trending technology: a comprehensive review. e-Prime-Adv Electr Eng Electron Energy 4:100183

Kramer IJ, Sargent EH (2011) Colloidal quantum dot photovoltaics: a path forward. ACS Nano 5(11):8506–8514

Kumar DS (2024) Essentials and applications of solar energy technology. Academic Guru Publishing House

Lee YH (2024) Beyond the Shockley-Queisser limit: exploring new frontiers in solar energy harvest. Science 383(6686):eado4308

Lee MM, Teuscher J, Miyasaka T, Murakami TN, Snaith HJ (2012) Efficient hybrid solar cells based on meso-superstructured organometal halide perovskites. Science 338(6107):643–647

Li H, Zhang W (2020) Perovskite tandem solar cells: from fundamentals to commercial deployment. Chem Rev 120(18):9835–9950

Lin FS, Priyanka P, Fan MS, Vegiraju S, Ni JS, Wu YC, Li YH, Lee GH, Ezhumalai Y, Jeng RJ, Chen MC (2020) Metal-free efficient dye-sensitized solar cells based on thioalkylated bithiophenyl organic dyes. J Mater Chem C 8(43):15322–15330

Liu N, Wang L, Xu F, Wu J, Song T, Chen Q (2020) Recent progress in developing monolithic perovskite/Si tandem solar cells. Front Chem 8:603375

Liu P, Wang H, Niu T, Yin L, Du Y, Lang L, Zhang Z, Tu Y, Liu X, Chen X, Wang S (2024) Ambient scalable fabrication of high-performance flexible perovskite solar cells. Energy Environ Sci 17(19):7069–7080

Maka AO, Alabid JM (2022) Solar energy technology and its roles in sustainable development. Clean Energy 6(3):476–483

Malinkiewicz O, Yella A, Lee YH, Espallargas GM, Graetzel M, Nazeeruddin MK, Bolink HJ (2014) Perovskite solar cells employing organic charge-transport layers. Nat Photonics 8(2):128–132

Martínez-Rodríguez G, Fuentes-Silva AL, Baltazar JC (2021) Simultaneous production of solar thermal heat and power for industrial applications. Chem Eng Trans 88:505–510

Mathew S, Yella A, Gao P, Humphry-Baker R, Curchod BF, Ashari-Astani N, Tavernelli I, Rothlisberger U, Nazeeruddin MK, Grätzel M (2014) Dye-sensitized solar cells with 13% efficiency achieved through the molecular engineering of porphyrin sensitizers. Nat Chem 6(3):242–247

Maurer C, Cappel C, Kuhn TE (2017) Progress in building-integrated solar thermal systems. Solar Energy 154:158–186

Mekhilef S, Saidur R, Safari A (2011) A review on solar energy use in industries. Renew Sustain Energy Rev 15(4):1777–1790

Min H, Kim M, Lee SU, Kim H, Kim G, Choi K, Lee JH, Seok SI (2019) Efficient, stable solar cells by using inherent bandgap of α-phase formamidinium lead iodide. Science 366(6466):749–753
Mishra A (2024) A sustainability perspective. In: The resilience of traditional knowledge systems for a sustainable future: a focus on agriculture and food practices in the Himalayas, p 249
Müller-Steinhagen H, Trieb F (2004) Concentrating solar power. A review of the technology. Ingenia Inform QR Acad Eng 18:43–50
Nam-Chol O, Kim H (2019) Towards the 2 C goal: achieving sustainable development goal (SDG) 7 in DPR Korea. Resour Conserv Recycl 150:104412
Olabi AG, Elsaid K, Obaideen K, Abdelkareem MA, Rezk H, Wilberforce T, Maghrabie HM, Sayed ET (2023) Renewable energy systems: comparisons, challenges and barriers, sustainability indicators, and the contribution to UN sustainable development goals. Int J Thermofluids 20:100498
Østergaard PA, Duic N, Noorollahi Y, Kalogirou SA (2021) Recent advances in renewable energy technology for the energy transition. Renew Energy 179:877–884
Öz AK, Neven-du Mont S, Nikitina V, Romer P, von Kutzleben D, Latif NA, Forster J, Wellens C, Heinrich M, Neuhaus DH (2025) Pushing the boundaries: challenges that arise in manufacturing and designing photovoltaic modules for new application areas. Solar Energy Mater Solar Cells 291:113735
Pietryga JM, Park YS, Lim J, Fidler AF, Bae WK, Brovelli S, Klimov VI (2016) Spectroscopic and device aspects of nanocrystal quantum dots. Chem Rev 116(18):10513–10622
Pitz-Paal R (2020) Concentrating solar power. In: Future energy. Elsevier, pp 413–430
Saadat F, Alizadeh A, Roudgar-Amoli M, Shariatinia Z (2022) Exploring the influence of Zn_2SnO_4/ZIF-8 nanocomposite photoelectrodes on boosting efficiency of dye sensitized solar cells. Ceram Int 48(15):21812–21823
Sakthivadivel D, Balaji K, Rufuss DDW, Iniyan S, Suganthi L (2021) Solar energy technologies: principles and applications. In: Renewable-energy-driven future. Academic Press, pp 3–42
Saliba M, Matsui T, Seo JY, Domanski K, Correa-Baena JP, Nazeeruddin MK, Zakeeruddin SM, Tress W, Abate A, Hagfeldt A, Grätzel M (2016) Cesium-containing triple cation perovskite solar cells: improved stability, reproducibility and high efficiency. Energy Environ Sci 9(6):1989–1997
Sawin JL (2013) Renewables 2013. Global status report
Shamoon A, Haleem A, Bahl S, Javaid M, Garg SB, Sharma RC, Garg J (2022) Environmental impact of energy production and extraction of materials—a review. Mater Today: Proc 57:936–941
Sutherland BR, Hoogland S, Adachi MM, Kanjanaboos P, Wong CT, McDowell JJ, Xu J, Voznyy O, Ning Z, Houtepen AJ, Sargent EH (2014) Perovskite thin films via atomic layer deposition. Adv Mater 27(1):53-58
Swain RB, Karimu A (2020) Renewable electricity and sustainable development goals in the EU. World Dev 125:104693
Tasmin N, Farjana SH, Hossain MR, Golder S, Mahmud MP (2022) Integration of solar process heat in industries: a review. Clean Technol 4(1):97–131
Tetteh EK, Amankwa MO, Yeboah C (2021) Emerging carbon abatement technologies to mitigate energy-carbon footprint—a review. Clean Mater 2:100020
Tsai H, Nie W, Blancon JC, Stoumpos CC, Asadpour R, Harutyunyan B, Neukirch AJ, Verduzco R, Crochet JJ, Tretiak S, Pedesseau L (2016) High-efficiency two-dimensional Ruddlesden–Popper perovskite solar cells. Nature 536(7616):312–316
Utami RWN, Dewi T (2024) Bio-inspired solar power: natural dyes and the advancements of dye-sensitized solar cells. In: 2024 international conference on electrical and information technology (IEIT). IEEE, pp 42–47
Wallerand AS (2018) Integration of solar energy with industrial processes. Doctoral dissertation, Ecole Polytechnique Fédérale de Lausanne
Wang YC, Cho CP (2017) Application of TiO_2-graphene nanocomposites to photoanode of dye-sensitized solar cell. J Photochem Photobiol A: Chem 332:1–9

Wang Q, Ito S, Grätzel M, Fabregat-Santiago F, Mora-Seró I, Bisquert J, Bessho T, Imai H (2006) Characteristics of high efficiency dye-sensitized solar cells. J Phys Chem B 110(50):25210–25221

Wang Y, Wu T, Barbaud J, Kong W, Cui D, Chen H, Yang X, Han L (2019) Stabilizing heterostructures of soft perovskite semiconductors. Science 365(6454):687–691

Weinstein LA, Loomis J, Bhatia B, Bierman DM, Wang EN, Chen G (2015) Concentrating solar power. Chem Rev 115(23):12797–12838

Wu T, Wang Y, Dai Z, Cui D, Wang T, Meng X, Bi E, Yang X, Han L (2019a) Efficient and stable $CsPbI_3$ solar cells via regulating lattice distortion with surface organic terminal groups. Adv Mater 31(24):1900605

Wu H, Xie X, Mei Y, Ren Y, Shen Z, Li S, Wang P (2019b) Phenalenothiophene-based organic dye for stable and efficient solar cells with a cobalt redox electrolyte. ACS Photonics 6(5):1216–1225

Wu J, Gao M, Chai Y, Liu P, Zhang B, Liu J, Ye L (2021) Towards a bright future: the versatile applications of organic solar cells. Mater Rep Energy 1(4):100062

Xiao K, Lin R, Han Q, Hou Y, Qin Z, Nguyen HT, Wen J, Wei M, Yeddu V, Saidaminov MI, Gao Y (2020) All-perovskite tandem solar cells with 24.2% certified efficiency and area over 1 cm^2 using surface-anchoring zwitterionic antioxidant. Nat Energy 5(11):870–880

Yang WS, Noh JH, Jeon NJ, Kim YC, Ryu S, Seo J, Seok SI (2015) High-performance photovoltaic perovskite layers fabricated through intramolecular exchange. Science 348(6240):1234–1237

Yella A, Lee HW, Tsao HN, Yi C, Chandiran AK, Nazeeruddin MK, Diau EWG, Yeh CY, Zakeeruddin SM, Grätzel M (2011) Porphyrin-sensitized solar cells with cobalt (II/III)–based redox electrolyte exceed 12 percent efficiency. Science 334(6056):629–634

Yogi Goswami D (1998) Solar thermal power technology: present status and ideas for the future. Energy Sources 20(2):137–145

Zanatta AR (2022) The Shockley–Queisser limit and the conversion efficiency of silicon-based solar cells. Results Opt 9:100320

Zhang Y, Park NG (2022) Quasi-two-dimensional perovskite solar cells with efficiency exceeding 22%. ACS Energy Lett 7(2):757–765

Zhu J, Xia J, Li Y, Li Y (2025) Perspective on flexible organic solar cells for self-powered wearable applications. ACS Appl Mater Interfaces 17(4):5595–5608

Chapter 9
Recycling of Polymer Nanocomposites in Renewable Energy Systems

Jyotirmoy Panda, Kaustav Mazumder, Soheli Biswas, Sk Mehebub Alom, Saista Parvin, Sayantan Sarkar, Divakar Pamanji, Bimal Das, and Deepshikha Datta

Abstract Polymer Nanocomposites (PNC) are the first choice materials that contribute majorly to the field of Renewable energy. The three primary applications (Solar energy, Wind Turbines, Fuel and Hydrogen cells), where PNCs are employed, ideally benefit from the High Mechanical Strength, Thermal Stability, and Conductivity properties of PNCs. Moreover, these specific properties not only make PNCs the first choice but also reduce the possibility of their recycling or easy degradation. The recyclability as well as the sustainability of PNC are some of the issues that result from the Complex composition and thermo-chemical stability, which subsequently become potential hazards to the environment. This article is about the importance of Polymer Nanocomposites (PNC) in Renewable energy and their associated issues and possible solutions of PNCs' Recyclability, Sustainability,

J. Panda
Department of Pharmaceutical Technology, Brainware University, Kolkata, India

K. Mazumder · S. Sarkar
Department of Chemical Engineering, National Institute of Technology Durgapur, Durgapur, India

S. Biswas
Department of Molecular Biology and Biotechnology, University of Kalyani, Kalyani, Nadia, India

S. M. Alom · S. Parvin
Department of Biotechnology, Brainware University, Barasat, India

D. Pamanji
Department of Civil Engineering, Mother Theresa Institute of Engineering and Technology, Palamaner, India

B. Das
Department of Chemical Engineering, National Institute of Technology Durgapur, Durgapur, India

D. Datta (✉)
Department of Basic Science & Humanities, Institute of Engineering and Management (IEM), Kolkata, University of Engineering and Management (UEM), Newtown, Kolkata, India

S. K. Verma et al. (eds.), *Sustainable and Natural Polymers*, Clean Energy Production Technologies, https://doi.org/10.1007/978-981-92-0092-4_9

and Waste Management. Moreover, this paper introduces in detail the different methods for recycling, including Mechanical Recycling, Chemical Recycling, Closed-loop Recycling, and the biodegradable approach. It also discusses Advanced material Science like self-healing PNCs, Recovery Techniques of Nanofillers, and Bio-based nanocomposites. This paper also offers all this information, along with a thorough study of the Economic and Environmental Impact of Nanocomposites plus their future trends and research directions.

9.1 Introduction

Nanoreinforced polymers or ones with nanofillers as a nanofiller well-dispersed in the bulk polymer matrix are called Polymer Nanocomposites (PNCs). Nanomaterials are those with at least one dimension on the nanoscale, in most cases from 1 nm to 100 nm, and may have different physical properties such as a very strong local electromagnetic field or quantum confinement with quantum dots (QDs) (Zhang and Hoshino 2019) that can be used as a nanofiller (the fillers with particle sizes of 1 nm to 100 nm in Polymer Nanocomposites (PNCs)).

Since the fillers are of the nanoscopic scale, the improvement in functionality can be seen, e.g., when the size of nanofillers is reduced, the surface area gets larger, thus increasing the efficiency. The classification of polymer nanocomposites is possible from the nature of polymer matrix and structure, and the properties of nanocomposites. The polymer matrices can be thermoplastic, thermoset, or elastomer, based on which polymer nanocomposites are categorized (Bokobza 2023).

By introduction of nanoparticles significantly improves the base polymer properties, including mechanical strength, thermal stability, electrical conductivities, barrier properties, and flame retardancy. Polymer nanocomposites represent a versatile and scalable class of matter with unique properties, and their application ranges from aerospace to biomedical. Polymer Nanocomposites display superior tensile strength, Young's modulus, and toughness due to a high aspect ratio and strong interfacial bonding between the nanofiller and polymer matrix. Nanomaterials like Carbon Nanotubes (CNTs), Graphene, and metal oxides, also known as conductive nanofillers, convert insulating polymers into conductive polymers. Carbon Nanotubes reinforced epoxy nanocomposites demonstrate Electromagnetic interference shielding properties and act as an antistatic material. Some Polymer Nanocomposites restrict the permeation of gases, water vapor, and chemicals, making them ideal packing materials. Graphene oxide reinforced Polylactic acid demonstrates a significant reduction in the permeability of oxygen (Rhim and Ng 2007). Biocompatible and bioresorbable PNCs deliver is application in the field of drug delivery, tissue engineering, and biosensors.

Some general methods of recycling of Polymer Nanocomposites are like Mechanical recycling, which includes processes like shredding and melting, Chemical recycling, or Feedstock recycling, in which polymers break into monomers or fuels by processes like pyrolysis, hydrolysis, or solvolysis (Rizos et al.

2023), Biological recycling, which promotes recycling of biodegradable plastic and organic waste, and decomposing processes like composting and anaerobic digestion (Tamjid et al. 2024), Thermal recycling or energy recovery in which waste is incinerated to recover heat or electricity (Waste-to-Energy from Municipal Solid Wastes 2019). Other than this, the General method of recycling can be classified into two types based on the end product. If the product is the same as the original before processing, then the type of recycling is called closed-loop recycling; if different, then the type of recycling is called open-loop recycling (Kasmi et al. 2023).

Even when recycling takes place, there are some challenges in performing such recycling process, which may negatively affect our environment. The challenges in recycling and their environmental impact are as follows:

1. Contamination of recyclables (due to mechanical recycling, some particles of polymer nanocomposites may be of micro or nanoscale, may enter our environment, and become toxic to our ecosystem);
2. Downcycling in recycling (in which polymer quality degrades upon recycling (Chen et al. 2025));
3. Economic viability (in which high collection, sorting, and processing costs make recycling less profitable);
4. Lack of infrastructure in developing countries;
5. Microplastic and Nanoplastic (due to leaching from the disposal of polymer as landfill, resulting in contamination of micro and nano plastic formation (Carpanez et al. 2025).
6. Energy consumption (recycling is a highly energy-consuming process, which makes it both less profitable and environmentally unsafe).
7. E-waste and Composite materials, (in e-waste and composite materials, polymers are hard to disassemble, resulting in problems in the recycling process).

Thus, this chapter describes the various challenges in recycling polymer composites, elaborating on the sustainable recycling methods. The work also highlights the applications of polymer nanocomposites in renewable energy with innovative approaches towards sustainability. The economic, environmental impact and the regulatory policies are also discussed so as to provide a detailed understanding of the feasibility of the process.

9.2 Challenges in Recycling Polymer Nanocomposites

While recycling polymer nanocomposites, several economic and environmental issues are faced, such as Complex Composition and Heterogeneity, Thermal and Chemical stability issues, and Toxicity and impact on the environment (Fig. 9.1).

Fig. 9.1 Key challenges in recycling polymer nanocomposites

9.2.1 Complex Composition and Heterogeneity of Polymer Nanocomposites

One of the main issues is that nanofillers used in polymer nanocomposites (PNCs) are usually highly incompatible with a standard mechanical recycling process, and this is because the nanofillers have strong interfacial bonding with the polymer matrix. The compatibility impairs the melting process in which the nanofillers either agglomerate or are unevenly dispersed in the recycling, resulting in lower quality and performance of the new material. The agglomeration of the nanofillers is caused by their high surface energy and small size, which, after the melting process, results in non-uniform dispersion that affects mechanical integrity. The presence of mixed polymer waste and additives not only makes recycling more difficult but also this mixture often leads to phase separation and poor interfacial adhesion between the recycled phases (Rasilainen et al. 2024).

Physically separating nanoparticles from the polymer matrix during recycling is challenging because of the nanoscale dimensions and high surface energy of the nanoparticles. The likes of carbon nanotubes, graphene, and metal oxide particles are often so tightly embedded in the polymer matrix that chemical or mechanical methods are usually not efficient in freeing them. The compact combination complicates the process of its reuse and may trigger the uncontrolled release of nanoparticles into the environment (Bijanu et al. 2024). Besides that, the removal of nanoparticles will face more difficulties due to their tendency to form agglomerates as a result of a recycling process that leads to thermodynamic instability.

9.2.2 Thermal and Chemical Stability Issues

Research has proven that low-density polyethylene and polystyrene weaken remarkably in tensile strength, elongation, and fracture after several reprocessing cycles. The existing studies dealing with this issue have focused on a solvent-assisted low-temperature chemical recycling method to effectively limit thermal degradation. Moreover, the use of stabilizing additives and antioxidants is extended to the area of thermally sensitive PNCs that are recycled and to which the degradation reactions are slowed down as a result (Patel et al. 2024).

On the other hand, repeated extrusion or thermal cycling may cause these interfacial bonds to gradually deteriorate and consequently lead to the loss of the reinforcement effect of the nanomaterials as well as poor adhesion and loss of interfacial integrity, especially in CNT and nanoclay-based composites (Rasilainen et al. 2024).

9.2.3 Toxicity and Impact on Environment

Nanoparticles, due to their extremely small size and high surface energy, can pass easily through the biological barriers and interact with the cellular components, such as generating reactive oxygen species (ROS), which causes oxidative stress and may results into several health related issue such as inflammatory response in tissue and may also results in damage in DNA (Kumah et al. 2023). Inhalation or dermal exposure to metal oxide nanoparticles such as ZnO or TiO_2 has been associated with respiratory inflammation, neurotoxicity, as well as cytotoxic effects at the cellular level. Chronic exposure to some engineered nanomaterials (ENMs) can result in their bioaccumulation in vital organs and results in increasing the risk of long-term toxicity (Kumah et al. 2023). Nanoparticles are highly reactive and tend to attract or carry toxic pollutants on their surface; thus, on their release in the environment may result in affecting the ecosystem by disrupting microbial communities, inhibiting plant growth, and accumulating in aquatic organisms. Recycling methods with their costs, Feasibility, Efficiency and Scalability are given in Table 9.1.

Table 9.1 Comparison of recycling methods for polymer nanocomposites

Recycling method	Cost	Feasibility	Efficiency	Scalability	Reference
Mechanical	Low	Common, but risk of property loss due to nanofiller agglomeration	Moderate—Degradation of mechanical properties over cycles	High	Hopewell et al. (2009)
Chemical (Depolymerization)	Moderate–High	Effective on PET, PU, PC with proper catalysts	High–monomer recovery is possible	Moderate	Ügdüler et al. (2020)
Biological (Enzymatic/ Microbial)	Low–Moderate	Limited to biodegradable/ bio-based polymers (e.g., PLA, PHAs)	Low–Moderate—Very slow rates	Low	Zhang et al. (2024)
Thermal (Pyrolysis)	High	Useful for mixed or contaminated waste with nanofillers (e.g., CNTs, MMT)	High—Produces oils, gases, carbon black	Moderate	Majka et al. (2016)
Solvent-based (Dissolution/ Precipitation)	Moderate	Efficient for multilayer or colored PNCs	High—Separates polymer from filler with minimal degradation	Moderate	Lee et al. (2024)
Microwave-assisted	Moderate–High	Fast and energy-efficient; experimental	High—Improves depolymerization speed	Low	Cafaro et al. (2025)
Electrochemical Recycling	Moderate–High	Suitable for conductive nanofillers (e.g., graphene, CNTs)	Moderate—Emerging; selective degradation pathways	Low–Moderate	Zhang et al. (2024)
Supercritical Fluid Recycling	High	Requires supercritical CO_2 or water; experimental	High—Excellent solubility and polymer-filler separation	Low	Goto (2009)

9.3 Sustainable Recycling Methods

9.3.1 Mechanical Recycling

Mechanical Recycling involves physical reprocessing of polymer nanocomposites, such as sorting, shredding, melting, and re-extrusion, without altering their chemical structure. Due to its simplicity and cost-effectiveness, it is widely accepted in industries where chemical changes are not favourable. Mechanical recycling demonstrates feasibility in the recycling of 3D printed wastes.

Repeating the recycling cycle for mechanical recycling may cause chain scission and oxidative degradation, resulting in mechanical property degradation and diminished mechanical integrity. Moreover, when associated with nanofillers, the process is further complicated.

9.3.2 Chemical Recycling

Chemical Recycling operates by breaking down polymers into their monomeric or oligomeric form, which enables the recovery of raw materials suitable for repolymerization or new polymer synthesis. With the principles of green chemistry, the selective depolymerization of plastic can be done using chemical recycling methods.

Solvent-assisted chemical recycling allows separating nanomaterials in the polymer nanocomposites from their polymer matrix.

Chemical recycling of polymer nanocomposites faces significant challenges during the scale-up process, such as depolymerization, nanoparticle recovery, and purification (Martínez-Narro et al. 2024).

9.3.3 Sustainable Recycling of Biodegradable Polymer Nanocomposites

With the ability to preserve the essential properties of polymer materials and still be biodegradable, biodegradable polymer nanocomposites are becoming an excellent alternative as eco-friendly polymer nanocomposites of improved function. The key point of such polymer nanocomposite is to combine the biodegradability of the polymer with the enhanced thermal, mechanical and barrier properties by using suitable nanofillers (Lors et al. 2024). However, the matter of the end-of-life of these materials must be handled with great care in order to ensure that the recyclability of the materials is sustainable. An essential factor in the successful sustainable recycling of biodegradable polymer nanocomposites is the innovation and improvement of green nanofillers and bio-based polymer matrices, which have become an effective approach to limit the environmental impact of the process. The production of polymers from renewable resources, including polylactic acid, polyhydroxyalkanoates, and bio-based epoxy resins, served as the major platform for sustainable recycling. Polylactic acid is one of the most widely utilized compostable bio-based polymers that are manufactured from sugarcane or corn starch, and it involves process of recycling biodegradable polymer nanocomposites and thus making environmentally friendly recycling possible, besides providing suitable mechanical properties to polymer nanocomposites (Wu et al. 2023). Similarly, polyhydroxyalkanoates and bio-based epoxy resins are completely biodegradable products and thus can be recycled into the environment without causing any harm to the environment as both are obtained from natural sources.

9.3.4 Closed-Loop Recycling Approaches

One of the methods by which closed-loop recycling of polymer nanocomposites can be carried out is depolymerization catalysed by reversible crosslinking agents that alter the polymer to one that is considered a starting material for the synthesis of the subsequent materials without affecting its main components. Qin et al. (2022) suggested the use of dynamic maleic acid tertiary amide bonds to introduce the character of composites having high mechanical strength and thereby offering the recyclability aspect. Very significantly, when fully dissociated into the original monomers, this reversibility process attains a recovery rate that exceeds 94%. Apart from that, another promising material recovery method is the use of magnetic or electrostatic separation. To give an example, one may implant magnetic nanoparticles, e.g., FeO in a composite, so that during recycling the nanoparticles can be pulled out using an external magnet thus allowing less contamination of the surface of the other materials. Thermal, redox, pH etc. bonds, linkages or adducts are employed in the retrieval of materials from the nanocomposites, and the instances of such crosslinkings are Diels-Alder adducts, Disulfide bonds, Boronic ester linkages, Hydrogen bonding, etc. (Chen et al. 2002).

9.4 Applications of Polymer Nanocomposites in Renewable Energy

Increased tensile strength, impact strength, thermal resistance, and water uptake have continued to make composite and nanocomposite materials a material of preference. Among the many properties that can be desired, these first three are the most sought-after latter. For materials applied in energy, the above-stated properties and some others, such as optical character, antireflection, conductivity, etc., are considered to be of great importance. Certain polymers possess an aspect of conductive nature, whereas certain other ones would need to be modified with other polymers or nanoparticles. Adding the nanoparticles into the polymer matrix has been said to provide the desired characteristics required for electrical use. The various applications of polymer nanocomposites in renewable energy systems are described in Table 9.2, Fig. 9.2.

9.4.1 Solar Energy Systems

Solar energy is one of the most universal forms of renewable energy sources available copiously in nature. In order to reap solar energy, systems need the use of big solar collectors and wide-open spaces. Solar energy has applications in various industrial as well as domestic applications like water heating, air heating, desalination, generation of power, etc. The performance of solar thermal systems is

Table 9.2 Nanocomposites in renewable energy applications

Application area	Specific nanocomposite	Key function	References
Solar cells	Graphene/polymer nanocomposites	Enhanced charge transport, mechanical strength, and light absorption	Ibrahim et al. (2019), Wan et al. (2016)
	Conjugated polymer/nanocrystal (CdSe, PbS, TiO_2, ZnO) composites	Improved photon harvesting and efficient charge separation	Ates et al. (2020), Ibrahim et al. (2019)
	Graphitic carbon nitride-based nanocomposites	Photocatalytic water splitting and pollutant degradation	Jawaid and Khan (2018)
Wind turbines	Carbon nanotube (CNT)/polymer nanocomposites	Increased blade strength, fatigue resistance, and reduced weight	Rahman (2019)
	Nanoclay or silica-reinforced epoxy composites	Enhanced durability and environmental resistance	Ibrahim et al. (2019)
Fuel cells	Polymer/graphene oxide nanocomposites	Improved proton conductivity and membrane stability	Antolini (2015)
	Poly(2,2′-(m-phenylene)-5′-bibenzimidazole)/nano-calcium titanate	High-temperature operation and enhanced ionic conductivity	Korniejenko (2024)
Batteries	CNT/polymer nanocomposites	Enhanced electrode conductivity and cycling stability	Croce et al. (1998)
	Nanoclay/polymer electrolytes	Improved ionic transport and safety	Ibrahim et al. (2019)
	PZT nanowire/PVDF composites	Increased energy density for capacitors and batteries	Wan et al. (2016)
Supercapacitors	Polyaniline/graphite oxide (GO) nanocomposites	High capacitance, long cycle life, and rapid charge/discharge	Wan et al. (2016)
	Metal oxide/graphene nanocomposites	Enhanced energy and power density	Ibrahim et al. (2019)

primarily based on the exposure of solar collectors/panels to solar radiation and their absorptance properties. To improve the energy absorption capability and efficiency of the solar collector and other solar thermal systems, different coatings and nanocomposite films are deposited on surfaces.

9.4.1.1 Polymer-Based Photovoltaic Cells (Organic Solar Cells)

The shortage of energy can be seen as one of the main problems of global warming that the world is facing, and so it has provoked the desire for more energy. Different solar technologies had been unfolded until now, but dye-sensitised solar cells (DSCs)

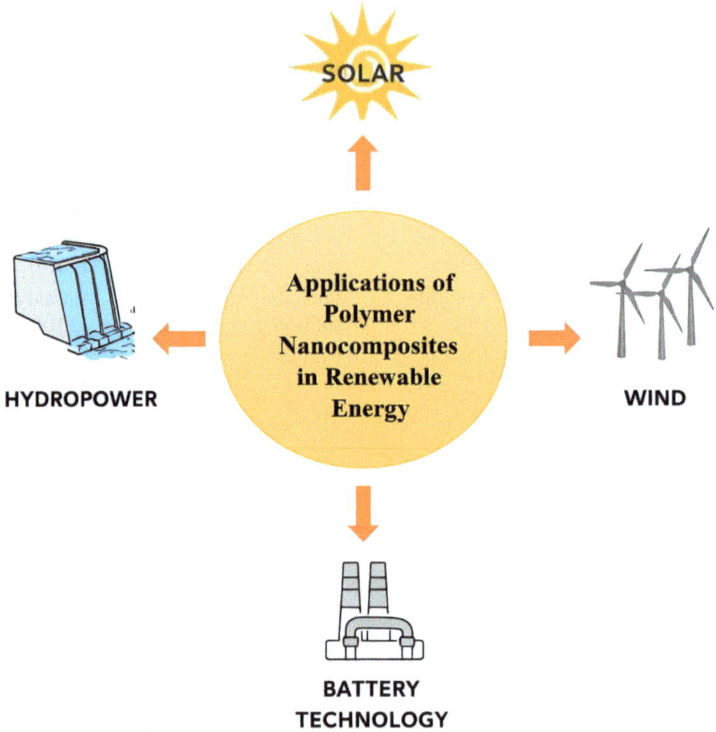

Fig. 9.2 Renewable energy applications of nanocomposites

seem to be the most advantageous due to their eco-friendly character and cost-effectiveness, which makes them the ideal solution for the increase of the world's energy needs, moreover, the use of nanocomposite coatings on their surfaces can also improve to a great extent the performance of these solar cells (Ibrahim et al. 2019). These materials are one of the most promising due to their antireflective properties, ability to provide surface passivation, and their suitable bandgap characteristics.

9.4.1.2 Nanocomposite Coatings for Efficiency Enhancement

The use of nanocomposite coatings will bring great benefits to the installation of solar panels, as they will be able to have a multifunctional coating that will not only absorb better the light but will also reflect less from the surface and be more resistant to environmental degradation. For example, titaniasilicalanthanum phosphate nanocomposite coatings are designed to show both photoactive and hydrophobic roles, maintaining the cleaning and cooling functions of the panel through the resistance against dirt, water, and UV degradations, and high thermal stability, as well as self-cleaning activity. Besides, the films can reduce the influence of solar heat and reflection losses, which are important efficiency shortfalls; hence, the UV-blocking polymer/ZnO nanocomposite films can play the role of preventing heat buildup (Smitha et al. 2016).

9.4.1.3 Thermal Stability and Weather Resistance

Nanocomposite films can extend the weather resistance and thermal stability of solar energy systems, thus, ensuring their operation for long periods under severe outdoor conditions. For instance, the titaniasilicalanthanum phosphate (TSL) nanocomposites are stable anatase with a wide temperature bracket and no notable changes in lattice structures or photoactivity, which is very important for a solar panel that is fluctuating in temperature and receiving intense sunlight (Ibrahim et al. 2019). These hydrophobic and self-cleaning materials could be combined in solar panels to address the problems of maintenance and efficiency loss due to atmospheric contamination. Coated with high-performance nanocomposite absorber, such as aluminium titanium oxynitride, these materials have been verified for a long-term stability performance with thermal cycling to up to 700 °C and their degradation has not been detected, thus they are considered potential materials for the next generation of concentrated solar power plants.

9.4.2 Wind Energy Systems

One major example of the rising necessity of polymer nanocomposites is in wind energy technology, where their use has expanded widely due to the fabrication of wind turbine blades and other crucial parts. By incorporating nanomaterials such as nanoclay, carbon nanotubes (CNTs), and nano-silica into polymer matrices, the mechanical properties, fatigue life, and environmental resistance of the composites for turbine blades are improved by several times. When one or both of the components that make up a material are engineered at the nanoscale, this is called nanoengineering; this process is carried out to obtain the desired characteristics, such as stiffness, strength, and damage tolerance needed to meet the strict operational requirements of wind turbines, which are subjected to cyclic loads and often harsh environments. By using these nanocomposites, the wear resistance of gears and other turbine components could be enhanced to reduce the required maintenance.

9.4.2.1 Lightweight Composite Materials for Wind Turbine Blades

The cost-effectiveness and efficiency of wind turbines are directly related to the weight of the blades because lighter blades consume less energy to turn and can capture more wind power. To improve the strength-to-weight ratio of materials, steel and aluminum can primarily be substituted by polymer-based materials, including carbon fiber and glass fiber reinforced polymers, as these have better corrosion resistance besides that. Carbon fiber composites, in offer enhanced mechanical strength, reduced weight, and superior durability, making them highly suitable for advanced wind turbine blade applications where improved performance and longer service life are critical.

9.4.2.2 Fatigue Resistance and Longevity Improvements

Wind turbine blades are under constant cyclic loading due to the variability of the wind, and hence fatigue resistance is a very important feature if the long-term reliability of the system is the goal. One of the most effective ways to achieve the optimization of fatigue resistance as well as toughness is by introducing carbon nanotubes (CNTs) into polymer matrices. The composites with CNT insertions are characterised by high strength, stiffness, and good interfacial adhesion, thus avoiding the generation and expansion of microcracks under cyclic loading (Jespersen and Mikkelsen 2017). The improvement of fatigue life does not only extends the lifetime of the blades but also reduces the downtime and the maintenance cost.

9.4.2.3 Nanocomposites in Wind Turbine Gears and Components

Aside from blades, polymer nanocomposites are increasingly being used in wind turbine components, including gears, hubs, and nacelles. These find advantages with the increased wear resistance, friction reduction, and strengthened mechanical properties that nanofiller-reinforced polymers have to offer. The addition of nanoparticles such as boron nitride or nano-silica to thermoplastic matrices yields gears and moving parts with less need for lubrication and more endurance, thus reducing maintenance requirements and downtime during operation (Thomas and Ramachandra 2018). The light weight of these nanocomposites also helps improve overall system efficiency by minimizing the load on the turbine structure.

9.4.3 Fuel Cells and Hydrogen Storage

The term fuel cell is used to describe an electrochemical device that produces on-the-spot clean power by converting chemical energy from hydrogen and/or other fuels by electrochemical reactions with oxygen and/or other oxidants. Unlike batteries, which operate without a constant supply of fuel, fuel cells only expend the available chemical energy of an oxidant to produce water or carbon dioxide as by-products, making them indispensable in sustainable energy systems and hydrogen storage developments.

9.4.3.1 Proton Exchange Membrane Fuel Cells (PEMFCs)

Proton Exchange Membrane Fuel Cells (PEMFCs) represent a prominent fuel cell technology on account of high efficiency, low temperature operation, and quick startup. The key material of a PEMFC is the polymer electrolyte membrane, which allows protons to pass through from the anode to the cathode, at the same time keeps electrons isolated, which are forced through the external circuit to generate electricity. The membrane and electrode materials have an indispensable role in the functioning of PEMFCs. Nanomaterials were implemented in the polymers to meet the requirements of proton conductivity, mechanical and thermal stability, as well as to maintain their individual properties in combination. For instance, the use of nanocalcium titanate (nanoCTO) in poly(2,2′-(m-phenylene)-5′-bibenzimidazole) (PBI) membranes has shown drastic conductivity and power density improvements, with values reaching as high as 32.7 mS/cm and 419 mW/cm^2 at high temperatures (Antolini 2015). Hybrid nanocomposites of chitosan with polyaniline/nano-silica fillers have also exhibited high proton conductivity, at 8.39×10^{-3} S/cm at 80 °C.

9.4.3.2 Nanomaterials for Electrolytes and Electrodes

Fuel cells have been revolutionised by the use of nanomaterials in electrodes and electrolytes. These changes have been made possible through enhanced ionic conductivity, catalytic activity, and durability. Nanostructured electrolytes, such as polymer matrices doped with nanoparticles (e.g., nano-silica, nano-titania, or carbon nanotubes), not only provide a larger surface area but also facilitate improved proton transport channels. Nanomaterials like carbon–carbon, metal–carbon, and polymer–carbon composites are used in electrodes to develop anode and cathode catalysts, yielding better catalytic efficiency and operational performance than in the case of traditional materials. Nanostructured catalysts enhance the active surface area for electrochemical reactions, minimizing the use of precious metals and overall costs.

9.4.3.3 Hydrogen Storage Using Polymer Nanocomposites

Polymer nanocomposites for hydrogen storage resolve central issues of safety, capacity, and reversibility by immersing complex hydrides in nanoporous carbon or wrapping hydride nanoparticles with polymer layers. For instance, combining metal hydride nanoparticles in nanocarbon scaffolds generates materials in which hydrogen is finely dispersed in nanoscale cavities, suppressing agglomeration and facilitating rapid absorption and release at modest temperatures and pressures. Layer-by-layer self-assembled polymer coatings on hydride particles also improve stability and prevent oxidation, extending the life and safety of storage systems.

Agglomeration is prevented, hydrogen uptake and release at reasonable temperatures and pressures are maximized, and stability against oxidation is improved (Zhu et al. 2017).

9.4.4 Energy Storage Devices (Batteries and Supercapacitors)

Batteries and supercapacitors are the key pillars of modern energy storage technologies. Lithium-ion batteries are useful for their high energy density storage and long-term power delivery. On the other hand, supercapacitors provide fast charging/discharging and better durability. Polymer nanocomposites are useful here due to their increasing property of conductivity, mechanical stability, and electrochemical performance, which enables progress in next-gen electronics and sustainable energy systems.

9.4.4.1 Polymer-Based Nanocomposites in Lithium-Ion Batteries

Lithium-ion batteries have some serious concerns, like ionic conductivity loss and safety issues, which can be optimised using polymer nanocomposites. These nanocomposites work by improving the electrode performance using polymer binder-stabilized silicon anodes, which reduces the volume expansion during cycling, by which their life cycle get extended. Clay nanoparticles, dispersed within polymer matrices, endow the solid electrolyte with enhanced stability and ionic transport capability compared with conventional liquid electrolytes. Polymer nanocomposites are also a key component of the separator and binder, which closely link battery architecture with cell cycle stability, via polymer porosity and adhesion to the electrodes. Lithium-sulfur batteries utilize conductive polymer-sulfur composites to mitigate polysulfide shuttling, enhancing capacity retention.

9.4.4.2 Electrochemical Performance of Nanocomposites

Nanocomposites exhibit greatly improved electrochemical parameters for energy storage. For lithium-ion batteries, silicon-polyaniline nanocomposites realize >2000 mAh/g capacity with 80% retention at 500 cycles to overcome silicon expansion issues. For supercapacitors, polyaniline/graphene composites on stainless steel fabric realize 1506.6 mF/cm^2 specific capacitance and 92% retention after 5000 cycles. Flexible graphene/polymer hybrid-based supercapacitors exhibit 90% capacitance retention on mechanical bending, suitable for wearable electronics3. Metal-organic framework (MOF)-polymer composites, such as UiO-66/polyaniline, have 1015 F/g at 1 A/g with 91% cycling stability. The materials also have low equivalent series resistance (ESR), allowing fast charge/discharge and high power density (Shao et al. 2018).

9.4.4.3 Role of Conductive Polymers in Supercapacitors

Conductive polymers like polyaniline (PANI), polypyrrole (PPy), and polythiophene (PTh) are central to supercapacitor innovation due to their pseudocapacitive charge storage and flexibility. PANI/graphene composites achieve 1506.6 mF/cm^2 capacitance with 92% retention over 5000 cycles, leveraging graphene's conductivity and PANI's redox activity. Similarly, PPy/MOF composites enhance surface area and ion accessibility, yielding 647 F/g at 1 A/g (Ibrahim et al. 2019). Such polymers facilitate light, flexible devices: stainless steel fabric-based supercapacitors preserve 95.8% capacitance upon 1000 bending cycles, essential in wearable technology. Conducting polymers also reduce self-discharge and increase cycle life in comparison to carbon-only electrodes.

9.5 Innovative Approaches for Sustainability

Advances in Sustainable Nanocomposites is a newly launched Special Issue of Sustainability, which will publish review and original papers on the emerging new scientific achievement of sustainable nanocomposite materials driven by zero-carbon policy targets. While the world continues to become more conscious of carbon emission-induced environmental issues, there has been greater demand for sustainable nanocomposite materials. Nanocomposites are showing promise as novel replacements for conventional materials due to their characteristic properties, making them viable candidates for environmentally compatible technological uses.

9.5.1 Development of Self-Healing Nanocomposites

Self-healing synthetic materials are the next generation of intelligent components that will repair themselves independently and automatically, just as all living organisms continue to repair their own structure. Polymer composites widely receive attention as potential material candidates in different applications for improving efficiency and sustainability among a range of materials, polymer nanocomposites could be considered as ideal candidates, which can show improved tensile strength, stiffness, cost-effectiveness thermal stable, lighter and environmentally acceptable material components. These also suffer from the drawbacks such as mesoscopic damage (e.g., micro-cracks and cavitation), aging, and wear scars (Thakur and Kessler 2015). The integration of self-healing principles into polymer nanocomposites provides new avenues towards sustainable, safer and more durable materials for multifunctional applications, such as electronics, energy, transport and coatings to mention just a few.

9.5.1.1 Smart Polymers for Damage Repair

Smart polymers in self-healing nanocomposites are designed to heal structural damage autonomously or non-autonomously, mimicking biological systems. These materials utilize various healing mechanisms, including intrinsic methods—where reversible chemical bonds like quadruple hydrogen bonding, Diels–Alder reactions, or metal–ligand coordination within the polymer matrix allow the material to reform its structure upon damage—and extrinsic methods, which involve encapsulated healing agents released upon crack development. Inclusion of nanomaterials such as carbon nanotubes, graphene oxide, or halloysite nanotubes greatly increases the healing effectiveness, mechanical durability, and heat stability of the composite. As an example, addition of mere 0.5–1 wt.% of graphene oxide into epoxy nanocomposites can register up to 88% efficiency in self-healing, and polyurethane/halloysite nanocomposites exhibit efficiencies of up to 90% healing (Jamil et al. 2024). Healing can be initiated by external forces like heat, light, or pressure, or can happen spontaneously at room conditions, depending on the involved chemistry.

9.5.1.2 Extending Product Lifespan to Reduce Waste

Incorporating self-healing polymer (SHP) nanocomposites represents (Fig. 9.3) an environmentally sustainable approach to extending product lifespans, as these composites improve mechanical strength, corrosion resistance, and anti-penetration capabilities, while also reducing waste generation.

9.5.2 AI & Machine Learning for Sorting and Recycling

AI and machine learning (ML) transform polymer nanocomposite recycling into a process with automated sorting and optimal separation through intelligent identification and control (Fig. 9.4). ML technologies improve efficiency in recycling by effectively sorting multiform waste streams by polymer content and nanofiller level. AI-based systems regulate processing conditions to maximize recovery yield and purity of recycled nanocomposites and preserve their advantageous properties for recycling use. This method minimizes waste, reduces energy usage, and facilitates a circular economy through the ability to recover precious nanomaterials.

9.5.2.1 Automated Identification of Polymer Nanocomposites

Artificial Intelligence (AI) and Machine Learning (ML) are revolutionizing the automated detection of polymer nanocomposites in recycling streams. Polymer nanocomposites, which consist of polymers blended with nanoscale fillers, pose special challenges for detection because of their intricate compositions and the

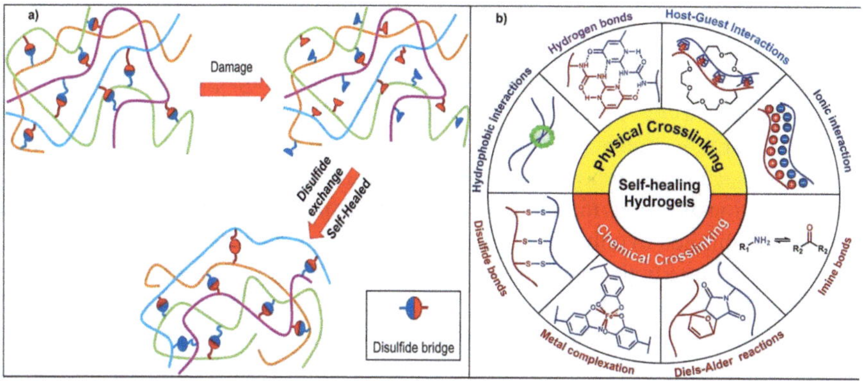

Fig. 9.3 (**a**) Self healing mechanism in nanocomposites and (**b**) Bonds work in self healing mechanism

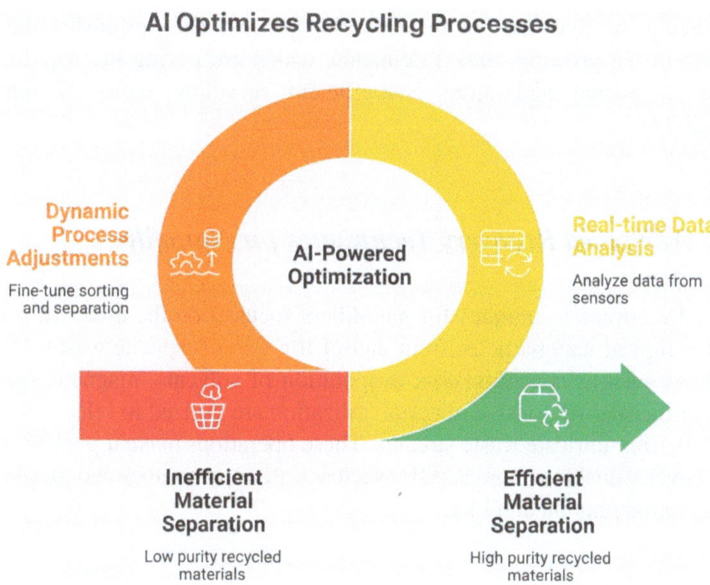

Fig. 9.4 AI-powered sorting and recycling system

minute differences between different formulations. AI-based computer vision technology, powered by deep learning, can scan high-resolution images captured by cameras and sensors to identify distinguishing characteristics like color, texture, and shape on conveyor belts. These systems are trained using massive datasets, allowing them to identify even challenging-to-classify nanocomposite materials that are easily missed by conventional optical sorting techniques (Farghali and Osman 2024).

9.5.2.2 AI-Based Separation and Processing Optimization

AI-based separation and processing technologies are changing the traditional recycling process using optimised and intelligence-driven operations. AI-powered machine learning-driven systems are analysing the recyclable materials using various sensors, cameras, robotic arms, etc. This real-time analysis helps in the sorting and separation of various materials, including polymer nanocomposites. Involvement of AI is not limited to only the sorting of the materials, rather it also generates various data using predictive analysis, which can provide crucial information about the optimisation of the collection route, distribution channel, recycling trends, etc., which not only reduces the operational costs but also reduces the environmental footprints. Quality and risk of contamination are a grave concern for the recyclable materials, as they can only be marketed when contamination is below the acceptable quality levels. Apart from AI, use of smart bins is also a solution for the waste collection of recyclable materials, as it can predict and notify the weight and level of contamination present in the bin (Tariq et al. 2022). Cumulatively, all of these efforts are changing the dynamics of the recycling process and optimising the system according to the growing market demands, which are paving the way for implementing a greener and more cost-efficient recycling sector (Farghali and Osman 2024).

9.5.3 Advanced Recovery Techniques for Nanofillers

Advanced recovery techniques for nanofillers focused on the effective separation and extraction of nanoparticles from end-of-life nanocomposite wastes (Fig. 9.5). Techniques such as centrifugation, evaporation of solvents, magnetic separation, and pH or thermal-responsive material utilization are utilized to selectively recover nanofillers from intricate waste streams. These operations make it possible to selectively extract valuable nanomaterials (such as carbon nanotubes and graphene with minimal contamination and loss.

9.5.3.1 Selective Extraction and Reuse of Nanoparticles

Selective recovery of nanoparticles from waste materials is performed using a mixture of physical and chemical separation processes. Centrifugation uses differences in density and particle size to recover nanofillers from polymer matrices, whereas solvent evaporation dissolves the polymer, with nanoparticles being left behind for harvesting. Magnetic separation works best for nanofillers that have magnetic properties, facilitating fast and non-destructive recovery.

After extraction, the recovered nanoparticles can be refined and reutilized during nanocomposite preparation. In doing so, raw materials are conserved in addition to

Fig. 9.5 AI-driven recycling framework

cutting environmental footprints through avoidance of sending nanowaste to the landfills, as well as by curbing production from virgin nanomaterial.

9.5.3.2 Sustainable Strategies for Carbon Nanotubes and Graphene

Carbon Nanotubes (CNTs) are incorporated with the polymer nanocomposites to increase the functional performance along with sustainable resource efficiency. Such CNTs incorporated nanocomposites not only enhance the thermomechanical and self-healing capacity but also achieve shape fixity and recovery rates of up to 100% and 95%, respectively, when activated by electrical heating. Furthermore, hybrid composites that combine CNTs with other nanofillers, such as montmorillonite nanoclay or halloysite nanotubes, exhibit superior recovery efficiency and mechanical strength, with electrical activation offering up to tenfold faster recovery compared to conventional thermal methods. Additionally, CNTs act as nucleating agents, enhancing the polymer's crystallinity and cross-link density, thereby improving both its recyclability and overall durability (Jamil et al. 2024). Graphene:

Graphene oxide (GO) or reduced graphene oxide (rGO) nanofillers are also pivotal in sustainable polymer nanocomposites by conferring outstanding shape memory and mechanical characteristics. Blending GO or rGO into polymers such as poly(vinyl alcohol) greatly enhances the glass transition temperature (from 81.1 °C to well above 100 °C) and allows for high shape fixity and recovery (to 95%) (Shahsavari et al. 2023). Graphene nanoplatelets (GNPs) are highly capable of increasing the material's strength and self-healing capacity of the material, not only that, it can also increase the tensile strength by up to 30% and increase stiffness by around 365% in comparison to regular polymers. However, their effectiveness highly relies on the even dispersion of the materials; only then, their durability and recovery are enhanced; otherwise, clumping leads to a drop in the material's performance.

9.5.4 Role of Bio-based Nanocomposites

Currently, Bio-based nanocomposites are getting attention due to their eco-friendly and sustainable approach, as they are composed of renewable polymers and nanofillers, which offer high performance as well as biodegradability. This can effectively reduce the consumption of fossil fuels and thus lower the environmental pollution. Studies also indicate that it may offer better thermal, mechanical, and physical barrier properties over the conventional Plastics. As the uses of Packaging, biomedicine, water purification, etc., are growing, environmental impacts are declining along with the extensive use of the circular economy. Bio-based nanocomposites provide a good answer to the issues faced by traditional non-biodegradable materials through utilizing natural resources and advanced processing techniques.

9.5.4.1 Sustainable Synthesis of Biodegradable Nanocomposites

The approach towards the synthesis of biodegradable nanocomposites is centrally focused on the use of biodegradable and renewable polymers such as polysaccharides (starch, chitosan, cellulose), biodegradable polyesters (PLA, PHB, PCL), and natural fibres, etc., and these are incorporated with the nanofillers to achieve high performance. To fabricate biodegradable nanocomposites, methods like solvent casting, in situ polymerization, and melt blending are used along with a green chemistry approach to reduce the environmental impact. Recent developments encompass the incorporation of novel catalysts and interface promoters, including dendritic MXene@CNT heterostructures, that enhance substantially dispersion, mechanical strength, and reprocessability of final products (Wang et al. 2025).

Several approaches are also emerging in which waste products like agro-industrial wastes are used as a source for polymers and nanofillers to not only save on raw materials but also prevent landfilling of organic waste, which additionally improves the environmental sustainability. Nanocomposites prepared using these approaches are also exhibiting improved properties such as increased strength, toughness, and physical barrier performance, etc., along with maintaining biodegradability. In addition, the nanocomposites can be designed for certain end-of-life applications, including compostability or recyclability, by selection of the appropriate degradable bonding agents and designed polymer matrices (Menossi et al. 2022).

9.5.4.2 Potential Applications in Eco-Friendly Renewable Energy

Bio-based and biodegradable nanocomposites are gaining popularity in the renewable energy field, offering greener and more sustainable alternatives to traditional materials. These materials exhibit excellent mechanical, thermal, and barrier characteristics, making them suitable for use in components of solar cells, wind turbine blades, and various energy storage systems. In the context of energy storage, biodegradable nanocomposites are increasingly utilised as separators or electrolytes in batteries and supercapacitors, where their tailored conductivity and stability not only boost device performance but also ensure environmentally safe degradation once the device reaches the end of its service life (Prasad and Bueno 2025). Such a move promotes global sustainable development objectives and encourages the formation of sophisticated, environmentally friendly technology for clean energy production and storage.

9.6 Economic and Environmental Impact

9.6.1 Cost-Benefit Analysis of Recycling Methods

9.6.1.1 Capital Investment Analysis

Capital investment analysis is a budget evaluation tool that can be used by investors to assess the potential risk and profitability of possible waste management techniques. The aim of this analysis is to identify the initial investment cost for starting the process. The cost of raw material, utility cost and factory site preparation were considered for analysis of capital expenditure. The capacity of factories corresponding to each fiber reclamation technique was designed based on the optimum expected recycling function with an annual capacity of 6000, 6000, 4000, 2000, 1500, 3000 and 3000 tonne per year for landfill, incineration, mechanical recovery, thermal recycling, fluidized bed, solvolysis and electrochemical recycling, respectively (Wei and Hadigheh 2022).

9.6.1.2 Benefit Analysis

Physical characteristics (e.g. length, alignment, and fiber surface condition) and mechanical properties (e.g. tensile and modulus of elasticity) of recovered fibers dictate the value of recycled product. The benefit analysis conducted in the study evaluated the worth of recycled fibres by assessing their length and mechanical properties achieved through different recycling methods (Dong et al. 2018). Pyrolysis and oxidation method is also a promising method for earning high economic returns, as more secondary products and energy could be recovered via thermal degradation of the matrix.

9.6.2 Market Trends in Recycled Polymer Nanocomposites

In recent years, the market for recycled polymer nanocomposites (RPNCs) has witnessed significant growth, primarily driven by increasing environmental concerns, legislative pressures on plastic waste management, and the growing demand for sustainable materials in sectors like automotive, packaging, construction, and renewable energy systems. The integration of nanotechnology with recycled polymers not only enhances mechanical, thermal, and barrier properties but also facilitates high-value reuse of plastic waste—an essential goal in achieving circular economy targets. The global recycled polymer market is projected to grow at a compound annual growth rate (CAGR) of 6.5% from 2022 to 2030, with nanocomposites emerging as a key sub-sector due to their superior performance characteristics. In the renewable energy sector, RPNCs are gaining traction in applications such as wind turbine blades, photovoltaic encapsulants, and battery enclosures, where lightweight and high-strength materials are critical (George et al. 2022).

9.6.3 Life Cycle Assessment (LCA) of Polymer Nanocomposites

LCA is a well-developed evaluation method is to investigate the impact of energy consumption on environment (Fig. 9.6). For an in-depth and comprehensive environmental evaluation, approaches such as global warming potential (GWP) impact have been developed. Energy consumption analysis can furnished the precise forecast of energy demands of FRP recycling methods. It could also be used as a framework by industry for decision making on selecting recycling technique by considering energy consumption intensity, especially in relation to non-renewable energy use (NREU) for each treatment method. The cumulative energy demand analysis is a step to provide a comprehensive summary of potential energy supply, which also could be used as an energy-related screening-indicator for other environmental impact assessments (George et al. 2022). Global warming potential impact analysis evaluates the influence of produced gases on global warming.

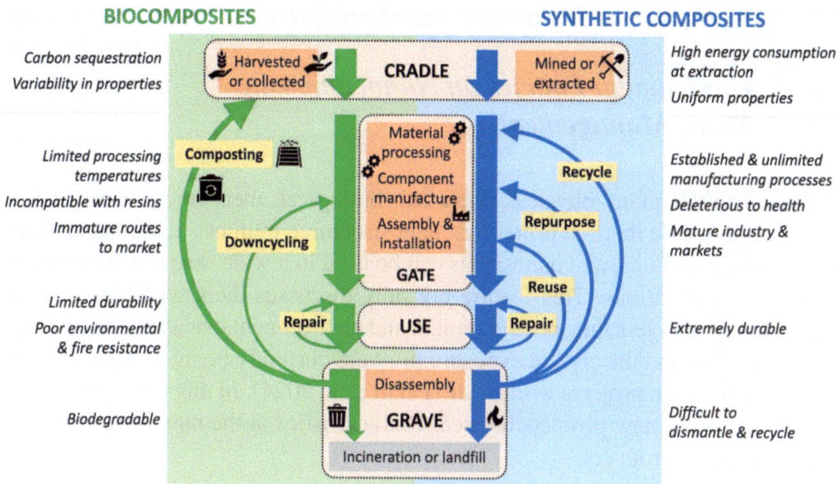

Fig. 9.6 LCA of polymer nanocomposite

9.6.4 *Energy Consumption of Virgin Production Versus Recycling Polymer Nanocomposites*

The total energy consumption of a product and its respective life cycle impact can be accessed using the CML2001 assessment criteria. Studies reported that the energy consumption for the recycling of polymer nanocomposites is significantly lower than the manufacturing of the product from virgin raw materials. Though it is true that the energy demand and consumption can vary based on the different parameters and their influence over the recycling process (Wei and Hadigheh 2022).

9.6.5 *Environmental Impact of CFRP Thermal Recycling*

The environmental impact of the Carbon Fibre Reinforced Polymer (CFRP) Thermal Recycling is closely connected with the greenhouse gas emissions and overall energy consumption by the process. Global Warming Potentials (GWPs) are one of the key indicators to evaluate these impacts, and it also quantify how much heat a greenhouse gas traps in the environment compared to Carbon dioxide. Thus, a lower GWP number signifies a lesser carbon footprint and lower environmental impact.

According to Chen's research, when CFRP goes through thermal recycling, heat reaches up to 615 °C and which releases carbon monoxide, carbon dioxide, and various aromatic hydrocarbons (Chen et al. 2018). This emission contributes greatly to environmental pollution.

So, considering this, it is high time to consider these environmental concerns and need to optimize the recycling process.

9.7 Regulatory and Policy Considerations

9.7.1 *Government Policies on Nanocomposite Waste Management*

Because of their many physical, chemical, biological, mechanical, electrical, and static and dynamic thermal properties, nanostructured entities such as nanospheres, nanotubes, nanowires, and nanosheets can be used in a wide range of structural and functional applications. This feature not only discusses their application but also highlights the dangers they pose to human and environmental health. The toxicity of these materials and the critical exposure limit are currently being extensively investigated by research projects worldwide (Dhall et al. 2024). In this way, materials can exhibit altered or new physicochemical characteristics at the nanoscale, which can help create new products.

9.7.2 *International Sustainability Standards and Guidelines*

Standards usually do not serve as the main channel for disseminating new research, as they are meant to reflect a consolidated understanding of existing knowledge. In most cases, the costs associated with developing these standards are covered by the organisations or developers involved (Allan et al. 2021). Regulation-recognized techniques for evaluating the safety of chemicals, including nanomaterials, are the OECD Test Guidelines (TG) (OUTLOOK 1990). Test results obtained after an OECD TG are legally acceptable in OECD member nations and other followers because they are part of the Mutual Acceptance of Data in the Assessment of Chemicals (MAD) system. For industrial chemicals, insecticides, biocides, food and feed additives, and cosmetics, MAD was founded. It seeks to minimize trade barriers and eliminate redundant testing and animal use. The scientific and regulatory approval of a TG in many countries is supported by a thorough awareness of the regulatory context; it is crucial to be aware of the regulatory requirements prior to beginning data collection.

9.8 Future Trends and Research Directions

9.8.1 *Emerging Technologies in Sustainable Nanocomposites*

9.8.1.1 Innovation in Green Synthesis of Nanocomposites

Today's scientific and engineering wonders include composite science, material science, and nanotechnology. The materials of today's scientific revival are nanocomposites. The greatest demands of humanity now are green engineering and green sustainability. To create sustainable materials, a variety of natural and manufactured green polymers and green nanofillers have been used. Green polymeric nanocomposites embody the unique qualities of eco-reinforcement and sustainable polymers. Biomedical, automotive, energy, and packaging industries have all made use of green nanocomposites reinforced with environmentally friendly nanofillers (Ying et al. 2022). This paper systematically discusses the properties and concepts of eco-polymers, eco-nanofillers, and green nanocomposites. The trends, challenges, and possibilities of green nanocomposites have been discussed in addition to their technical applications. The same materials can function better when handled in nanoscale dimensions and with a high surface-to-volume ratio than when handled at the micrometric scale. In this way, integrating nanometric components with larger-scale materials, or the two together, opens up a world of possibilities and impacts day-to-day existence. Moreover, nanotechnology and sustainability are closely related concepts (da Silva Freitas et al. 2021). Less material is needed to provide excellent performance as compared to conventional micrometre systems green synthesis and Sustainable methods have been applied to increase efficiency and cut waste levels. Carbon nanotubes, graphene, metals, oxides, double lamellar hydroxides, phyllosilicates, lamellar phosphates, and more are being created as nanomaterials that can be used alone or in polymeric nanocomposites and nanostructured materials.

9.8.1.2 Next-Generation Recyclable Polymers

The accumulation of polymer materials and their evolution in addition to causing major environmental issues, the environment also wastes resources. Common methods for handling garbage that contains nanoparticles (NPs) include recycling, landfilling, and incineration. However, there are currently no specialized techniques or infrastructure in place to effectively catch and recycle the NPs discharged into waste streams. Innovative methods like green synthesis of nanocomposites, producing biodegradable nanocomposites using renewable sources as raw materials, and magnetic nanoparticles have stimulated researchers' curiosity to help synthesize and manufacture sustainable Using nanocomposites to stop the landfilling of non-recyclable nanomaterials, particularly of man-made engineered nanomaterials (ENMs), and to counteract their detrimental effects on the environment (Hao et al. 2024) (Fig. 9.7).

Along with the substantial amount of plastic waste produced by food packaging, this emphasizes the need for high-performance materials. Lightweight packaging materials. To overcome these obstacles, organic–inorganic hybrids with excellent gas barrier qualities have been created, such as inorganic-material-coated plastic films and inorganic-filler-reinforced polymer nanocomposites. A continuous polymer matrix and at least one scattered component (filler) in sub-micrometric sizes—typically in the nanoscale—make up polymer composites. Because of their large surface area, inorganic fillers, including clays, metal nanoparticles (NPs), and carbon nanomaterials, significantly enhance the physical characteristics of the polymer at modest loadings (Peng et al. 2022).

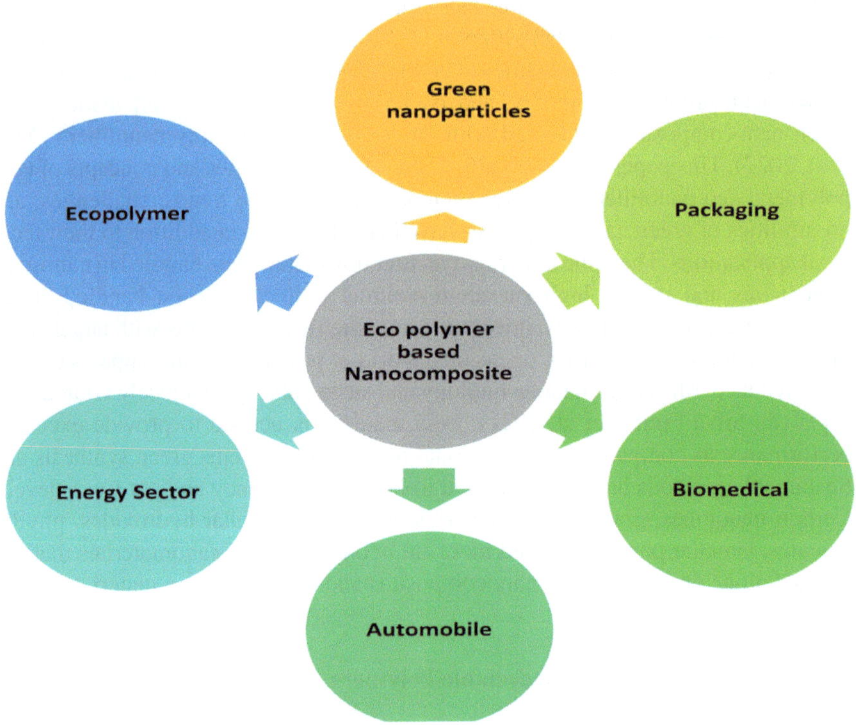

Fig. 9.7 Eco-polymer-based nanocomposite

9.8.2 Innovation in Green Synthesis and Recycling

9.8.2.1 Use of Natural Catalysts for Recycling Under Atmospheric Pressure

A laboratory fluidized bed reactor was utilized to operate at 360 °C with a 2:1 catalyst to polymer feed ratio and at 450 °C with a 6:1 catalyst to polymer feed ratio to catalytically degrade high-density HDPE (polyethylene) under nitrogen (Hao et al. 2024). Pyrolysis is a promising technology that can be used to turn plastic waste into chemicals and fuels. The pyrolysis mechanism and product distribution are the main topics of discussion when discussing the characteristics of common plastics, including polyethylene terephthalate (PET), polyethylene (PE), polyvinyl chloride (PVC), polypropylene (PP), and polystyrene (PS) and their blends (Peng et al. 2022).

Kaolin was moistened with increasing volumes of $AgNO_3$ and of ELE, at RT or 50 °C, and the conditions under which the experiment was conducted are described. Displays the photocatalytic IC degradation results using different AgNP/kaolin composites. Increasing the amounts of $AgNO_3$ and ELE, thus fabricating more AgNP, the photocatalytic performance increases, with 58% and 77% IC degradation for AgNP/kaolin 20:20 RT and AgNP/kaolin 50:50 RT, respectively. However, among the synthesized photocatalysts, AgNP/ kaolin100:100 synthesized with 100 mL of $AgNO_3$, and 100 mL of ELE showed high decolorization of IC dye under visible irradiation, regardless of the temperature used in the synthesis.

9.8.2.2 Low-Energy Processing Method

Unquestionably, one new trend in sustainable and green chemistry is the production of low carbon H_2 and its incorporation into existing chemical processes to reduce their carbon footprint (Papanikolaou et al. 2024). introducing low-carbon H_2 ways, electrifying processes, and defossilizing the chemical industry to achieve carbon neutrality. In certain respects, they symbolize unique facets of the same enormous task. Cement hydration rate accelerators and strength enhancers, including nano-$CaCO_3$ powders, can be made from CO_2 from cement manufacture. Utilizing ethanol and chloride ions, high-pure monticellite ceramic ($CaMgSiO_4$) nanoparticles were created utilizing a unique sol–gel technique to speed up the polycondensation process (Kalantari et al. 2017). In order to track the structural development of the finished samples, the gel made from metal alkoxide and metal salt precursors was first dried at 100 °C and then calcined at various temperatures. According to Scherrer's equation, the synthesized samples' grain size at 1200 °C was around 28 nm. The majority of the particles were around 85 nm in size, according to dynamic light scattering spectroscopy. The samples' Additionally, lattice strain was calculated. Using the Williamson-Hall formula (Papanikolaou et al. 2024).

Functional nanomaterials with low waste levels were created during the nanomaterial manufacturing process using the green chemistry principles (GCP). Green methods for producing nanomaterials include using sugars, vitamins, and plants or agricultural waste that can be used as a biodegradable capping and reducing agent. Using d-glucose as a reducing agent and starch as a capping agent to create starch silver nanoparticles (AgNPs). Additionally, AgNPs were made using natural polysaccharides such heparin, gum Arabic, chitosan, gum. When creating a green method for producing NP, the stabilizing agent, reducing agent, and solvent should all be taken into account (Papanikolaou et al. 2024).

9.9 Circular Economy Model for Renewable Energy Materials

According to the European Commission's 7th Environmental Action Programme, a Circular Economy (Fig. 9.8) is one in which biodiversity is preserved, natural resources are handled responsibly, and nothing is wasted. The present patterns of global resource consumption, waste generation, and emissions are pushing developing nations towards severe climate change and environmental decline. For instance, it is striking to note that extractive industries alone account for nearly half of global

Fig. 9.8 The circular economy integrated resource flow diagram

carbon emissions, while around 90% of greenhouse gas emissions are causing significant harm to biodiversity and intensifying water stress, posing even greater risks to the climate and natural life-support systems than previously understood (Olabi 2019).

The circular economy model for renewable energy materials can be sustainably translated for the lifecycle of polymer nanocomposite materials. The process started with the manufacturing of the polymer nanocomposite-based durable and lightweight renewable energy components like solar panel, wind turbine etc. The prepared materials can then be introduced in the use phase, where it generates clean energy for society or consumers according to their life span. After that, closed-loop circular economy model dictates, that instead of discarding the components, it needs to collect and process for the next stage. Then, these collected components can be processed through physical and chemical recycling, to recover valuable polymer and nanofillers, which can be reused as a feed for the next manufacturing batch. Using this closed-loop sustainable recycling methods, generation of waste can be reduced, as well as clean economical gain can be achieved. Overall, this environmentally friendly sustainable process helps society to reduce the carbon footprint of the material.

9.10 Conclusion

Polymer nanocomposites emerged as a critical material for advancing the rapidly growing renewable energy technologies. As they offer better mechanical and thermal superiority in the precision application-based renewable energy sector. Their popularity can be witnessed from the growing use of polymer nanocomposites from solar panel to blades of wind turbine and even in energy storing systems. But their complex composition and nano-dispersed structure, which makes them suitable for this renewable technology, also creates challenges to manage their end-of-life discard, which raises the concern of sustainability, as these materials tend to produce hurdles like material heterogenicity, environmental toxicity and thermal degradation etc. So, the recycle and recovery of the economically high value polymer and nanofillers, become challenging.

This chapter provides a comprehensive overview of the benefits and challenges of the polymer nanocomposites. It also discussed the emerging solutions in this field which addressing these key challenges of the recycling of polymer nanocomposite-based renewable energy system. Alternative advance methods like chemical, biological, solvent-based recycling methods are also discussed widely along with their scalability, cost, and industry integration issues. As the research progresses, a sustainable shift is also happening in this field, where new generation tools like AI and machine learning models are taking part in the process of sorting and recycling of the discarded materials. Novel developments like life-span enhancing self-healing nanocomposites, biodegradable polymer nanocomposites are also paving the path of sustainable and closed-loop life management of the polymer nanocomposites. So, from the rising trends of sustainable research advancements, soon the hurdles of recycling will be overcome using circular economy model, and the environmental impacts will be greatly reduced.

References

Allan J, Belz S, Hoeveler A, Hugas M, Okuda H, Patri A, Rauscher H, Silva P, Slikker W, Sokull-Kluettgen B, Tong W (2021) Regulatory landscape of nanotechnology and nanoplastics from a global perspective. Regul Toxicol Pharmacol 122:104885

Antolini E (2015) Composite materials for polymer electrolyte membrane microbial fuel cells. Biosens Bioelectron 69:54–70

Ates B, Koytepe S, Ulu A, Gurses C, Thakur VK (2020) Chemistry, structures, and advanced applications of nanocomposites from biorenewable resources. Chem Rev 120(17):9304–9362

Bijanu A, Rajak G, Paulose R (2024) Challenges of nanofillers in environment, legislation, health, and safety. In: Handbook of nanofillers, pp 1–25. https://doi.org/10.1007/978-981-99-3516-1_158-1

Bokobza L (2023) Elastomer nanocomposites: effect of filler–matrix and filler–filler interactions. Polymers 15(13):2900. https://doi.org/10.3390/POLYM15132900

Cafaro F, Ferrari F, Carallo GA, Greco A, Maffezzoli A (2025) A sustainable microwave-assisted process for chemical recycling and the reuse of epoxy resin matrices. Polymers 17(7):989. https://doi.org/10.3390/POLYM17070989/S1

Carpanez TG, Castro LMC, Amaral MCS, Moreira VR (2025) Occurrence and environmental consequences of microplastics and nanoplastics from agricultural reuse of wastewater and biosolids in the soil ecosystem: a review. Sci Total Environ 980:179538. https://doi.org/10.1016/J.SCITOTENV.2025.179538

Chen X, Dam MA, Ono K, Mal A, Shen H, Nutt SR, Sheran K, Wudl F (2002) A thermally re-mendable cross-linked polymeric material. Science 295(5560):1698–1702. https://doi.org/10.1126/SCIENCE.1065879/SUPPL_FILE/1065879S3_THUMB.GIF
Chen R, Xu X, Lu S, Zhang Y, Lo S (2018) Pyrolysis study of waste phenolic fibre-reinforced plastic by thermogravimetry/Fourier transform infrared/mass spectrometry analysis. Energy Convers Manag 165:555–566
Chen S, Wang M, Huang FW, Chang CH, Lin JJ (2025) Downcycling in circular production through sustainable insurance under cap-and-trade regulation and carbon tariffs. Energy Econ 142:108209. https://doi.org/10.1016/J.ENECO.2025.108209
Croce F, Appetecchi GB, Persi L, Scrosati B (1998) Nanocomposite polymer electrolytes for lithium batteries. Nature 394(6692):456–458
da Silva Freitas DDF, da Silva PSRC, de Paula Amantes B, Cestari SP, Albitres GAV, de Mattos Mariano D et al (2021) Sustainable routes and green synthesis for nanomaterials and nanocomposites production. In: Handbook of greener synthesis of nanomaterials and compounds. Elsevier, pp 637–650
Dhall S, Nigam A, Harshavardhan M, Mukherjee A, Srivastava P (2024) A comprehensive overview of methods involved in nanomaterial production and waste disposal from research labs and industries and existing regulatory guidelines for handling engineered nanomaterials. Environ Chem Ecotoxicol 6:269–282. https://doi.org/10.1016/j.enceco.2024.06.002
Dong PAV, Azzaro-Pantel C, Cadene AL (2018) Economic and environmental assessment of recovery and disposal pathways for CFRP waste management. Resour Conserv Recycl 133:63–75
Farghali M, Osman AI (2024) Revolutionizing waste management: unleashing the power of artificial intelligence and machine learning. In: Advances in energy from waste. Woodhead Publishing, pp 225–279
George SC, Haponiuk JT, Thomas S, Reghunath R, Sarath PS (eds) (2022) Tribology of polymers, polymer composites, and polymer nanocomposites. Elsevier
Goto M (2009) Chemical recycling of plastics using sub- and supercritical fluids. J Supercrit Fluids 47(3):500–507. https://doi.org/10.1016/J.SUPFLU.2008.10.011
Hao LT, Kim S, Lee M, Park SB, Koo JM, Jeon H, Park J, Oh DX (2024) Next-generation all-organic composites: a sustainable successor to organic–inorganic hybrid materials. Int J Biol Macromol 269(P2):132129. https://doi.org/10.1016/j.ijbiomac.2024.132129
Hopewell J, Dvorak R, Kosior E (2009) Plastics recycling: challenges and opportunities. Philos Trans R Soc B: Biol Sci 364(1526):2115–2126. https://doi.org/10.1098/RSTB.2008.0311;PAGEGROUP:STRING:PUBLICATION
Ibrahim ID, Jamiru T, Sadiku ER, Hamam Y, Alayli Y, Eze AA (2019) Application of nanoparticles and composite materials for energy generation and storage. IET Nanodielectr 2(4):115–122
Jamil H, Faizan M, Adeel M, Jesionowski T, Boczkaj G, Balčiūnaitė A (2024) Recent advances in polymer nanocomposites: unveiling the frontier of shape memory and self-healing properties—a comprehensive review. Molecules 29(6):1267
Jawaid M, Khan MM (eds) (2018) Polymer-based nanocomposites for energy and environmental applications. Woodhead Publishing
Jespersen KM, Mikkelsen LP (2017) Three dimensional fatigue damage evolution in non-crimp glass fibre fabric based composites used for wind turbine blades. Compos Sci Technol 153:261–272
Kalantari E, Naghib SM, Reza Naimi-Jamal M, Mozafari M (2017) Green solvent-based sol–gel synthesis of monticellite nanoparticles: a rapid and efficient approach. J Sol-Gel Sci Technol 84(1):87–95. https://doi.org/10.1007/s10971-017-4461-5
Kasmi N, Bäckström E, Hakkarainen M (2023) Open-loop recycling of post-consumer PET to closed-loop chemically recyclable high-performance polyimines. Resour Conserv Recycl 193:106974. https://doi.org/10.1016/j.resconrec.2023.106974
Korniejenko K (2024) Future scope of synthetic fibers and their composites. In: Synthetic and mineral fibers, their composites and applications. Woodhead Publishing, pp 657–677
Kumah EA, Fopa RD, Harati S, Boadu P, Zohoori FV, Pak T (2023) Human and environmental impacts of nanoparticles: a scoping review of the current literature. BMC Public Health 23(1):1–28. https://doi.org/10.1186/S12889-023-15958-4

Lee N, Ahn B, Lee J (2024) Solvent-based recovery of acrylate polymer from display film waste. Energy Environ 35(5):2296–2306. https://doi.org/10.1177/0958305X231151681/ASSET/1DFB311D-4832-431B-AAD7-DA4620CA560D/ASSETS/IMAGES/LARGE/10.1177_0958305X231151681-FIG6.JPG

Lors C, Leleux P, Park CH (2024) State of the art on biodegradability of bio-based plastics containing polylactic acid. Front Mater 11:1476484. https://doi.org/10.3389/FMATS.2024.1476484/XML/NLM

Majka TM, Bartyzel O, Raftopoulos KN, Pagacz J, Leszczyńska A, Pielichowski K (2016) Recycling of polypropylene/montmorillonite nanocomposites by pyrolysis. J Anal Appl Pyrolysis 119:1–7. https://doi.org/10.1016/J.JAAP.2016.04.005

Martínez-Narro G, Hassan S, Phan AN (2024) Chemical recycling of plastic waste for sustainable polymer manufacturing—a critical review. J Environ Chem Eng 12(2):112323. https://doi.org/10.1016/J.JECE.2024.112323

Menossi M, Casalongué C, Alvarez VA (2022) Bio-nanocomposites for modern agricultural applications. In: Handbook of consumer nanoproducts. Springer Nature, Singapore, pp 1201–1237

Olabi AG (2019) Circular economy and renewable energy. Energy 181:450–454. https://doi.org/10.1016/j.energy.2019.05.196

OUTLOOK OE (1990) OECD Economic Outlook

Papanikolaou G, Centi G, Perathoner S, Lanzafame P (2024) Green synthesis and sustainable processing routes. Curr Opin Green Sustain Chem 47(385):100918. https://doi.org/10.1016/j.cogsc.2024.100918

Patel AD, Schyns ZOG, Franklin TW, Shaver MP (2024) Defining quality by quantifying degradation in the mechanical recycling of polyethylene. Nat Commun 15(1). https://doi.org/10.1038/s41467-024-52856-8

Peng Y, Wang Y, Ke L, Dai L, Wu Q, Cobb K, Zeng Y, Zou R, Liu Y, Ruan R (2022) A review on catalytic pyrolysis of plastic wastes to high-value products. Energy Convers Manag 254:115243. https://doi.org/10.1016/j.enconman.2022.115243

Prasad R, Bueno J (2025) Bioactive containing plant based waste for (nano)-biocomposites: applications in biomedicine, health, and bioremediation. Front Chem 13:1575200

Qin B, Liu S, Huang Z, Zeng L, Xu JF, Zhang X (2022) Closed-loop chemical recycling of cross-linked polymeric materials based on reversible amidation chemistry. Nat Commun 13(1):7595. https://doi.org/10.1038/S41467-022-35365-4

Rahman AS (2019) Design of cost-effective and efficient fiber-reinforced composite blades for wind turbines. Reinf Plast 63(1):21–25

Rasilainen I, Lahtela V, Kärki T (2024) A review of plastic waste nanocomposites: assessment of features and applications. Discover Nano 19(1):1–22. https://doi.org/10.1186/S11671-024-04062-0/METRICS

Rhim JW, Ng PKW (2007) Natural biopolymer-based nanocomposite films for packaging applications. Crit Rev Food Sci Nutr 47(4):411–433. https://doi.org/10.1080/10408390600846366;REQUESTEDJOURNAL:JOURNAL:BFSN20

Rizos V, Urban P, Righetti E, Kassab A, CEPS (2023) Chemical recycling of plastics [Report]. https://cdn.ceps.eu/wp-content/uploads/2023/07/CEPS-In-depth-analysis-2023-11_Chemical-recycling-of-plastics.pdf

Shahsavari E, Ghasemi I, Karrabi M, Azizi H (2023) Starch/polycaprolactone/graphene nanocomposites: shape memory behavior. Iran Polym J 32(6):763–772

Shao L, Wang Q, Ma Z, Ji Z, Wang X, Song D et al (2018) A high-capacitance flexible solid-state supercapacitor based on polyaniline and Metal-Organic Framework (UiO-66) composites. J Power Sources 379:350–361

Smitha VS, Vidya K, Jayasankar M, Mohamed AP, Hareesh US, Warrier KGK (2016) Energy revamping of solar panels through titania nanocomposite coatings; influence of aqueous silica precursor. RSC Adv 6(37):31114–31121

Tamjid E, Najafi P, Khalili MA, Shokouhnejad N, Karimi M, Sepahdoost N (2024) Review of sustainable, eco-friendly, and conductive polymer nanocomposites for electronic and thermal applications: current status and future prospects. Discov Nano 19(1). https://doi.org/10.1186/s11671-024-03965-2

Tariq W, Ali F, Arslan C, Nasir A, Gillani SH, Rehman A (2022) Synthesis and applications of graphene and graphene-based nanocomposites: conventional to artificial intelligence approaches. Front Environ Chem 3:890408

Thakur VK, Kessler MR (2015) Self-healing polymer nanocomposite materials: a review. Polymer 69:369–383

Thomas L, Ramachandra M (2018) Advanced materials for wind turbine blade—a review. Mater Today Proc 5(1):2635–2640

Ügdüler S, Van Geem KM, Denolf R, Roosen M, Mys N, Ragaert K, De Meester S (2020) Towards closed-loop recycling of multilayer and coloured PET plastic waste by alkaline hydrolysis. Green Chem 22(16):5376–5394. https://doi.org/10.1039/D0GC00894J

Wan S, Peng J, Jiang L, Cheng Q (2016) Bioinspired graphene-based nanocomposites and their application in flexible energy devices. Adv Mater 28(36):7862–7898

Wang H, Ding J, Zhao H, Chu Q, Zhu J, Wang J (2025) Robust and reprocessable biorenewable polyester nanocomposites in situ catalyzed and reinforced by dendritic MXene@ CNT heterostructure. Nano-Micro Lett 17(1):161

Waste-to-Energy from municipal solid wastes (2019) Office of Energy Efficiency and Renewable Energy. U.S. Department of Energy (DOE). https://www.energy.gov/eere/bioenergy/articles/waste-energy-municipal-solid-wastes-report. Accessed 1 Jul 2025

Wei Y, Hadigheh SA (2022) Cost benefit and life cycle analysis of CFRP and GFRP waste treatment methods. Constr Build Mater 348:128654

Wu Y, Gao X, Wu J, Zhou T, Nguyen TT, Wang Y (2023) Biodegradable polylactic acid and its composites: characteristics, processing, and sustainable applications in sports. Polymers 15(14):3096. https://doi.org/10.3390/POLYM15143096

Ying S, Guan Z, Ofoegbu PC, Clubb P, Rico C, He F, Hong J (2022) Green synthesis of nanoparticles: current developments and limitations. Environ Technol Innov 26:102336. https://doi.org/10.1016/j.eti.2022.102336

Zhang JXJ, Hoshino K (2019) Nanomaterials for molecular sensing. In: Molecular sensors and nanodevices, pp 413–487. https://doi.org/10.1016/B978-0-12-814862-4.00007-7

Zhang H, Zheng T, Wang Y, Li T, Chi Q (2024) Multifaceted impacts of nanoparticles on plant nutrient absorption and soil microbial communities. Front Plant Sci 15:1497006. https://doi.org/10.3389/FPLS.2024.1497006/XML/NLM

Zhu J, Deng H, Zhu B, Dong W, Zhang W, Li J, Bao X (2017) Polymer-assistant ceramic nanocomposite materials for advanced fuel cell technologies. Ceram Int 43(7):5484–5489

Chapter 10
Resorbable Polymer Nanocomposites: Current Developments and Future Perspectives

Trideep Saikia, Sandip Bordoloi, and Nirmala Devi

Abstract This chapter thoroughly reviews the polymeric resorbable nanocomposites (PRNs) and their development towards a more eco-friendly and clean energy technology. The chapter first presents the major groups of naturally occurring and artificial resorbable polymers that are the basis of these composites. It then proceeds to discuss the very concept of nanocomposites and the reasons behind their positive impact on material properties. After that, the chapter exposes the primary methods of synthesis and fabrication critically, showing the vital dependency between the mechanical properties, the microstructure, and the processing route. Lastly, the chapter reports on the recent advancements and utilization of PRNs in the clean energy sector with particular emphasis on fuel cells, supercapacitors, and biodegradable energy storage systems. In this way, the chapter ends by signposting future research paths and, at the same time giving a futuristic vision on the capacity of these new materials to contribute to the technological sustainability of the Earth.

Keywords Polymers • Nanocomposites • Resorbable nanocomposites • Energy • Sustainability

---

T. Saikia
School of Pharmaceutical Sciences, Girijananda Chowdhury University, Guwahati, Assam, India

Department of Pharmaceutical Sciences, Dibrugarh University, Dibrugarh, Assam, India

S. Bordoloi
Department of Electrical Engineering, Girijananda Chowdhury University, Guwahati, Assam, India

N. Devi (✉)
Department of Chemistry, Girijananda Chowdhury University, Guwahati, Assam, India

S. K. Verma et al. (eds.), *Sustainable and Natural Polymers*, Clean Energy Production Technologies, https://doi.org/10.1007/978-981-92-0092-4_10

10.1 Introduction

The twenty-firs century faces two global imperatives that are converging: the need for an immediate transition to a sustainable and circular materials economy, which will help in eliminating environmental problems associated with plastic waste, and on top of that, an ongoing demand for a new class of materials with exceptional performance to support the development of clean-energy technologies. In some cases, materials design has been rooted in the idea of long-lasting and permanent materials that more or less initiated their own linear take-make-dispose journey, thus contributing to the ever-increasing e-waste and pollution problem (Tamjid et al. 2024). However, materials science is now redefining the whole concept by coming up with designs that are not only functionally useful throughout their use but also ready for responsible disposal at the end of their life. The resorbable nanocomposites of the polymers (PRNs) are playing a key role in this transformation as they are among the very few materials that can fulfill the requirements of both high performance and sustainability.

10.1.1 Definition and Significance of Resorbable Polymeric Nanocomposites

Nanobiodegradable polymeric composites are defined as multiphase materials consisting of a polymeric matrix that degrades and gets absorbed into the natural environment, plus some amounts of nanofiller in reinforcement, where one of the filler dimensions is within the nanoscale (1–100 nm) (Tamjid et al. 2024). The synergetic nature of the composite structure is the key aspect. The resorbable polymeric matrix, typically derived from a renewable biological source, serves as a framework for the material's designed, short-term existence, thereby addressing concerns related to end-of-life environmental issues. The simultaneous impregnation of nanofillers at very low wt% values—generally less than 5 wt% of the total polymer weight—confers extraordinary properties onto its host matrix, a phenomenon known as "nano-effect" (Al-Mutairi et al. 2022). The large surface-area-to-volume ratio of the nanoparticles is the main reason for the effect, which results in a significant interfacial region between the filler and the polymer matrix. This interface dominates the properties of the material, and some remarkable improvements are made in mechanical strength, thermal stability, electrical conductivity, and barrier properties that cannot be achieved by either the neat polymer or the conventional micro-scale composite (Tamjid et al. 2024). Thus, PRNs are not just disappearing plastics; they are sophisticated and engineered materials for specific applications that have a high demand.

The most important thing about PRNs is their multifaceted application in the direction of greener and benign technologies. On one hand, PRNs follow green chemistry principles by using renewable resources and reducing reliance on petrochemicals (Guragain and Vadlani 2021). On the other hand, since these polymers are resorbable, they provide a partial mitigation for the mounting problem of electronic waste being generated due to the rapid obsolescence of portable gadgets (Kumar et al. 2023; Nunes et al. 2023). Rather than spending hundreds of years in landfills, PRNs could be smartly designed to minimize contamination by breaking down into environmentally friendly components. The existence of this capability implies a shift in design philosophy from managing waste after production to planning waste-free existence ahead of its creation. Hence, following this notion, transitory functional systems can be designed-diegetic devices performing temporary functions for a specified duration and resorbing back into nature without causing harm. The idea lends itself well to long-term material involvement in single-use environmental sensors, disposable medical diagnostic tools, and temporary power supplies because not only is material persistence unwelcome in the first place, but it might even lead to consequences higher than being unwanted (Tamjid et al. 2024).

10.1.2 Overview of Resorbable Polymers

Polymeric resorbable nanocomposites utilize their matrix as the base, which serves as the continuous phase embedding the nanofillers and whose dissolution determines the nanocomposite's life expectancy in the environment. Selection of the polymer herein is an important design consideration, as it is the polymer that controls the degradation profile and biocompatibility, as well as the base mechanical and thermal properties of the composite (Bharadwaz and Jayasuriya 2020). Resorbable polymers are generally divided into two main categories according to their origin: natural polymers, which are derived/isolated directly from biological sources, and synthetic polymers, which are formed through a controlled chemical polymerization. Both classes present different sets of pros and cons, especially from the perspective of their shift from traditional biomedical roles to the harsh environments of clean energy applications (Kumar et al. 2019).

10.1.3 Classification: Natural Vs. Synthetic Resorbable Polymers

Natural Polymers Natural polymers are macromolecules synthesized by living beings, obtained from renewable resources, such as plants, animals, and microbes (Balaji et al. 2017). In this context, the category further splits into polysaccharides (e.g., chitosan, cellulose, alginate) and proteins (e.g., collagen, gelatin, silk) (Satchanska et al. 2024). Natural polymers have the advantage of being inherently biocompatible, bioactive, and structurally similar to biological macromolecules, allowing for favourable interactions within biological systems (Amiri et al. 2022). Generally, their degradation products are nontoxic and easily metabolized. On the contrary, some restrictions to their application are high batch-to-batch variability, a tendency to cause immunogenic reactions and microbial contamination, and, commonly, decreased or uncontrollable mechanical properties and degradation rates vis-à-vis synthetic polymers (Reddy et al. 2021).

Synthetic Polymers This class of polymers represents a true range of chemical synthesis, controlling every aspect of their properties. Key characteristics such as molecular weight, crystallinity, biodegradation rate, and mechanical strength may be very precisely controlled to suit the needs of a particular application. These unique properties of predictability and reproducibility endear them most to the engineering friction. The major family of synthetic (including semi-synthetic and biobased) resorbable polymers is formed by the aliphatic polyesters, including poly(lactic acid) (PLA), poly(glycolic acid) (PGA), poly(ε-caprolactone) (PCL), and their copolymers. Another biobased vital set are the polyhydroxyalkanoates (PHA), which are produced naturally by microorganisms but are mostly treated with synthetics due to the controlled fermentation and processing needed for their production. The worst pitfalls for synthetic polymers include eliciting foreign body reactions, which, in general, have a lack of intrinsic bioactivity or specific cell-adhesion sites and may have to undergo chemical modification in special situations.

This transitioning of these materials from their older ancestries in medicine to newer ones in energy opens up a fundamental conflict in material requirements. A property that is desired for a biomedical scaffold is for degradation to take place in a warm aqueous neutral-pH physiological environment. This, however, would be considered a property that is undesirable and even counter to a useful energy device. An energy storage would require electrochemical stability against electrolytes, thermal stability to withstand generated heat in operation, and resistance to environmental degradation within its useful lifetime. As for example, hydrolytic degradation that makes PLA an excellent material for resorbable sutures would be a catastrophic failure mechanism in a moisture-sensitive battery (Dalwadi et al. 2023). The successful implementation of resorbable polymers in the clean energy field, therefore, is not a mere technology transfer; it is a matter of fundamentally engineering a new material system. That is just where the nanocomposite route becomes crucial. The nanofiller may be treated to enhance any single property. Its function is in fact to confer several energy-relevant functionalities for conductivity, thermal stability, or barrier properties, which should in turn protect the resorbable matrix from the harsh operating environment for its assigned lifespan, while, of course, finally permitting its degradation into benign form at the end-of-life (Chiong et al. 2021).

10.1.4 *Key Examples and Properties*

A select group of natural and synthetic polymers has emerged as the most promising candidates for PRN matrices. Their properties are summarized in Table 10.1 and detailed below.

10.1.4.1 Synthetic Aliphatic Polyesters

Poly(Lactic Acid) (PLA)

PLA is a thermoplastic polyester produced through the fermentation of renewable resources such as corn starch or sugarcane (Benatti et al. 2019). It degrades by hydrolysis of the ester linkages between lactic acids, with lactic acid being a natural metabolite in the human body, and is finally eliminated as carbon dioxide and water (Yang et al. 2024). Its mechanical properties include a high tensile modulus (3 GPa) comparable to those of some commodity plastics, although it tends to be brittle (DeStefano et al. 2020). The degradation is relatively slow, taking anywhere from months to years, depending on crystallinity and molecular weight (Yang et al. 2024). With its processability and renewability, it is considered the foremost candidate toward sustainable electronics, with research undertaken in using it as a dielectric material for capacitors, among other applications, including as a binder (Skorda et al. 2024).

Poly(Glycolic Acid) (PGA)

As an aliphatic linear polyester with a simple structure, PGA is highly crystalline and, thus, has a high melting point and high tensile strength, with the downside of being hardly soluble in most organic solvents (Ciocîlteu et al. 2024). The hydrolytic degradation is much faster than that of PLA (weeks to months), which makes it suitable for applications requiring rapid resorption. The fastest degradation and cumbersome processing hinder its application as a homopolymer, whereas it is prevalently used in copolymers with PLA (PLGA) to tune degradation rates (Bamyacı et al. 2025).

Poly(ε-Caprolactone) (PCL)

PCL is a semi-crystalline polyester characterized by high toughness and flexibility, with exceptionally slow degradation rates (2–4 years) (Balaji et al. 2017). Since it has a rather low melting point (60 °C), it is easy to process and blend with other polymers (Suryani et al. 2024). In energy-related contexts, its high stability and compatibility allow it to serve as a matrix for long-type devices with conductive fillers.

Table 10.1 Comparative properties and energy applications of key resorbable polymers

Polymer	Type (natural/synthetic)	Key mechanical property (e.g., tensile modulus)	Degradation mechanism & kinetics	Key advantages for energy apps	Key limitations for energy apps	Specific role in clean energy	References
PLA	Synthetic	High Modulus (3 GPa)	Hydrolysis (months to years)	Renewable source, processability	Hydrolytic instability, insulating	Dielectric in capacitors, binder	DeStefano et al. (2020), Klimczuk et al. (2024)
PGA	Synthetic	High Modulus, crystalline	Hydrolysis (weeks to months)	High strength	Rapid degradation, poor solubility	Limited direct use; copolymer use	Bamyacı et al. (2025), Casalini et al. (2019)
PCL	Synthetic	Tough, flexible	Hydrolysis (2–4 years)	Slow degradation, blend-compatible	Low melting point, insulating	Host for conductive fillers	Buyuksoy-Fekraoui et al. (2022), Chrysafi et al. (2021)
PHAs	Synthetic	Variable (brittle to elastic)	Enzymatic (slow)	Biosynthesized	Brittle, thermal instability	Research in biodegradable electronics	Dalton et al. (2022), Turco et al. (2021)
Collagen	Natural	Low mechanical strength	Enzymatic	Porous structure after carbonization	Poor mechanical/ thermal stability	Porous carbon electrode for supercapacitors, battery additive	Das et al. (2023), Xiao et al. (2024)
Chitosan	Natural	Poor mechanical strength	Enzymatic (lysozyme)	Film-forming, active functional groups	Low ionic conductivity, swells in water	Anion/proton exchange membranes (fuel cells), separator/binder (batteries)	Amiri et al. (2022), Xiao et al. (2024)

Polyhydroxyalkanoates (PHAs)

PHAs are a family of polyesters biosynthesized by various microorganisms. Their properties can be tailored by selecting the bacterial strain and setting the fermentation conditions, with an array of possibilities spread from stiff and brittle ones like poly(3-hydroxybutyrate) (PHB) to flexible and elastomeric ones (Turco et al. 2021). They degrade enzymatically in soil and in water. However, these changes in properties, such as brittleness during processing and thermal instability during film processing, decrease the potential for their commercial adoption (Zhou et al. 2023).

10.1.4.2 Natural Polymers

Collagen

Being the most abundant protein in the mammalian system, collagen essentially gives tensile strength to the connective tissue (Prete et al. 2023). It can undergo enzymatic breakdown by collagenases. While it is weak in mechanical strength by nature, the unique porous fibrous nature of collagen may be exploited. Carbonization followed by the use of collagen bio-scaffold electrodes may be utilized as high-surface-area porous carbon electrodes for supercapacitors. Their functional groups can also trap polysulfides, thus serving as useful additives in lithium-sulfur batteries for enhanced cycle life (Dalwadi et al. 2023).

Chitosan

Chitosan is a cationic polysaccharide obtained through the deacetylation of chitin, which is obtainable from crustacean shells (Amiri et al. 2022). It is biodegradable, biocompatible, and has reactive amine and hydroxyl groups, which allow easy chemical modification (Ye et al. 2012). While it exhibits low mechanical attributes and ionic conductivity, and tends to swell in aqueous media, the film-forming ability and functional groups of chitosan make it an attractive material for energy-related purposes. It is widely researched for anion and proton exchange membranes in fuel cells because its structure can be modified to facilitate ion transport (Periayah et al. 2016). It also works as a proper binder and separator for zinc-ion and lithium-sulfur batteries to suppress dendritic formation with off-shuttling effects (Lalarukh et al. 2024).

10.2 Nanocomposites: Fundamentals and Rationale

The transition from a simple resorbable polymer into a high-performance polymeric resorbable nanocomposite happens with the careful incorporation of nanofillers. This combination creates a new class of materials whose properties are not an average of those of its constituents but rather are peculiar to the physics that emerge at the nanoscale (Bikiaris 2022). The whole point of creating these nanocomposites is to overcome the intrinsic limitations of the polymer matrix, such as low strength, poor conductivity, or high gas permeability, while still retaining the positive characteristics of being low-weight, processable, and resorbable. Basic understanding of nanocomposite design thus becomes an important aspect in engineering advanced materials for clean energy applications (Bhattacharya 2016).

10.2.1 The Nanocomposite Concept and the "Nano-Effect"

A polymer nanocomposite is a hybrid material in which a polymer matrix is reinforced with fillers having a dimension between 1 and 100 nm (Tamjid et al. 2024). The fundamental attribute that differentiates nanocomposites from their conventional micro-scale counterparts is the profound influence of what has been termed the "nano-effect (Paul and Robeson 2008)." So, as we reduce all filler dimensions to the nanoscale, a sharp increase is observed in the surface-area-volume ratio (Huang et al. 2022). This results in an interfacial zone of enormous area between the nanofiller and the polymer matrix. This interfacial zone possesses properties that differ from those of both the bulk polymer and the filler and can be several nanometers thick. In a nanocomposite, the volume fraction of the interfacial region can be so significant that it overwhelms the macroscopic properties of the material (Mittal 2009).

With superior interface characteristics, the extraordinary transformation produced in the material properties requires certainly low filler concentration ranges, typically 1–5 wt%, while conventional composites may need 20–30 wt% filler loading to bring about an equivalent change in properties (Pielichowska 2022). The low filler concentration is indeed an advantage in that it supports the lightness and processability of the base polymer, properties that are more often compromised in heavily filled systems (Le et al. 2020). The end result is a material that is strong, light in weight, conductive, and resorbable-all the properties that rarely come together in other systems.

10.2.2 Classification and Role of Nanofillers

In large measure, the properties of a nanocomposite are determined by the type, geometry, and concentration of the nanofiller. Nanofillers are generally classified according to their dimensionality (Table 10.2), which determines their chief function in the composite (Fu et al. 2019).

Table 10.2 Properties and functions of common nanofillers for clean energy applications

Nanofiller	Dimensionality	Key intrinsic property	Primary function in composite	Example clean energy application	References
Carbon Nanotubes (CNTs)	1D	High Electrical Conductivity (106 S/m), High Young's Modulus (1 TPa)	Electrical percolation, Mechanical reinforcement	Conductive electrodes (supercapacitors, batteries), EMI shielding	Bhattacharya (2016), Biswas et al. (2023)
Graphene	2D	High Electrical Conductivity (107 S/m), Excellent Barrier Properties	Electrical percolation, Gas/liquid barrier, Thermal dissipation	Electrodes, Proton exchange membranes, Thermal management	Guo et al. (2012)
Nano clays (MMT)	2D	High Aspect Ratio, Platelet Structure	Mechanical reinforcement, Flame retardancy, Barrier improvement	Electrolyte membranes (reduced fuel crossover), Encapsulation materials	Kausar et al. (2022a)
Metal Oxides (e.g., TiO_2, $BaTiO_3$)	0D, 1D	High Dielectric Constant, Photocatalytic Activity	Dielectric enhancement, Energy conversion	High-k dielectrics for energy storage capacitors, Photocatalysts	Wang and Zhu (2011)
Cellulose Nanocrystals (CNCs)	1D	High Stiffness (150 GPa), Renewable	Mechanical reinforcement, Rheology modifier, Dispersion aid for other fillers	Structural components, "Green" filler in hybrid systems, Separators	Das et al. (2023), Mukherjee and Koller (2023)

10.2.2.1 Zero-Dimensional (0D) Nanofillers

Those are iso-dimensional particles, spherical nanoparticles or cubical nanoparticles, having all three dimensions in the nanometer scale. These include silica nanoparticles and metal oxides, such as TiO_2, ZnO, $BaTiO_3$, and quantum dots (Fu et al. 2019). Their main objective is often the imparting of specific functional properties rather than mechanical reinforcing. For example, high-dielectric-constant ceramic nanoparticles such as $BaTiO_3$ serve to increase the energy storage capacity of polymer-based capacitors, whereas TiO_2 serves as a photocatalyst (Wang and Zhu 2011).

10.2.2.2 One-Dimensional (1D) Nanofillers

Biaxial dimensions can be smaller than a nanometer; one dimension is, however, considerably larger in nanomaterials such as nanotubes and nanofibers, with an exceptionally high aspect ratio. The most famous example is the CNTs, which have exceptional mechanical properties (a Young's modulus value of around 1 TPa) and conductivity properties (Lee et al. 2023). Cellulose nanocrystals are another example of this 1D filler system and are perhaps the best choice for mechanical reinforcement of polymeric matrices due to their renewable nature (Bhattacharya 2016). A high aspect ratio enables 1D reinforcers to perform superbly during mechanical reinforcement and in the formation of electrically or thermally conductive networks in the polymer matrix.

10.2.2.3 Two-Dimensional (2D) Nanofillers

They represent sheet- or platelet-like materials with one dimension in the nanometer range and two much larger dimensions. Key examples are graphene, a single atomic carbon layer having exceptional strength, conductivity, and barrier properties, and nano clays such as montmorillonite (MMT), which are layered aluminosilicates (Espíndola et al. 2023). Because of their planar shape and huge aspect ratio, 2D fillers are great at generating a tortuous path for diffusing molecules. They thus are wonderful candidates for gas and liquid barrier enhancement as well as flame retardancy (Kim et al. 2018).

Advanced PRN design for more demanding energy applications will be a textbook example of multi-objective optimization. One particular nanofiller may be excellent in enhancing one property, but may not be so good for other properties. For instance, a biodegradable electrode for a flexible supercapacitor should have good electrical conductivity, with excellent mechanical integrity to tolerate bending, and with a high electrochemically active surface area (Kulkarni et al. 2024; Sonika et al. 2022). Graphene, in contrast, is very conductive but tends to agglomerate, which can weaken the mechanical properties (Li et al. 2016). In contrast, CNCs provide an excellent reinforcing effect but are electrically insulating in character. This problem has led to the development of hybrid nanocomposites with two or more distinct types of nanofillers to provide synergistic property combinations. For example, CNCs can be used to exfoliate and stabilize graphene sheets in a PLA matrix, thereby forming a composite that is strong, conductive, and sustainable, which single-filler systems cannot do (Uyor et al. 2023). This method of synergistic hybridization forms the pivotal enabling step in the creation of next-generation multifunctional materials.

10.2.3 Property Enhancement in Nanocomposites

The incorporation of nanofillers improves the properties (Table 10.2) of the polymer matrix through several distinct physical mechanisms:

10.2.3.1 Mechanical Reinforcement

The mechanism principally at play here is the efficient transfer of an applied mechanical load from the soft, flexible polymer matrix to the stiff and strong nanofillers (Vemuganti et al. 2024). This, however, requires efficient interfacial adhesion between the two phases. Fillers of high aspect ratios, such as CNTs and CNCs, are considered the best candidates. Other mechanisms, however, include crack deflection, whereby cracks are able to draw tortuous paths for propagation, and crack pinning, which increases the energy for fracture (Jamil et al. 2024).

10.2.3.2 Electrical and Thermal Conductivity

The conductive fillers, such as CNTs and graphene, improve conductivity in an insulating polymer matrix by forming an appropriate percolation network (Kadkhodazadeh et al. 2025). With the increased concentration of the filler, the particles either touch or come so close that hopping of electrons occurs between the particles, thus providing an uninterrupted conductive path through the material (Kalaitzidou et al. 2010). The concentration at which this formation takes place is called the percolation threshold and is usually quite low for a filler of very high aspect ratio (Payandehpeyman and Mazaheri 2025).

10.2.3.3 Barrier Properties

Barrier performance is mainly enhanced by 2D nanofillers such as graphene or nano clays. When the impermeable platelets are well-dispersed and oriented perpendicular to the direction of diffusion, gas or liquid molecules have to travel a much longer and more tortuous path through the composite, which effectively reduces the permeability of the material (Kim et al. 2018; Paduvilan et al. 2021).

10.2.3.4 Thermal Stability and Flame Retardancy

Thermal stability enhancement in polymers through nanofillers is achieved by preventing polymer chains from moving at high temperatures. During burning, on the other hand, a char layer can be formed by fillers like nano clay if they have gathered on the polymer surface. Such a char layer then constitutes a barrier to the transmission of heat, along with confining the release of flammable volatile gases (Xu et al. 2022).

Nanocomposites are worth the extra effort as they offer various advantages over traditional composites, including a better performance-to-weight ratio, the opportunity to introduce multifunctionality (e.g., a material that provides structure and conduction), and improved processability due to lower viscosity at the same performance level.

10.3 Synthesis, Fabrication and Modification Techniques

There is an absolute dependence on synthesis and fabrication methods for transforming the nominal promise of polymeric resorbable nanocomposites into materials of real utility. Processing methodology is not a mere detail; it stands as the supreme factor in determining the final composite nanoscale architecture. The architecture, in turn, depending upon the degree of nanofiller dispersion, orientation of the fillers, and quality of interfacial adhesion between filler and polymer matrix, governs the macroscopic properties and hence the performance of the material. A perfectly made composite would thus fail to realize the expected enhancements if it were fabricated in such a way as to form large filler agglomerations or weak interfaces. Therefore, a thorough understanding of all the techniques available, along with their advantages and disadvantages, is required to develop PRNs for clean energy and other high-tech applications successfully (Harito et al. 2019).

Most processing techniques aims to overcome the strong van der Waals attraction forces between nanoparticles and attain a state in which they are dispersed uniformly and well within the polymer matrix (Rong et al. 2006). Ideally, the structure should comprise filler particles exfoliated either as individual particles or as small bundles, each fully wetted by the polymer. Nevertheless, an unregulated process produces large agglomerates, consequently disrupting barrier properties and breaking conductive pathways, which affects overall properties of the material (Müller et al. 2017; Hiremath et al. 2021).

10.3.1 Principal Fabrication Methodologies

Several principal methodologies are employed to fabricate PRNs (Table 10.3), each operating on a different physical principle and presenting a unique set of trade-offs between dispersion quality, scalability, and cost.

10.3.1.1 Solvent Casting

This has always been a laboratory-scale method that satisfied the need for polymer-metal oxide nanocomposites and consisted of three primary stages: (1) dissolution of the polymer in a volatile solvent that would evaporate fast, (2) preparation of the nanofillers in the same or a compatible solvent, with the help of ultrasonication or high-shear mixing, and (3) combining both solutions, spreading the resultant on a surface, and letting the solvent evaporate slowly (Mathew and Oksman 2014). The high fluidity leads to breakage of agglomerates and therefore, uniform dispersion can be controlled with great ease (Marras et al. 2010). On the other hand, this method has some serious drawbacks that make it unsuitable for the manufacture of industrial products. First, the use of large amounts of organic solvents is not only

Table 10.3 Comparison of fabrication techniques for polymeric resorbable nanocomposites

Technique	Basic principle	Typical dispersion quality	Interfacial adhesion potential	Scalability/ throughput	Key advantage	Major limitation	References
Solvent Casting	Co-dissolving/ dispersing polymer and fillers in a solvent, then evaporating the solvent	Excellent	Moderate (can be improved with surface modification)	Poor/Low	High degree of dispersion control in the lab	Use of volatile organic compounds; difficult to scale	Hiremath et al. (2021), Mathew and Oksman (2014)
Melt Compounding	Mixing polymer and fillers in the molten state under high shear	Fair to Poor (agglomeration is common)	Poor to Moderate (highly dependent on shear and compatibility)	Excellent/ High	Solvent-free, cost-effective, industrially mature	Difficult to break down agglomerates; potential for thermal degradation	Musa et al. (2025), Tanahashi (2010)
Electrospinning	Drawing a polymer solution/melt into nanofibers using an electric field	Good (within fibers)	Good (high surface contact)	Poor/Low	Creates unique nanofibrous structures with high surface area	Low production rate; limited to fiber-based architectures	Huang et al. (2003)", Maran et al. (2024)
In Situ Polymerization	Polymerizing monomers in the presence of dispersed nanofillers	Good to Excellent	Excellent (potential for covalent bonding)	Fair/ Medium	Creates very strong interfacial bonds	Complex chemistry; limited to specific monomer/filler systems	Zhao et al. (2018), Aljibori et al. (2025)

costly but also causes environmental and health issues; second, there is the possibility of plasticization of the final composites due to the presence of residual solvent, and finally, the process is slow and has a low throughput which makes scaling up very difficult and costly (Jouault et al. 2014).

10.3.1.2 Melt Compounding

Melt compounding is the industrial version of polymer processing in manufacturing. It directly incorporates nanofillers into the superheated polymer by means of seldom-used high-shear equipment like a twin-screw extruder (Tanahashi 2010). The mixing produces forces that are both intensive and elongational, which help to break the clusters of minerals down until they are evenly distributed across the polymer melt. The primary benefit of melt compounding is that it is an environmentally friendly, continuous, and highly scalable process; thus, it is directly compatible with all subsequent manufacturing processes, injection molding, or film extrusion, for example (Müller et al. 2017). Herein, high viscosity of the polymer melt acts as a resisting force that counteracts the attractive forces holding the fine particles together (Musa et al. 2025). Care needs to be taken for thermally sensitive resorbable polymer or PRNs because the high temperature and shear forces used in the process may affect and damage the polymer or PRN.

10.3.1.3 Electrospinning

In this processing method electric field is used that helps the polymer solution or melt to be drawn into continuous fibers of micron to sub-micron scale range (Xue et al. 2019). Addition of nanofillers in the polymer solution before electrospinning produces fibres with entrapped nanofillers. This special technique can also be tailored for making fibres with high-surface-area, interconnected porosity, and adjustable fiber alignment (Maran et al. 2024). Because of such interesting features, it is one of the fascinating techniques for designing materials specially for purposes like battery separators, fuel cell membranes, and supercapacitor electrodes (Pham et al. 2006). The most significant disadvantage of electrospinning technique lies in its slow production rate, which limits its application to high-value, low-volume markets rather than the bulk production of materials (Huang et al. 2003).

10.3.1.4 In Situ Polymerization

In this technique, nanofillers are initially dispersed in a monomer liquid or solution of monomer. Subsequently, the polymerization reaction initiates, and polymer chains start growing directly from or about the surface of the nanofillers (Tanahashi 2010). This has the special advantage of developing strong covalent chemical bonds between the polymer matrix and the nanofillers, leading to superior interfacial adhesion and the efficient transfer of loads (Yang et al. 1998). It produces excellent dispersion because the fillers are mixed into the low-viscosity monomer before the polymer is formed with a higher viscosity (Zhao et al. 2018). However, the chemistry involved can be rather complex and is not applicable to every polymer-filler system, and controlling reaction kinetics to avoid unwanted side reactions may pose a severe challenge (Aljibori et al. 2025).

10.3.2 The Critical Role of Surface Modification

A controversy arises in the fabrication of PRNs due to the chemical incompatibility between the mostly hydrophilic inorganic nature of the nanofiller's surface (such as silica or nano clays) and the organic polymer matrix, which is usually hydrophobic (Rong et al. 2006). This incompatibility causes poor wetting, weak interfacial adhesion, and therefore, nanofillers do not disperse but agglomerate. To circumvent it, surface modifications of the nanofillers have become a widely accepted and often necessary step. The purpose behind this modification would be to change the surface chemistry of the filler so as to render it more compatible with the polymer matrix. This can be achieved by way of:

10.3.2.1 Physical Modification

This involves the use of surfactants or dispersing agents that physically adsorb another layer onto the filler surface through non-covalent interactions like van der Waals or electrostatic forces. Usually, a surfactant molecule has a "head" that is attracted to the filler surface and a "tail" that is compatible with the polymer matrix, acting as a bridge between the two phases (Kamiya and Iijima 2010; Mariano et al. 2017).

10.3.2.2 Chemical Modification

This provides a more robust and permanent solution by the formation of strong covalent bonds between the filler and a modifying agent. This can include grafting of small functional molecules or even entire polymer chains onto the surface of the filler. This method improves the compatibility, dispersion, and also the interface strength for better mechanical properties of the composite (Johnson et al. 2023; Poggetto et al. 2025).

10.3.3 Overarching Challenges in Fabrication: The Fabrication Trilemma

A set of three continuous, interrelated problems is keeping away the profitable commercialization of PRNs: dispersion, interfacial adhesion, and scalability (Musa et al. 2025). One can think of these challenges as a Fabrication Trilemma, where it becomes challenging to simultaneously maximize high performance (requiring good dispersion and adhesion), scalability (industrial), and low cost.

Dispersion Getting the materials dispersed uniformly at the nanoscale level, and preventing their re-agglomeration during processing, is known to be the most critical parameter (Šupová et al. 2011).

Interfacial Adhesion Even with good dispersion, a weak interface will not transfer load or energy effectively, negating the benefits of the nanofiller (Bagchi et al. 2018).

Scalability and Cost Techniques effective for final properties in the laboratory scale (solvent casting, LBL assembly) are often economically not feasible for large-scale production, whereas scale-up, such as melt compounding, always poses a compromising threat to ultimate performance (Goda et al. 2024).

The way into the future in PRN manufacturing is not to attempt to discover the "perfect" way but, rather, to find ways that can push the limit of the trilemma. Advanced surface modification essentially acts as an enabler. It essentially makes nanofillers inherently more dispersible and compatible with polymer melts, thus affording the use of scalable and low-cost methods such as melt compounding to deliver levels of performance previously achieved purely through laboratory techniques. Further investigations will, however, probably focus on novel "process-agnostic" nanofillers and hybrid manufacturing approaches that would somehow bring together the advantages of opposing methodologies to surpass the present trade-offs (Bagchi et al. 2018).

10.4 Applications in Energy Sectors

An up-and-coming field in energy applications that is still rapidly growing is that of these polymers. While synthetic polyesters such as PLA are tested as biodegradable dielectric films in capacitors with their insulating properties improved with ceramic nanoparticles of extremely high dielectric constants, natural polymers, particularly chitosan, represent the forefront for developing fuel cell membranes (Fan et al. 2015). Modified chitosan membranes, meanwhile, seemed to offer a possibly inexpensive and greener solution to replace the rather expensive Nafion® Membranes either for proton or anion exchange (Chotiradsirikun et al. 2022). For numerous applications related to the manufacture of batteries/supercapacitors, natural and synthetic resorbable polymers can be exploited. They can be applied as binders that hold active materials in the electrodes. These biodegradable porous separators allow ions to move through and prevent shorts, and polymer electrolytes, whether gel or solid (Mecerreyes et al. 2024). Carbonization of biopolymers such as cellulose and collagen allows a very environmentally friendly method for producing high-performance porous carbon electrodes for supercapacitors by taking advantage of the inherent nano-architecture of the biologically derived precursor. Conscious application of the polymers in concrete energy devices rests upon the aforementioned nanocomposite strategies, which bestow the necessary performance features on these sustainable matrices (Das et al. 2023).

10.5 Relevance to Clean Energy Applications

While resorbable polymers are traditionally applied in the biomedical field, their combination with functional nanofillers has opened a new window in clean energy. The intrinsic property possessed by most polymers is that they act as thermal and electrical insulators, thus limiting their use in energy conversion and storage devices (Uyor et al. 2023). Yet, the nanocomposite route serves to lift these restrictions so that the materials can be modified and tailored to the complex requirements of energy systems (Tamjid et al. 2024). By judiciously selecting nanofillers, it is possible to prepare PRNs whose electrical conductivity, dielectric permittivity, thermal stability, and ion transport can be tuned to the extent that makes them interesting for many clean energy-based applications (Bera et al. 2023; Bikiaris 2022).

Key areas where PRNs are poised to make a significant impact include:

10.5.1 Energy Storage Systems

Biodegradable energy storage devices, such as biopolymer supercapacitors and batteries, are still considered to be the most crucial goal in sustainable electronics. PRNs offer multiple functions, such as acting as electrodes, electrolytes, or separators, thus unifying good performance with non-toxicity at the end of life (Amalraj et al. 2023). For example, a desorbable matrix can be used to encapsulate conductive nanofillers like carbon nanotubes or graphene, thus generating flexible electrodes with supercapacitors having a large surface area (Mahajan and Sharma 2024).

10.5.2 Fuel Cells

Proton exchange membranes (PEMs), being the fundamental subject in fuel cells, are offering a very nice option to conventional perfluorinated materials. Among various suggested alternatives, nano-composite polymers based on chitosan and other biopolymers are very promising materials because the methods of preparation can be designed to hold the conductive protons while reducing the fuel crossover and thus lowering the operating costs and environmental impact of the fuel cell technologies (Paul and Robeson 2008).

PEMs are the main and pivotal parts of fuel cells that allow the protons to move from anode to cathode thus separating the chemicals reacting together (Mahajan and Sharma 2024). Traditional PEMs, like Nafion, encounter problems that include inadequate high-temperature and low-humidity performance and at the same time giving way to methanol diffusion. In order to overcome these problems, the scientific community has shifted its research work towards polymer nanocomposite membranes (Pandey 2018; Kausar et al. 2022b; Mahajan and Sharma 2024), where the polymers are augmented with inorganic additives, resulting in higher proton conductivity, better mechanical stability, and longer life span. Research done lately has focused on mixing different nano-additives such as silica, titanium dioxide, graphene oxide, and metal-organic frameworks (MOFs) into polymers like Nafion, sulfonated polybenzimidazole, and biopolymers such as chitosan (Hickner 2010). These nanocomposite membranes have increased water absorption ability even at high temperatures and high ionic conductivity through the additives having water absorption properties. Furthermore, MOFs have started to be regarded for their high proton conductivity, durability, and polymer material compatibility, thus making them suitable for PEM application (Satishkumar et al. 2023).

The development of nanocomposite PEMs has led to several advantages. Improved proton conductivity is one among many. The inclusion of proton-conducting inorganic additives enhances the transport of protons, leading to better fuel cell performance. Overall better mechanical stability and reduced methanol permeability are also advantageous for direct methanol fuel cells. Thus, these advantages have paved the way for the commercialization of more sustainable and high-performance PEMs, particularly those based on biodegradable polymers (Mahajan and Sharma 2024; Hickner 2010; Satishkumar et al. 2023).

10.5.3 Energy Harvesting and Management

Although the material is lightweight and flexible, it is still being used in the production of nanogenerators that capture the mechanical energy of the surroundings and in thermal management parts of electronic devices, which are in dire need of improved thermal conductivity to get rid of heat effectively. The creation of biodegradable energy-storage systems is a significant milestone in the quest for sustainability in energy applications. The systems are intended to have the least possible negative impact on the environment as they are made from biopolymers and natural nano-fillers that can readily biodegrade at the end of their life cycle.

10.6 Future Perspectives and Conclusion

The application of PRNs for clean energy applications is a rapidly growing area with enormous potential. Future research in the field is promising, with a focus on the scalability, cost-effectiveness, multidisciplinary associations, and environmental impact assessment. Scaling up the production of PRNs along with maintaining their performance at reduced cost is a major challenge. Designing advanced PRNs with tailored and value-added properties and assessing the long-term environmental benefits to ensure their sustainability also needs to be considered for large scale production and direct applications.

Rising demand for sustainable and eco-friendly energy solutions has driven significant advancements in the PRNs. These materials, which combine mostly bio-based polymers with nano-scale fillers, are being extensively explored for their potential in clean energy applications such as proton exchange membranes (PEMs) in fuel cells, supercapacitors, and energy storage systems. The PRNs thus possess great potential to reform energy generation and storage technologies for a sustainable future.

References

Aljibori HS, Al-Amiery A, Isahak WRW (2025) Unveiling the potential of halloysite nanotubes: insights into their synthesis, properties, and applications in nanocomposites. Starch 77(3):70003. https://doi.org/10.1002/star.202400080

Al-Mutairi NH, Mehdi AH, Kadhim BJ (2022) Nanocomposites materials definitions, types and some of their applications: a review. Eur J Res Dev Sustain 3(2):102–108

Amalraj J, Lakshmanan P, Amarnath CA, Pyarasani RD, Saravanan C (2023) Biodegradable polymer blend nanocomposites for energy storage application. In: Sabu Thomas S, Ajitha AR, Jaroszewski M (eds) Polymer blend nanocomposites for energy storage applications. Elsevier eBooks, pp 175–202. https://doi.org/10.1016/B978-0-323-99549-8.00009-1

Amiri H, Aghbashlo M, Sharma M, Gaffey J, Manning L, Basri SMM, Kennedy JF, Gupta VK, Tabatabaei M (2022) Chitin and chitosan derived from crustacean waste valorization streams can support food systems and the UN Sustainable Development Goals. Nat Food 3(10):822–828. https://doi.org/10.1038/s43016-022-00591-y

Bagchi S, Harpale A, Chew HB (2018) Interfacial load transfer mechanisms in carbon nanotube-polymer nanocomposites. Proc R Soc Math Phys Eng Sci 474(2216):20170705. https://doi.org/10.1098/rspa.2017.0705

Balaji AB, Pakalapati H, Khalid M, Walvekar R, Siddiqui H (2017) Natural and synthetic biocompatible and biodegradable polymers. Elsevier eBooks, pp 3–32

Bamyacı MC, Çetin D, Cengiz C, Belen SN, Mert O, Cengiz U, Mert S (2025) Synthesis, hydrolytic degradation behavior, and surface properties of poly(alkyl glycolide)-polyglycolide copolymers. ACS Omega. https://doi.org/10.1021/acsomega.4c10768
Benatti ACB, Pattaro AF, Rodrigues AA, Xavier MV, Kaasi A, Barbosa MIR, Jardini AL, Filho RM, Kharmandayan P (2019) Bioreabsorbable polymers for tissue engineering: PLA, PGA, and their copolymers. Elsevier eBooks, pp 83–116. https://doi.org/10.1016/B978-0-12-816901-8.00004-3
Bera S, Singh M, Thantirige R, Tiwary SK, Shook BT, Nieves E, Raghavan D, Karim A, Pradhan NR (2023) 2D-nanofiller-based polymer nanocomposites for capacitive energy storage applications. Small Sci 3(7). https://doi.org/10.1002/smsc.202300016
Bharadwaz A, Jayasuriya AC (2020) Recent trends in the application of widely used natural and synthetic polymer nanocomposites in bone tissue regeneration. Mater Sci Eng C 110:110698. https://doi.org/10.1016/j.msec.2020.110698
Bhattacharya M (2016) Polymer nanocomposites—a comparison between carbon nanotubes, graphene, and clay as nanofillers. Materials 9(4):262. https://doi.org/10.3390/ma9040262
Bikiaris DN (2022) Nanocomposites with different types of nanofillers and advanced properties for several applications. Appl Nano 3(3):160–162. https://doi.org/10.3390/applnano3030012
Biswas PK, Omole O, Peterson G, Cumbo E, Agarwal M, Dalir H (2023) Carbon and cellulose based nanofillers reinforcement to strengthen carbon fiber-epoxy composites: processing, characterizations, and applications. Front Mater 9. https://doi.org/10.3389/fmats.2022.1089996
Buyuksoy-Fekraoui K, Lacoste C, Pucci MF, Lopez-Cuesta J-M, Perrin D (2022) Characterization of optimized ternary PLA/PHB/organoclay composites processed through fused filament fabrication and injection molding. Materials 15(9):3398. https://doi.org/10.3390/ma15093398
Casalini T, Bassas-Galia M, Girard H, Castrovinci A, De Carolis A, Brianza S, Zinn M, Perale G (2019) A systematic experimental and computational analysis of commercially available aliphatic polyesters. Appl Sci 9(16):3397. https://doi.org/10.3390/app9163397
Chiong JA, Tran H, Lin Y, Zheng Y, Bao Z (2021) Integrating emerging polymer chemistries for the advancement of recyclable, biodegradable, and biocompatible electronics. Adv Sci 8(14). https://doi.org/10.1002/advs.202101233
Chotiradsirikun S, Guo R, Bhalla AS, Manuspiya H (2022) Biodegradable nanocomposite derived from PLA/PBAT blends with transition metal cation-doped porous clay heterostructures for flexible capacitor applications. Ferroelectrics 586(1):41–59. https://doi.org/10.1080/00150193.2021.2014259
Chrysafi I, Ainali NM, Bikiaris DN (2021) Thermal degradation mechanism and decomposition kinetic studies of poly(lactic acid) and its copolymers with poly(hexylene succinate). Polymers 13(9):1365. https://doi.org/10.3390/polym13091365
Ciocîlteu M-V, Gabriela R, Amzoiu MO, Amzoiu DC, Pisoschi CG, Poenariu BA-M (2024) PLGA-the smart biocompatible polimer: kinetic degradation studies and active principle release. PubMed 49(3):416–422. https://doi.org/10.12865/chsj.49.03.15
Dalton B, Bhagabati P, De Micco J, Padamati RB, O'Connor K (2022) A review on biological synthesis of the biodegradable polymers polyhydroxyalkanoates and the development of multiple applications. Catalysts 12(3):319. https://doi.org/10.3390/catal12030319
Dalwadi S, Goel A, Kapetanakis C, La Cruz DS-D, Hu X (2023) The integration of biopolymer-based materials for energy storage applications: a review. Int J Mol Sci 24(4):3975. https://doi.org/10.3390/ijms24043975
Das AK, Islam MN, Ghosh RK, Maryana R (2023) Cellulose-based bionanocomposites in energy storage applications—a review. Heliyon 9(1):e13028. https://doi.org/10.1016/j.heliyon.2023.e13028
DeStefano V, Khan S, Tabada A (2020) Applications of PLA in modern medicine. Eng Regen 1:76–87. https://doi.org/10.1016/j.engreg.2020.08.002
Espíndola SP, Zlopasa J, Picken SJ (2023) Systematic study of the nanostructures of exfoliated polymer nanocomposites. Macromolecules 56(18):7579–7586. https://doi.org/10.1021/acs.macromol.3c00575

Fan Y, Huang X, Wang G, Jiang P (2015) Core–shell structured biopolymer@$BaTiO_3$ nanoparticles for biopolymer nanocomposites with significantly enhanced dielectric properties and energy storage capability. J Phys Chem C 119(49):27330–27339. https://doi.org/10.1021/acs.jpcc.5b09619
Fu S, Sun Z, Huang P, Li Y, Hu N (2019) Some basic aspects of polymer nanocomposites: a critical review. Nano Mater Sci 1(1):2–30. https://doi.org/10.1016/j.nanoms.2019.02.006
Goda I, Padayodi E, Raoelison RN (2024) Enhancing fiber/matrix interface adhesion in polymer composites: mechanical characterization methods and progress in interface modification. J Compos Mater 58(29):3077–3110. https://doi.org/10.1177/00219983241283958
Guo F, Silverberg G, Bowers S, Kim S-P, Datta D, Shenoy V, Hurt RH (2012) Graphene-based environmental barriers. Environ Sci Technol 46(14):7717–7724. https://doi.org/10.1021/es301377y
Guragain YN, Vadlani PV (2021) Renewable biomass utilization: a way forward to establish sustainable chemical and processing industries. Clean Technol 3(1):243–259. https://doi.org/10.3390/cleantechnol3010014
Harito C, Bavykin DV, Yuliarto B, Dipojono HK, Walsh FC (2019) Polymer nanocomposites having a high filler content: synthesis, structures, properties, and applications. Nanoscale 11(11):4653–4682. https://doi.org/10.1039/c9nr00117d
Hickner MA (2010) Proton exchange membrane nanocomposites. In: ACS symposium series, pp 155–170. https://doi.org/10.1021/bk-2010-1034.ch011
Hiremath A, Murthy AA, Thipperudrappa S, Bharath KN (2021) Nanoparticles filled polymer nanocomposites: a technological review. Cogent Eng 8(1). https://doi.org/10.1080/23311916.2021.1991229
Huang Z-M, Zhang Y-Z, Kotaki M, Ramakrishna S (2003) A review on polymer nanofibers by electrospinning and their applications in nanocomposites. Compos Sci Technol 63(15):2223–2253. https://doi.org/10.1016/s0266-3538(03)00178-7
Huang J, Zhou J, Liu M (2022) Interphase in polymer nanocomposites. JACS Au 2(2):280–291. https://doi.org/10.1021/jacsau.1c00430
Jamil H, Faizan M, Adeel M, Jesionowski T, Boczkaj G, Balčiūnaitė A (2024) Recent advances in polymer nanocomposites: unveiling the frontier of shape memory and self-healing properties—a comprehensive review. Molecules 29(6):1267. https://doi.org/10.3390/molecules29061267
Johnson EC, Varlas S, Norvilaite O, Neal TJ, Brotherton EE, Sanderson G, Leggett GJ, Armes SP (2023) Adsorption of aldehyde-functional diblock copolymer spheres onto surface-grafted polymer brushes via dynamic covalent chemistry enables friction modification. Chem Mater 35(15):6109–6122. https://doi.org/10.1021/acs.chemmater.3c01227
Jouault N, Zhao D, Kumar SK (2014) Role of casting solvent on nanoparticle dispersion in polymer nanocomposites. Macromolecules 47(15):5246–5255. https://doi.org/10.1021/ma500619g
Kadkhodazadeh M, Pourmahdian S, Afshar-Taremi F (2025) Piezoresistive polyurethane nanocomposites of MWCNT-graphene with high modulus and thermomechanical stability. Polym Compos. https://doi.org/10.1002/pc.29528
Kalaitzidou K, Fukushima H, Drzal LT (2010) A route for polymer nanocomposites with engineered electrical conductivity and percolation threshold. Materials 3(2):1089–1103. https://doi.org/10.3390/ma3021089
Kamiya H, Iijima M (2010) Surface modification and characterization for dispersion stability of inorganic nanometer-scaled particles in liquid media. Sci Technol Adv Mater 11(4):044304. https://doi.org/10.1088/1468-6996/11/4/044304
Kausar A, Ahmad I, Maaza M, Eisa MH (2022a) State-of-the-art nanoclay reinforcement in green polymeric nanocomposite: from design to new opportunities. Minerals 12(12):1495. https://doi.org/10.3390/min12121495
Kausar A, Ahmad I, Maaza M, Eisa M, Bocchetta P (2022b) Cutting-edge green polymer/nanocarbon nanocomposite for supercapacitor—state-of-the-art. J Compos Sci 6(12):376. https://doi.org/10.3390/jcs6120376

Kim SJ, Kim TY, Kang BH, Lee G-H, Ju B-K (2018) Fabrication of graphene oxide/montmorillonite nanocomposite flexible thin films with improved gas-barrier properties. RSC Adv 8(68):39083–39089. https://doi.org/10.1039/c8ra08232d

Klimczuk B, Rudnicka A, Owczarek O, Puszkarz AK, Szparaga G, Puchalski M (2024) Investigation of the hydrolytic degradation kinetics of 3D-printed PLA structures under a thermally accelerated regime. Materials 17(5):1043. https://doi.org/10.3390/ma17051043

Kulkarni ND, Saha A, Kumari P (2024) Utilizing multicriteria decision-making approach for material selection in hybrid polymer nanocomposites for energy-harvesting applications. Polym Compos 45(7):6264–6277. https://doi.org/10.1002/pc.28194

Kumar VV, Balaganesan G, Lee JKY, Neisiany RE, Surendran S, Ramakrishna S (2019) A review of recent advances in nanoengineered polymer composites. Polymers 11(4):644. https://doi.org/10.3390/polym11040644

Kumar S, Darshna A, Ranjan D (2023) A review of literature on the integration of green energy and circular economy. Heliyon 9(11):e21091. https://doi.org/10.1016/j.heliyon.2023.e21091

Lalarukh N, Hussain SM, Ali S, Zahoor AF, Azmat H, Nazish N, Alshehri MA, Riaz D, Naeem E, Mahrukh N (2024) Innovation of advanced polymers from seafood waste: applications of chitin and chitosan. Polym Adv Technol 35(6). https://doi.org/10.1002/pat.6471

Le B, Khaliq J, Huo D, Teng X, Shyha I (2020) A review on nanocomposites. Part 1: Mechanical properties. J Manuf Sci Eng 142(10). https://doi.org/10.1115/1.4047047

Lee J, Kim JH, Han JT, Chae HG, Eom Y (2023) Achieving both ultrahigh electrical conductivity and mechanical modulus of carbon films: templating-coalescing behavior of single-walled carbon nanotube in polyacrylonitrile. Adv Sci 10(8). https://doi.org/10.1002/advs.202205924

Li B, Emami N, Gong G, Song W (2016) Applications of nanomaterials in multifunctional polymer nanocomposites. J Nanomater 2016:1. https://doi.org/10.1155/2016/5790194

Mahajan P, Sharma M (2024) A comprehensive review on developments and future perspectives of biopolymer-based materials for energy storage. Energy Storage 6(4). https://doi.org/10.1002/est2.634

Maran BAV, Jeyachandran S, Kimura M (2024) A review on the electrospinning of polymer nanofibers and its biomedical applications. J Compos Sci 8(1):32. https://doi.org/10.3390/jcs8010032

Mariano M, Pilate F, De Oliveira FB, Khelifa F, Dubois P, Raquez J-M, Dufresne A (2017) Preparation of cellulose nanocrystal-reinforced poly(lactic acid) nanocomposites through noncovalent modification with PLLA-based surfactants. ACS Omega 2(6):2678–2688. https://doi.org/10.1021/acsomega.7b00387

Marras SI, Zuburtikudis I, Panayiotou C (2010) Solution casting versus melt compounding: effect of fabrication route on the structure and thermal behavior of poly(l-lactic acid) clay nanocomposites. J Mater Sci 45(23):6474–6480. https://doi.org/10.1007/s10853-010-4735-6

Mathew AP, Oksman K (2014) Processing of bionanocomposites: solution casting. In: Materials and energy/materials and energy, pp 35–52. https://doi.org/10.1142/9789814566469_0018

Mecerreyes D, Casado N, Villaluenga I, Forsyth M (2024) Current trends and perspectives of polymers in batteries. Macromolecules 57(7):3013–3025. https://doi.org/10.1021/acs.macromol.3c01971

Mittal V (2009) Polymer layered silicate nanocomposites: a review. Materials 2(3):992–1057. https://doi.org/10.3390/ma2030992

Mukherjee A, Koller M (2023) Microbial polyhydroxyalkanoate (PHA) biopolymers—intrinsically natural. Bioengineering 10(7):855. https://doi.org/10.3390/bioengineering10070855

Müller K, Bugnicourt E, Latorre M, Jorda M, Sanz YE, Lagaron J, Miesbauer O, Bianchin A, Hankin S, Bölz U, Pérez G, Jesdinszki M, Lindner M, Scheuerer Z, Castelló S, Schmid M (2017) Review on the processing and properties of polymer nanocomposites and nanocoatings and their applications in the packaging, automotive and solar energy fields. Nanomaterials 7(4):74. https://doi.org/10.3390/nano7040074

Musa AA, Bello A, Adams SM, Onwualu AP, Anye VC, Bello KA, Obianyo II (2025) Nano-enhanced polymer composite materials: a review of current advancements and challenges. Polymers 17(7):893. https://doi.org/10.3390/polym17070893

Nunes A, Coelho L Junior, Abrahão R, Júnior ES, Simioni F, Rotella P Junior, Rocha L (2023) Public policies for renewable energy: a review of the perspectives for a circular economy. Energies 16(1):485. https://doi.org/10.3390/en16010485

Paduvilan JK, Velayudhan P, Amanulla A, Maria HJ, Saiter-Fourcin A, Thomas S (2021) Assessment of graphene oxide and nanoclay based hybrid filler in chlorobutyl-natural rubber blend for advanced gas barrier applications. Nanomaterials 11(5):1098. https://doi.org/10.3390/nano11051098

Pandey J (2018) Recent progresses in membranes for proton exchange membrane fuel cell (PEMFC) for clean and environmentally friendly applications. Advances in environmental engineering and green technologies book series, In, pp 308–343. https://doi.org/10.4018/978-1-5225-5745-6.ch013

Paul, Robeson LM (2008) Polymer nanotechnology: nanocomposites. Polymer 49(15):3187–3204. https://doi.org/10.1016/j.polymer.2008.04.017

Payandehpeyman J, Mazaheri M (2025) Micromechanical modeling of electrical conductivity and percolation in CNT/Polymer nanocomposites by considering morphology: homogeneous, agglomerated, and segregated structures. Polym Compos. https://doi.org/10.1002/pc.70042

Periayah MH, Halim AS, Saad AZM (2016) Chitosan: a promising marine polysaccharide for biomedical research. Pharmacogn Rev 10(19):39. https://doi.org/10.4103/0973-7847.176545

Pham QP, Sharma U, Mikos AG (2006) Electrospinning of polymeric nanofibers for tissue engineering applications: a review. Tissue Eng 12(5):1197–1211. https://doi.org/10.1089/ten.2006.12.1197

Pielichowska K (2022) Polymer nanocomposites: preparation, characterisation and applications. Nanomaterials 12(11):1900. https://doi.org/10.3390/nano12111900

Poggetto GD, D'Amora U, Ronca A, Raucci MG, Soriente A, D'Ayala GG, Laurienzo P (2025) Chemical modification of PLA for the design of 3D printed nanocomposite scaffolds with enhanced degradability for bone tissue engineering. Polym Compos. https://doi.org/10.1002/pc.29470

Prete S, Dattilo M, Patitucci F, Pezzi G, Parisi OI, Puoci F (2023) Natural and synthetic polymeric biomaterials for application in wound management. J Funct Biomater 14(9):455. https://doi.org/10.3390/jfb14090455

Reddy MSB, Ponnamma D, Choudhary R, Sadasivuni KK (2021) A comparative review of natural and synthetic biopolymer composite scaffolds. Polymers 13(7):1105. https://doi.org/10.3390/polym13071105

Rong MZ, Zhang MQ, Ruan WH (2006) Surface modification of nanoscale fillers for improving properties of polymer nanocomposites: a review. Mater Sci Technol 22(7):787–796. https://doi.org/10.1179/174328406x101247

Satchanska G, Davidova S, Petrov PD (2024) Natural and synthetic polymers for biomedical and environmental applications. Polymers 16(8):1159. https://doi.org/10.3390/polym16081159

Satishkumar P, Isloor AM, Farnood R (2023) Nanocomposite membranes for proton exchange membrane fuel cells, pp 73–110. https://doi.org/10.1002/9781119829553.ch5

Skorda S, Bardakas A, Vekinis G, Tsamis C (2024) Influence of SIC and ZNO doping on the electrical performance of polylactic acid-based triboelectric nanogenerators. Sensors 24(8):2497. https://doi.org/10.3390/s24082497

Sonika N, Verma SK, Samanta S, Srivastava AK, Biswas S, Alsharabi RM, Rajput S (2022) Conducting polymer nanocomposite for energy storage and energy harvesting systems. Adv Mater Sci Eng 2022:1–23. https://doi.org/10.1155/2022/2266899

Šupová M, Martynková GS, Barabaszová K (2011) Effect of nanofillers dispersion in polymer matrices: a review. Sci Adv Mater 3(1):1–25. https://doi.org/10.1166/sam.2011.1136

Suryani N, Rihayat T, Fitria N, Sariadi N, Yunus M, Hasanah U, Safitri A (2024) The impact of chitosan incorporation on the mechanical characteristics of biodegradable packaging based on PLA/PCL blend. E3S Web Conf 503:08004. https://doi.org/10.1051/e3sconf/202450308004

Tamjid E, Najafi P, Khalili MA, Shokouhnejad N, Karimi M, Sepahdoost N (2024) Review of sustainable, eco-friendly, and conductive polymer nanocomposites for electronic and thermal applications: current status and future prospects. Discov Nano 19(1). https://doi.org/10.1186/s11671-024-03965-2

Tanahashi M (2010) Development of fabrication methods of filler/polymer nanocomposites: with focus on simple melt-compounding-based approach without surface modification of nanofillers. Materials 3(3):1593–1619. https://doi.org/10.3390/ma3031593

Turco R, Santagata G, Corrado I, Pezzella C, Di Serio M (2021) In vivo and post-synthesis strategies to enhance the properties of PHB-based materials: a review. Front Bioeng Biotechnol 8. https://doi.org/10.3389/fbioe.2020.619266

Uyor UO, Popoola API, Popoola OM (2023) Flexible dielectric polymer nanocomposites with improved thermal energy management for energy-power applications. Front Energy Resarch 11. https://doi.org/10.3389/fenrg.2023.1114512

Vemuganti S, Soliman E, Taha MMR (2024) Manipulating interfacial bond for controlling load transfer in 3D printed fiber reinforced polymer composites. J Reinf Plast Compos. https://doi.org/10.1177/07316844241238502

Wang Q, Zhu L (2011) Polymer nanocomposites for electrical energy storage. J Polym Sci Part B Polym Phys 49(20):1421–1429. https://doi.org/10.1002/polb.22337

Xiao B-H, Xiao K, Li J-X, Xiao C-F, Cao S, Liu Z-Q (2024) Flexible electrochemical energy storage devices and related applications: recent progress and challenges. Chem Sci 15(29):11229–11266. https://doi.org/10.1039/d4sc02139h

Xu J, Liu X, Ren S, Kang C, Niu L (2022) Synergistic effect of montmorillonite and nano-Sb_2O_3/brominated polystyrene on the flame retardancy of polypropylene matrix composites. Polym Polym Compos 30. https://doi.org/10.1177/09673911221083769

Xue J, Wu T, Dai Y, Xia Y (2019) Electrospinning and electrospun nanofibers: methods, materials, and applications. Chem Rev 119(8):5298–5415. https://doi.org/10.1021/acs.chemrev.8b00593

Yang F, Ou Y, Yu Z (1998) Polyamide 6/silica nanocomposites prepared by in situ polymerization. Wiley Online Library. https://doi.org/10.1002/(SICI)1097-4628(19980711)69:2

Yang Z, Qiao C, Liu J, Qin T, Gao P, Chen J, Yin G (2024) The current status and prospects of poly(lactic acid) applications in clinical treatment. J Biomed Nanotechnol 20(7):1049–1061. https://doi.org/10.1166/jbn.2024.3826

Ye Y-S, Rick J, Hwang B-J (2012) Water soluble polymers as proton exchange membranes for fuel cells. Polymers 4(2):913–963. https://doi.org/10.3390/polym4020913

Zhao D, Jiang Y, Ding Y, Zhu G, Zheng J (2018) Polymer/carbon nanotubes nanocomposites: relationship between interfacial adhesion and performance of nanocomposites. J Mater Sci 53(14):10160–10172. https://doi.org/10.1007/s10853-018-2335-z

Zhou L, Zhang Z, Shi C, Scoti M, Barange DK, Gowda RR, Chen EY-x (2023) Chemically circular, mechanically tough, and melt-processable polyhydroxyalkanoates. Science 380(6640):64–69. https://doi.org/10.1126/science.adg4520

Chapter 11
Sustainable Materials for Energy Harvesting, Piezoelectric Applications, Sensors, and Fuel Cells

Swasti Saxena, Ankit K. Srivastava, and Ashish Kumar Kaushal

Abstract The creation of high-performance, reasonably priced, and environmentally benign materials is essential to the shift to sustainable energy systems. This chapter examines how sustainable materials, such as polymer compounds, nanomaterials, and bio-based substitutes, are essential to the development of clean energy technology. The materials' uses in fuel cells, sensors, energy harvesting equipment, and piezoelectric devices—where they improve efficiency, lessen environmental impact, and allow for scalability—are highlighted. Nanostructured substances like graphene and nanotubes made of carbon (CNTs) offer better electrical, catalytic, and mechanical capabilities, while biodegradable polymers like chitosan, cellulose, along with silk fibroin offer flexible and environmentally friendly devices for sensors and energy harvesting. Sustainable triboelectric and piezoelectric energy harvesting technologies show promise for supplying power to Internet of Things (IoT) sensors and low-energy electronics. Similar to this, the use of catalysts generated from biomass and non-platinum group substitutes in fuel cells marks a significant advancement in lowering costs and environmental impact. Sustaining material stability under operating circumstances, guaranteeing large-scale production, and striking a balance between biodegradability and performance continue to be difficult tasks. Sustainable materials offer a revolutionary route toward environmentally friendly, effective, and scalable energy solutions by fusing developments in material science with device engineering. This aids in the global shift toward a cleaner and greener future.

S. Saxena (✉)
Gandhinagar University, Khatraj Kalol Rd, Moti Bhoyan, Kalol, Gujarat, India

A. K. Srivastava
Indrashil University, Mehsana, Gujarat, India

A. K. Kaushal
Jindal Global Business School, O P Jindal Global University, Sonepat, Haryana, India

S. K. Verma et al. (eds.), *Sustainable and Natural Polymers*, Clean Energy Production Technologies, https://doi.org/10.1007/978-981-92-0092-4_11

11.1 Introduction

Researchers are looking at new methods for producing, storing, and using energy as a result of the escalating energy crisis, environmental damage, and the pressing need to cut greenhouse gas emissions. In addition to being limited, conventional energy sources like natural gas, coal, and oil are also largely to blame for the world's carbon emissions (Chu and Majumdar 2012). The usage of sustainable materials is therefore strongly trending worldwide in order to provide economical, renewable, and environmentally pleasant energy solutions. Materials that come from renewable feedstocks, biodegradable polymers, or abundant natural resources and that can be recycled with little harm to the environment are generally referred to as sustainable materials (Rajendran and Chinnaiyan 2020). Because of their potential to lessen dependency on costly, poisonous, or scarce materials like lead, platinum, and rare--earth metals, these materials are being utilized more and more in energy-related technologies (Yang et al. 2017). Additionally, the development of revolutionary nanomaterials (graphene, carbon nanotubes, nanobelts, and nanowires) has been accelerated by nanotechnology and material engineering. These materials have excellent mechanical, electrical, and thermal properties, making them perfect candidates for next-generation energy devices (Novoselov et al. 2012).

Among the many energy technologies, sustainable materials are being rapidly adopted in four key areas: fuel cells, sensors, piezoelectric devices, and energy harvesting. Energy harvesting technologies use composites based on polymers or nanomaterials that combine efficiency and environmental friendliness to transform ambient energies, such as solar radiation, thermal fluctuations, and mechanical vibrations, into electrical power (Wang 2017). Similar to this, lead-free piezoelectrics like barium titanate ($BaTiO_3$), potassium sodium niobate (KNN), and zinc oxide (ZnO) nanostructures, as well as polymeric substitutes like PVDF, are progressively replacing piezoelectric devices, which were previously dominated by lead zirconate titanate (PZT). These alternatives not only lessen toxicity but also enhance flexibility and sustainability (Panda 2009). Because sensors are used in wearable technology, healthcare, and environmental monitoring, there is a growing need for flexible, biodegradable, and environmentally friendly materials. Biodegradable sensors with a small environmental impact can be developed using green materials, such as silk fibroin, chitosan, and cellulose (Someya et al. 2016). Similarly, high-sensitivity gas, biosensor, and temperature sensors are being designed using graphene and carbon nanotube hybrid nanocomposites (Varghese et al. 2015).

Sustainable materials are also widely used in fuel cells, which are clean energy technologies. Nafion membranes and platinum-based catalysts, which are expensive and scarce, are essential components of conventional fuel cells. In order to assure long-term sustainability and reduce costs, research is now moving toward carbon catalysts made from biomass, transition metal oxides, and bio-polymer membranes such as cellulose acetate and chitosan (Zeng and Zhang 2010).

There are difficulties in implementing these sustainable materials in several energy-related fields. For them to be widely used, problems including scalability, stability over the long term, and the trade-off between biodegradability and performance must be resolved (Shapter et al. 2019). However, given the speed at which materials science, nanotechnology, and bioengineering are developing, sustainable materials are anticipated to be crucial in providing clean, dependable, and renewable energy sources in the future. We provide a thorough review of the function of sustainable materials in fuel cells, sensors, piezoelectric applications, and energy harvesting in this chapter. The choice of materials, their characteristics, design approaches, and possible uses are highlighted, along with the advancements and difficulties in reaching widespread implementation.

11.2 Sustainable Materials for Energy Harvesting

The process of transforming minute amounts of ambient energy—such as mechanical vibrations, temperature changes, solar radiation, or electromagnetic waves—into useful electrical power is known as energy harvesting. Because it makes self-powered gadgets possible and lessens reliance on traditional batteries, this technology is especially significant in the context of environmentally friendly and renewable power systems (Beeby et al. 2006). Incorporating sustainable materials is essential to enhancing these devices' functionality and environmental friendliness while maintaining their effectiveness, affordability, and environmental safety.

11.2.1 Polymer-Based Materials

Polymeric materials have drawn a lot of interest in energy harvesting because of its special blend of low manufacturing costs, flexibility, and lightweight nature. Polyvinylidene fluoride (PVDF) and its copolymers, including PVDF-TrFE, stand out among them. Strong ferroelectric and piezoelectric characteristics of PVDF enable it to efficiently transform mechanical strain into electrical energy (Martins et al. 2014). PVDF-based materials, in contrast to conventional inorganic piezoelectric ceramics, are mechanically flexible and non-toxic, which makes them appropriate for use in flexible sensors, biomedical implants, and wearable electronics (Kim et al. 2011). Additionally, they are appealing for the production of sustainable devices because of their biodegradability and capacity for large-scale, low--temperature processing.

11.2.2 Nanomaterials

Because of their remarkable electrical, thermal, & mechanical qualities, nanostructured materials are revolutionizing the efficiency of energy harvesting devices.

(a) **Carbon Nanotubes (CNTs):** CNTs can function as effective charge carriers in hybrid energy harvesters because of their strong electric conductivity, large aspect ratios, and outstanding mechanical strength (Cao and Rogers 2009). It has been demonstrated that their incorporation into polymer matrices greatly increases output power and durability.
(b) **Graphene:** Because of its two-dimensional arrangement, graphene has exceptional heat conductivity and electron mobility. Graphene decreases energy losses and improves charge collection efficiency when employed as an electrode material or additive (Novoselov et al. 2012).
(c) **Nanobelts and Nanowires:** High surface area-to-volume ratios are provided by materials like titanium dioxide (TiO_2) nanobelts and zinc oxide (ZnO) nanowires, which enhance sensitivity and efficiency in converting energy. They are extensively used in hybrid energy harvesting systems, triboelectric devices, and piezoelectric nanogenerators (Wang and Song 2006).

A design of an energy harvesting system using hybrid materials based on graphene and carbon nanotubes (CNTs) is shown in Fig. 11.1. It shows how a flexible substrate covered in CNT/graphene composites transforms mechanical, thermal, and solar energy inputs into electrical energy. The picture highlights these nanoparticles' excellent conductivity and versatility, demonstrating their potential for use in wearable and flexible energy harvesting systems. Frequently, these nanoparticles

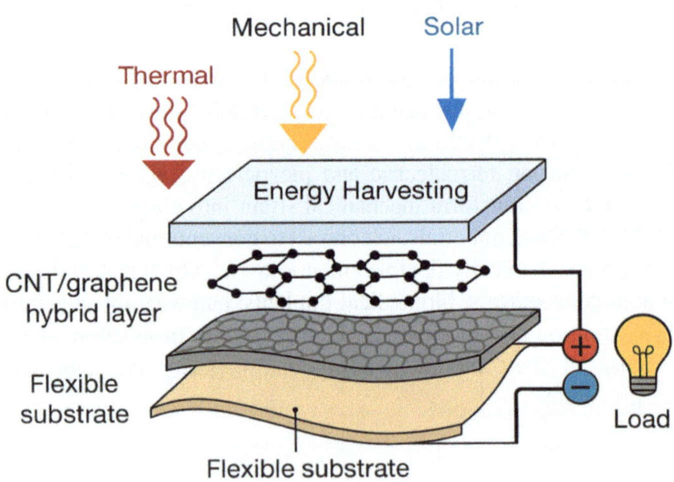

Fig. 11.1 Schematic representation of energy harvesting using CNT/graphene-based hybrid materials

are mixed with bio-based matrices or polymers to create hybrid energy harvesting systems that optimize sustainability, efficiency, and adaptability.

11.2.3 Bio-based Materials

In recent years, there has been increasing interest in using bio-derived and biodegradable materials for energy harvesting devices, in line with the principles of green technology. Cellulose, derived from natural plant fibers, exhibits piezoelectric properties when aligned at the nanoscale and can be used as a sustainable alternative to synthetic polymers (Csoka et al. 2011). Chitosan**,** a natural polysaccharide obtained from the shells of crustaceans, has excellent film-forming ability, biodegradability, and mechanical strength, making it suitable for use in proton-conducting membranes and flexible harvesters (Wu et al. 2018). Silk fibroin, a protein-based biopolymer obtained from silkworm cocoons, has recently been studied for its unique mechanical strength, biocompatibility, and ability to be engineered at the molecular level for energy conversion applications (Kim et al. 2019).

These bio-based materials not only provide a renewable alternative to petroleum--based polymers but also align with the goal of reducing the ecological footprint of energy technologies. Importantly, their biodegradable nature makes them particularly attractive for disposable or short-lifetime electronic devices.

11.3 Piezoelectric Applications

Piezoelectric materials possess the unique capability to generate an electric charge when subjected to mechanical stress, a property that has been harnessed in a wide range of energy harvesting and sensing devices. Traditionally, lead zirconate titanate (PZT) and other lead-based ceramics have dominated piezoelectric technology due to their high piezoelectric coefficients. However, the environmental concerns associated with lead toxicity and regulations such as the Restriction of Hazardous Substances (RoHS) directive have shifted research focus toward sustainable, lead--free alternatives (Shrout and Zhang 2007; Rödel et al. 2009). In this context, the exploration of environmentally benign, cost-effective, and high-performance materials has become essential for next-generation applications.

11.3.1 Lead-Free Piezoelectrics

Lead-free piezoelectric ceramics are increasingly being recognized as promising substitutes for conventional PZT-based materials. Among these, **barium titanate ($BaTiO_3$)** has been widely studied due to its high dielectric constant, strong

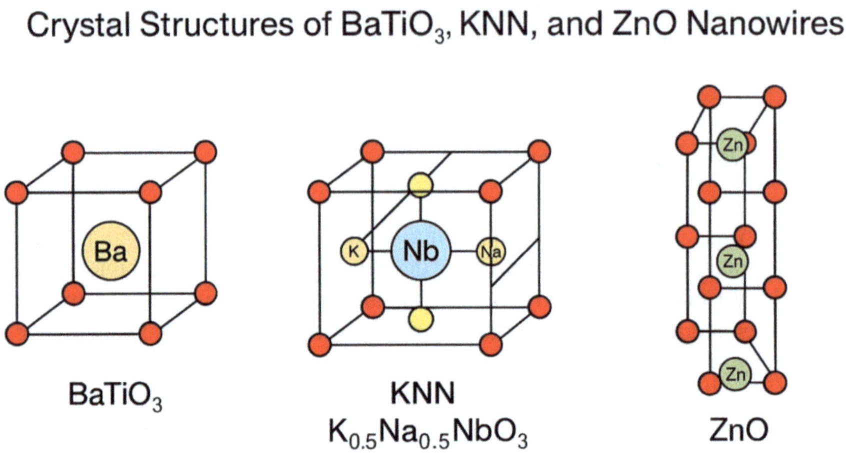

Fig. 11.2 Schematic illustration showing the crystal structures of $BaTiO_3$, $K_{0.5}Na_{0.5}NbO_3$ (KNN), and ZnO nanowires used in lead-free piezoelectric applications

ferroelectric properties, and excellent piezoelectric response (Safari and Akdoğan 2008). Another significant material, **potassium sodium niobate ($K_{0.5}Na_{0.5}NbO_3$, KNN)**, exhibits high Curie temperature, superior electromechanical coupling, and long-term stability, making it highly suitable for high-temperature and high-power applications (Saito et al. 2004). Furthermore, **zinc oxide (ZnO) nanostructures** such as nanowires and nanorods are particularly attractive due to their simple synthesis, nontoxic nature, and high anisotropy, which enhance energy conversion efficiency (Wang 2004). These lead-free alternatives are not only environmentally sustainable but also offer versatile functional properties for integration into modern devices.

This schematic Fig. 11.2 compares the crystal lattice structures of three lead-free piezoelectric materials: barium titanate ($BaTiO_3$), potassium sodium niobate ($K_{0.5}Na_{0.5}NbO_3$, KNN), and zinc oxide (ZnO) nanowires. By visualizing their atomic arrangements, the figure highlights the structural basis for their piezoelectric properties and underscores their role as sustainable alternatives to lead-based ceramics in energy applications.

11.3.2 Polymer-Based Piezoelectrics

In addition to ceramics, polymer-based piezoelectric materials have gained significant attention due to their **lightweight nature, mechanical flexibility, and ease of processing**. Among them, **polyvinylidene fluoride (PVDF)** and its copolymers (e.g., PVDF-TrFE) are widely studied for their high piezoelectric coefficients and excellent mechanical durability (Lovinger 1983).

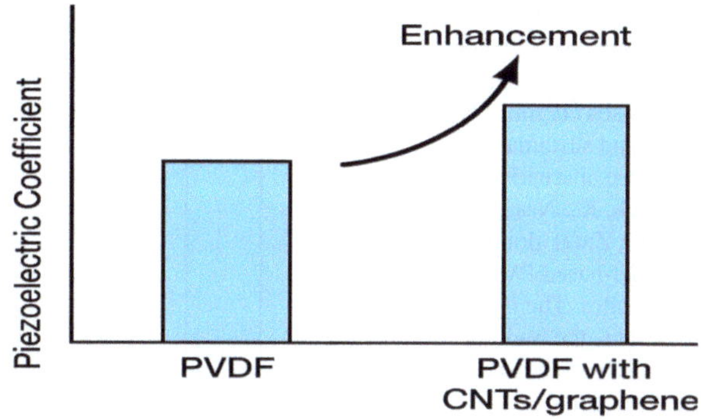

Fig. 11.3 Piezoelectric performance enhancement in PVDF before and after reinforcement with CNTs/graphene fillers

Incorporation of **nanofillers such as carbon nanotubes (CNTs) or graphene** into PVDF matrices further enhances the material's piezoelectric properties by improving polarization, dielectric constant, and charge transfer pathways (Li et al. 2016). These hybrid composites are particularly well-suited for **flexible and wearable energy harvesting devices**, where mechanical adaptability and miniaturization are crucial. Figure 11.3 shows a comparative bar chart of the piezoelectric coefficient (d_{33}) of pure PVDF versus PVDF reinforced with CNTs or graphene fillers. It demonstrates how nanomaterial reinforcement significantly boosts the piezoelectric response of the polymer, making it more suitable for flexible energy devices. The figure supports the narrative of hybrid material design for performance optimization.

11.3.3 Applications of Sustainable Piezoelectrics

The development of lead-free and polymer-based piezoelectrics has opened up new frontiers in **self-powered electronics and green technologies**. Some of the most notable applications include:

- **Self-powered sensors** for environmental monitoring, structural integrity analysis, and smart agriculture (Dagdeviren et al. 2014; Saxena and Srivastava 2020; Saxena and Verma 2013, 2014; Verma et al. 2011; Saxena et al. 2015, 2019; Srivastava et al. 2024).
- **Wearable electronics**, including energy-harvesting textiles and biomedical monitoring devices, which benefit from the lightweight and flexible nature of polymer composites.
- **Structural health monitoring devices** in civil and aerospace engineering, where piezoelectric sensors detect vibrations, cracks, and stress distributions without requiring an external power supply (Lynch and Loh 2006).

By integrating these materials into advanced devices, researchers are not only addressing sustainability concerns but also paving the way for innovative, eco-friendly technologies in the energy and sensor industry. Demonstration of

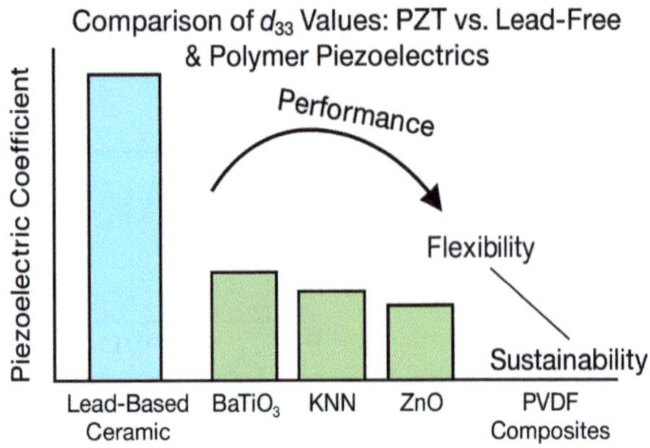

Fig. 11.4 Comparison of piezoelectric coefficients (d_{33} values) between lead-based ceramics (e.g., PZT) and sustainable lead-free alternatives ($BaTiO_3$, $K_{0.5}Na_{0.5}NbO_3$ (KNN), ZnO) along with polymer-based PVDF composites. The figure highlights the trade-off between performance, flexibility, and sustainability

piezoelectric applications: (a) wearable energy-harvesting patch, (b) self-powered wireless sensor node, (c) embedded piezoelectric device for structural health monitoring.

Figure 11.4 compares the piezoelectric coefficients of lead-based ceramics (e.g., PZT), lead-free alternatives ($BaTiO_3$, $K_{0.5}Na_{0.5}NbO_3$ (KNN), ZnO), and polymer-based PVDF composites. It visually communicates the trade-offs between performance, flexibility, and sustainability, helping readers understand why certain materials are chosen for specific applications despite lower piezoelectric output.

11.4 Sustainable Materials in Sensors

Sensors are crucial for modern smart devices and IoT applications. Sustainable materials improve their sensitivity, durability, and environmental impact. Sensors play a crucial role in energy harvesting and monitoring applications by converting physical, chemical, or biological stimuli into measurable electrical signals. Recent advances in materials science have enabled the design of highly sensitive and selective sensors based on sustainable and nanostructured materials.

11.4.1 Types of Sensors

A few types of sensors have been discussed here (Saxena and Srivastava 2020; Saxena and Verma 2013, 2014; Verma et al. 2011; Saxena et al. 2015, 2019; Srivastava et al. 2024):

(a) **Gas Sensors:** Gas sensors are widely employed in industrial safety, environmental monitoring, and medical diagnostics. Because of their superior gas adsorption and charge transfer capabilities, materials like conducting polymers (e.g., polyaniline, polypyrrole), graphene oxide (GO), and zinc oxide (ZnO) nanostructures can be used to detect volatile organic compounds (VOCs) and harmful gases like CO and NO_2. The high surface-to-volume ratio of ZnO nanowires in particular greatly increases the sensitivity of their gas-sensing capabilities.

(b) **Biosensors:** In order to detect biomolecules, biosensors combine electrical substrates with biological components (such as enzymes, antibodies, or DNA fragments). Carbon nanotube (CNT) composites and enzyme-immobilized biopolymers offer improved electron transfer efficiency and biocompatibility, which makes them perfect for environmental monitoring, illness diagnosis, and glucose sensing. Because of their excellent conductivity and mechanical resilience, CNT–polymer composites also enhance signal transmission (Saxena and Srivastava 2020).

(c) **Temperature Sensors:** Conducting polymer blends, including polyaniline/polyvinylene fluoride (PANI/PVDF), have robust resistance behavior that varies with temperature, which makes them appropriate for flexible and wearable temperature sensors. Because these materials are inexpensive, lightweight, and simple to produce, they can be used in flexible electronics and healthcare on a big scale.

11.4.2 Green Materials in Sensors

Researchers are investigating environmentally benign and biodegradable materials for the production of next-generation sensors in accordance with sustainability aims.

(a) **Biopolymer-based Sensors:** Biodegradable substrates and active layers for sensor devices are being created using materials including silk fibroin, cellulose, and chitosan. These biomaterials have a low environmental impact because of their inherent flexibility, biocompatibility, and degradability. For instance, in sensor applications, flexible substrates based on cellulose can take the place of polymers made from petroleum.

(b) **Flexible Nanomaterial-based Sensors:** Electronic sensors that are wearable and stretchable have been made possible by the incorporation of graphene, nanowires, and hybrid composites into flexible substrates. These sensors have outstanding mechanical robustness and can track physiological signals in real time, including breathing, sweat composition, and pulse. Performance is further improved without sacrificing sustainability by the incorporation of green polymers with nanoparticles (Fig. 11.5).

A flexible sensor constructed from biodegradable polymers like cellulose or chitosan combined with conductive nanostructures like graphene and nanowires is shown in this conceptual illustration. In keeping with the chapter's sustainability theme, the graphic highlights the possibilities of environmentally benign, disposable sensors in wearable electronics and environmental monitoring (Saxena and

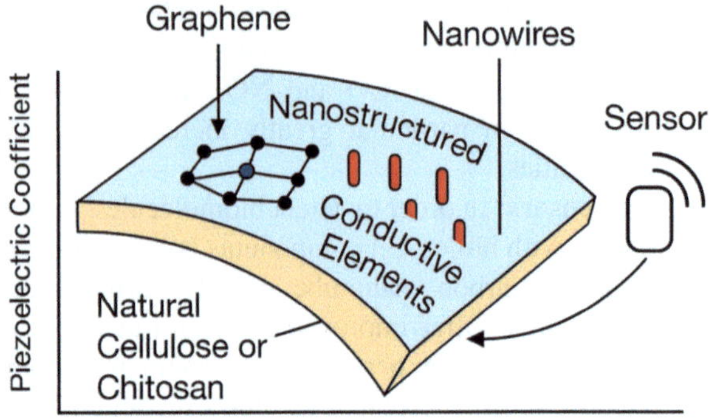

Fig. 11.5 Conceptual illustration of a biodegradable polymer--based sensor integrating natural cellulose or chitosan with nanostructured conductive elements such as graphene and nano-wires for flexible sensing applications

Srivastava 2020; Saxena and Verma 2013, 2014; Verma et al. 2011; Saxena et al. 2015, 2019; Srivastava et al. 2024).

11.5 Fuel Cells and Sustainable Materials

Because fuel cells transform chemical energy directly into electricity with great efficiency and little emissions, they are becoming one of the most promising sustainable energy conversion technologies. Costs can be reduced and efficiency is increased with sustainable materials. Fuel cells are an essential part of the shift to sustainable energy systems because, in contrast to conventional combustion engines, they run without emitting greenhouse gases (GHGs) when hydrogen or biofuels are utilized (Dutta et al. 2021). However, their dependency on rare materials like platinum, high component costs, and durability issues has prevented them from being widely commercialized. Researchers are concentrating more on sustainable materials that not only lower costs but also guarantee resource availability and environmental compatibility in order to overcome these constraints.

11.5.1 Polymer Electrolyte Membranes (PEM)

Because of its high power density, low operating temperature, and suitability for portable and automotive applications, polymer electrolyte membrane fuel cells, or PEMFCs, are the fuel cells that are the subject of the most research. Nafion, a perfluorinated polymer, has historically dominated PEMFC applications because of its superior mechanical stability and proton conductivity. Nevertheless, Nafion is costly, non-biodegradable, and its effectiveness declines in hot or cold temperatures.

Sustainable alternatives are being developed to address these issues. Because of their cost-effectiveness, tunable chemical characteristics, and biodegradability, natural polymers like cellulose acetate, sulfonated biopolymers, and chitosan (derived from chitin) have garnered interest. For instance, cellulose derivatives provide excellent thermal stability and environmental safety, while chitosan membranes doped with inorganic nanoparticles exhibit improved mechanical strength and proton conductivity. These findings imply that PEMs based on natural polymers may be able to maintain competitive performance while drastically lowering the environmental impact of fuel cells.

11.5.2 Catalysts

Another essential feature of fuel cells is the catalyst layer, which usually uses platinum (Pt) nanoparticles because of their superior catalytic activity for the oxygen reduction reaction (ORR). However, platinum's scalability is severely limited because of its high cost, scarcity, and susceptibility to carbon monoxide (CO) poisoning.

Alternatives for sustainable catalysts have been thoroughly studied in order to address these shortcomings. Iron oxide, manganese oxide, and cobalt oxide are examples of transition metal oxides that offer promise low-cost catalytic activity for ORR. Graphene, carbon nanotubes (CNTs), and carbons generated from biomass are examples of carbon-based nanostructures that are appealing as non-precious catalysts due to their great surface area, superior conductivity, and robust endurance. Specifically, nitrogen-doped graphene and CNT composites have ORR activities that are on par with Pt-based catalysts, although being substantially more affordable and readily available.

In addition to addressing waste management, the incorporation of carbons produced from biomass—such as lignin, coconut shells, or agricultural residues—allows for the creation of environmentally friendly, affordable, and scalable catalysts. These developments demonstrate how fuel cell technology is increasingly moving toward circular economy principles (Graph 11.1).

11.5.3 Applications

Fuel cells are finding uses in a variety of fields due to advancements in sustainable materials. Because they can be recharged more quickly and have a longer operational life than traditional lithium-ion batteries, small PEM fuel cells can be used in portable gadgets. Hydrogen-powered fuel cell vehicles (FCVs) are becoming more popular in the automobile industry as zero-emission substitutes for fossil fuel vehicles. Models like the Mirai and NEXO are already being sold by Toyota and Hyundai. Large-scale fuel cell-based stationary power generation is also being investigated for grid stability through integration with renewable energy sources like solar and wind, backup systems, and distributed energy networks.

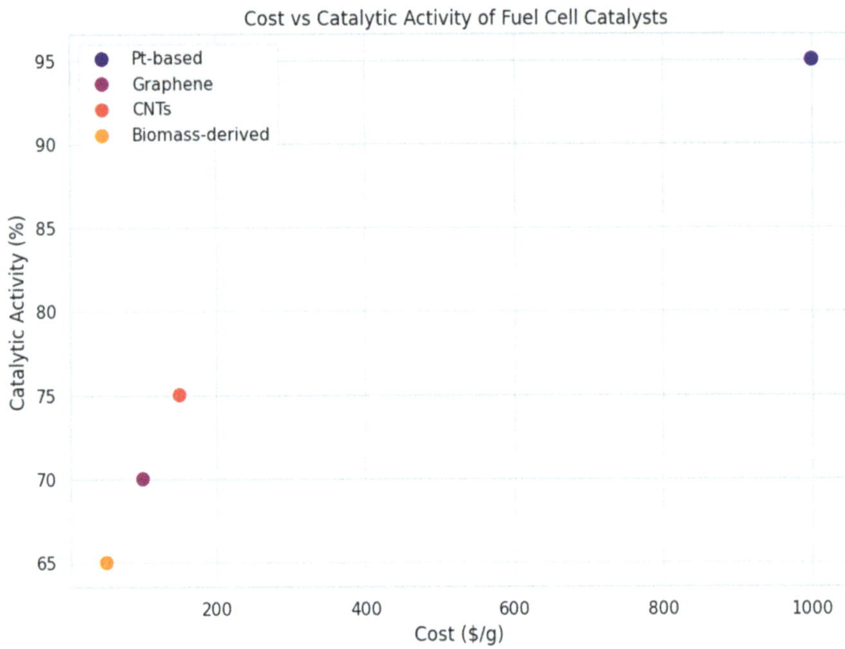

Graph 11.1 Cost vs catalytic activity of fuel cell

Fig. 11.6 Comparison of Pt-based and non-Pt-based catalysts in PEM fuel cells

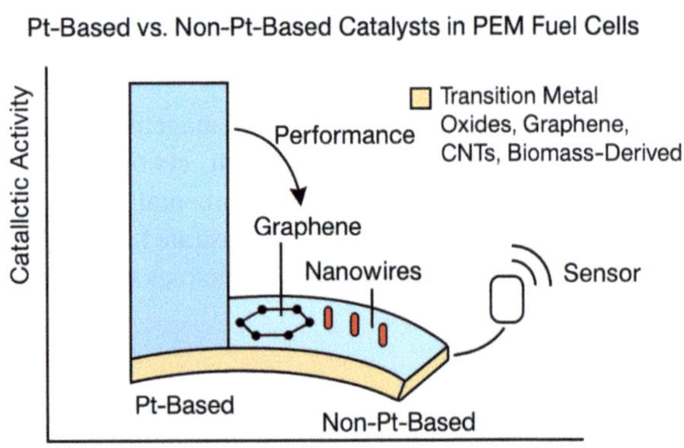

As a result, using sustainable materials in membranes and catalysts is not only cutting costs but also increasing the likelihood that fuel cells will be widely used commercially in the clean energy ecosystem (Fig. 11.6).

In fuel cell applications, the graphic emphasizes the stark difference between platinum-based catalysts and environmentally friendly substitutes. Platinum is still the gold standard for catalytic activity, but its scalability is severely constrained by its high cost, limited supply, and vulnerability to long-term deterioration. On the other hand, sustainable substitutes such as graphene, carbon nanotubes, biomass-derived carbons, and transition metal oxides are quickly becoming game-changers. These materials are excellent in terms of stability, durability, and cost-effectiveness in addition to providing competitive catalytic activity. They are far more suited for extensive deployment due to their abundance and reduced environmental impact. Non-platinum catalysts open a revolutionary path toward the widespread commercialization of fuel cells by combining high performance, affordability, and sustainability. This has ramifications for next-generation energy systems, portable electronics, and clean transportation (Dutta et al. 2021).

11.6 Challenges and Future Prospects

Even while sustainable materials for energy and sensor applications have advanced significantly, there are still a number of issues that need to be resolved before widespread commercialization is possible. Scaling up bio-based and nanomaterials is one of the main problems. Although high-quality materials can be produced using laboratory synthesis techniques like sol-gel processing, electrospinning, and hydrothermal growth, these technologies sometimes have issues with reproducibility, homogeneity, and cost-effectiveness when applied to industrial production. For sustainable materials to be widely used, scalable and environmentally friendly manufacturing methods must yet be developed.

A significant obstacle is striking a balance between biodegradability and performance. Though their long-term stability under environmental factors including humidity, temperature fluctuations, and mechanical wear still needs to be improved, several bio-based polymers and composites show promising mechanical and piezoelectric capabilities (Kumar et al. 2021)., To guarantee dependable device performance in practical applications, durability must be increased without sacrificing biodegradability.

The reliance on rare and hazardous materials is a third issue. Even while research on lead-free piezoelectrics, carbon-based catalysts, and natural polymers is ongoing, certain alternatives still call for the use of uncommon or hazardous precursors. For example, niobium, which is scarce and presents supply chain issues, is necessary for several niobate-based ceramics. Future studies must concentrate on cheap, readily available, and completely non-toxic basic materials. Furthermore, there are advantages and disadvantages of using sustainable materials into multipurpose hybrid systems. For instance, hybrid devices that combine thermoelectric, photovoltaic, and piezoelectric components may result in extremely effective self-powered systems. However, preventing cross-degradation, preserving interfacial stability, and attaining compatibility across several material classes are challenging technical issues.

Future developments in additive manufacturing (3D printing), nanotechnology, and artificial intelligence-driven material design should hasten the search for and improvement of sustainable materials. Researchers can anticipate material qualities, minimize trial-and-error methods, and accelerate innovation cycles by utilizing

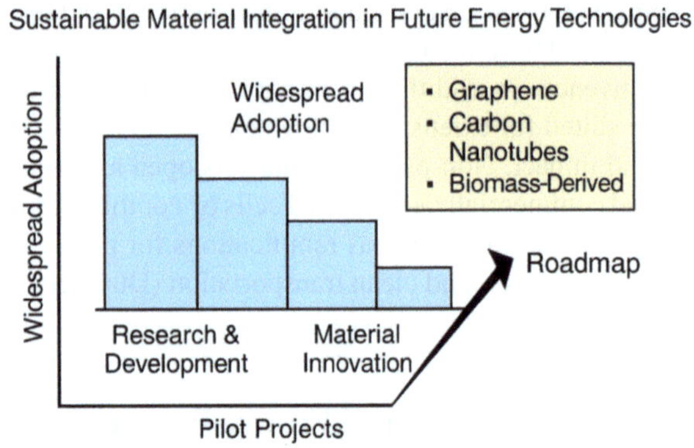

Fig. 11.7 Roadmap of sustainable material integration in future energy technologies

computer modeling, high-throughput tests, and machine learning. Sustainable materials have the potential to transform next-generation energy, healthcare, and environmental technologies while reducing their ecological impact, provided that interdisciplinary efforts are sustained.

A well-organized road map that charts the development of sustainable materials from early research to widespread commercialization in energy technologies is shown in Fig. 11.7. It starts with basic research and development, in which novel ideas and material characteristics are investigated in lab settings. Subsequently, potential alternatives are optimized for performance, efficiency, and environmental compatibility in the material innovation phase. These materials bridge the gap between theoretical promise and practical use by moving forward to pilot projects and prototype testing after validation. When sustainable materials are incorporated into commercial energy systems like fuel cells, energy harvesters, and sensors, the roadmap's goal is broad adoption. By outlining this methodical procedure, the figure emphasizes how crucial methodical development, teamwork, and scalability are to guaranteeing that innovative sustainable materials may leave the lab and provide significant, environmentally responsible answers to the world's energy problems.

11.7 Conclusion

Next-generation clean energy technologies now rely heavily on sustainable materials. They are essential enablers for meeting the world's increasing energy demands while minimizing environmental impacts because of their special blend of cost-effectiveness, eco-friendliness, and high functional performance. By incorporating nanomaterials and bio-based substitutes into gadgets like fuel cells, sensors, and piezoelectric harvesters, scientists are revolutionizing the production, storage, and use of energy.

Biopolymers and composite nanomaterials have made flexible, lightweight, and biodegradable energy solutions possible in the field of piezoelectric devices and energy harvesters. These solutions are appropriate for wearable electronics and self-powered systems (Kumar et al. 2021). The use of green materials in sensors, such chitosan, cellulose, and silk fibroin, has also accelerated the creation of flexible, biodegradable devices that preserve sensitivity and dependability while cutting down on long-term electronic waste.

The use of sustainable substitutes such as transition metal oxides, graphene, carbon nanotubes, and carbon catalysts generated from biomass is revolutionizing fuel cells, which have historically relied on expensive and rare platinum catalysts. In addition to lowering reliance on essential raw resources, these developments open the door for reasonably priced uses in stationary power generation, electric cars, and portable electronics (Zhao et al. 2022; Singh and Sharma 2023).

Notwithstanding these developments, there are still many obstacles to overcome, especially in increasing the production of bio-based and nanomaterials, enhancing their effectiveness without sacrificing biodegradability, and guaranteeing long-term stability in a variety of operating environments. Additionally, a crucial area for further study is the use of sustainable materials into multipurpose hybrid systems (Kumar et al. 2021).

This Fig. 11.8 provides a holistic view of how sustainable materials such as graphene, carbon nanotubes, and biodegradable polymers are being applied across diverse energy systems, including piezoelectric devices, sensors, and fuel cells. By illustrating the interconnected roles of these advanced materials, it reinforces the central theme of the chapter—that sustainable materials are not isolated innovations but integral components of a unified approach to advancing clean energy technologies. Their collective contribution demonstrates how material science is shaping a greener, more efficient, and scalable energy future.

In conclusion, sustainable materials are not merely substitutes for traditional counterparts but **disruptive enablers of innovation** in energy harvesting, storage, and conversion technologies. With continuous progress in materials science, coupled with strong interdisciplinary collaboration, the vision of a **greener, self-sustaining, and low-carbon future** appears increasingly attainable.

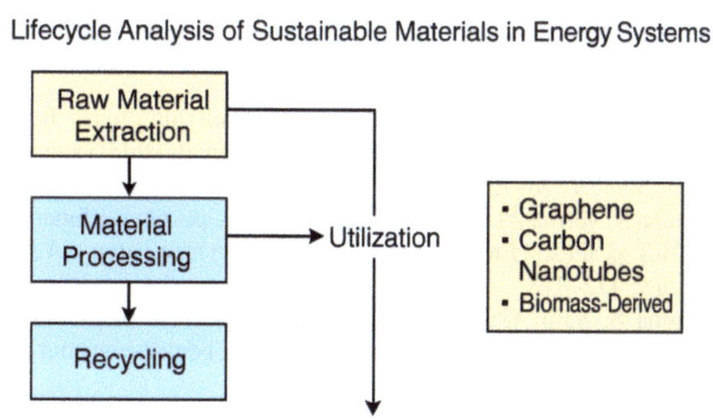

Fig. 11.8 Schematic representation of sustainable materials in energy systems driving future clean energy technologies across piezoelectric devices, sensors, and fuel cells

References

Beeby SP, Tudor MJ, White NM (2006) Energy harvesting vibration sources for microsystems applications. Meas Sci Technol 17(12):R175–R195

Cao Q, Rogers JA (2009) Ultrathin films of single-walled carbon nanotubes for electronics and sensors: a review. Adv Mater 21(1):29–53

Chu S, Majumdar A (2012) Opportunities and challenges for a sustainable energy future. Nature 488(7411):294–303

Csoka L et al (2011) Piezoelectric effect of cellulose nanocrystals thin films. ACS Macro Lett 1(1):45–48

Dagdeviren C et al (2014) Conformal piezoelectric energy harvesting and storage from motions of the heart, lung, and diaphragm. Proc Natl Acad Sci U S A 111(5):1927–1932

Dutta S et al (2021) Sustainable fuel cells: advances in materials and technology. Renew Energy 172:1384–1399

Kim HS, Kim JH, Kim J (2011) A review of piezoelectric energy harvesting based on vibration. Int J Precis Eng Manuf 12:1129–1141

Kim DH et al (2019) Silk fibroin-based bioelectronics for energy harvesting and sensing applications. Adv Funct Mater 29(3):1804910

Kumar S et al (2021) Lead-free and eco-friendly materials for energy applications. Mater Today Sustain 15:100073

Li L, Zhang M, Guo Y (2016) Enhanced piezoelectric properties of PVDF composites with CNT and graphene fillers. Compos Sci Technolgy 127:68–75

Lovinger AJ (1983) Ferroelectric polymers. Science 220(4602):1115–1121

Lynch JP, Loh KJ (2006) A summary review of wireless sensors and sensor networks for structural health monitoring. Shock Vib Digest 38(2):91–128

Martins P, Lopes AC, Lanceros-Mendez S (2014) Electroactive phases of poly(vinylidene fluoride): determination, processing and applications. Prog Polym Sci 39(4):683–706

Novoselov KS et al (2012) A roadmap for graphene. Nature 490(7419):192–200

Panda PK (2009) Review: environmental friendly lead-free piezoelectric materials. J Mater Sci 44:5049–5062

Rajendran S, Chinnaiyan SK (2020) Sustainable materials and green nanotechnology for energy applications. J Mater Sci: Mater Electron 31(20):17135–17150

Rödel J, Jo W, Seifert KTP, Anton EM, Granzow T, Damjanovic D (2009) Perspective on the development of lead-free piezoceramics. J Am Ceram Soc 92(6):1153–1177

Safari A, Akdoğan EK (2008) Piezoelectric and acoustic materials for transducer applications. Springer

Saito Y, Takao H, Tani T, Nonoyama T, Takatori K, Homma T et al (2004) Lead-free piezoceramics. Nature 432(7013):84–87

Saxena S, Srivastava AK (2020) Carbon nanotube-based sensors and their application. In: Nano-optics. Elsevier, pp 265–291

Saxena S, Verma AL (2013) Metal-tetraphenylporphyrin functionalized carbon nanotube thin films as sensor for detection of nitrogen dioxide vapors. AIP Conf Proc Ame Inst Phys 1536(1):1298–1299

Saxena S, Verma AL (2014) Metal-tetraphenylporphyrin functionalized carbon nanotube composites as sensor for benzene, toluene and xylene vapors. Adv Mater Lett Int Assoc Adv Mater 5(8):472–478

Saxena S, Saini GSS, Verma AL (2015) Co-TPP functionalized carbon nanotube composites for detection of nitrobenzene and chlorobenzene vapours. Bull Mater Sci Indian Acad Sci 38(2):443–449

Saxena S, Srivastava AK, Srivastava R, Kheraj V (2019) Metal-phthalocyanine functionalized CNTs sensor for chloroform series. Eur J Eng Sci Technol 2:71–80

Shapter JG et al (2019) Challenges and opportunities in sustainable nanomaterials for energy applications. Adv Mater 31(47):1903022

Shrout TR, Zhang SJ (2007) Lead-free piezoelectric ceramics: alternatives for PZT? J Electroceram 19(1):113–126
Singh R, Sharma P (2023) Biopolymer-based sustainable energy devices: challenges and future outlook. J Clean Prod 380:135152
Someya T, Bao Z, Malliaras GG (2016) The rise of plastic bioelectronics. Nature 540(7633):379–385
Srivastava AK, Saxena S, Yadav SK, Ashok P (2024) Smart sensing solutions for the growth of agriculture. In: Next-generation smart biosensing. Academic Press, pp 43–66
Varghese SS et al (2015) Recent advances in graphene-based gas sensors. Sens Actuat B: Chem 218:160–183
Verma AL, Saxena S, Saini GSS, Gaur V, Jain GSSVK (2011) Hydrogen peroxide vapor sensor using metal-phthalocyanine functionalized carbon nanotubes. Thin Solid Films 519(22):8144–8148
Wang ZL (2004) ZnO nanostructures: growth, properties and applications. J Phys: Condens Matter 16(25):R829–R858
Wang ZL (2017) On Maxwell's displacement current for energy and sensors: the origin of nanogenerators. Mater Today 20(2):74–82
Wang ZL, Song J (2006) Piezoelectric nanogenerators based on zinc oxide nanowire arrays. Science 312(5771):242–246
Wu Q et al (2018) Chitosan-based biomaterials for sustainable energy applications. Sustain Energy Fuels 2(4):822–839
Yang S et al (2017) Sustainable materials for renewable energy devices: a review. Renew Sustain Energy Rev 72:87–110
Zeng K, Zhang D (2010) Recent progress in alkaline anion exchange membrane fuel cells. J Power Sources 196(1):22–31
Zhao Y et al (2022) Advances in green nanomaterials for energy applications. Renew Energy 185:145–160

Chapter 12
Renewable Energy Resource: Current Status, Future Prospect and Their 3D Printing Technology

Gopikishan Sabavath, Mansi Sharma, Sarojni Rai, Sonika, and Ashutosh Kumar

Abstract Energy is essential for economic expansion, industrial advancement, and human development. Fossil fuels, including coal, oil, and natural gas, have traditionally been the main energy sources. But an over-reliance on these non-renewable resources has resulted in major geopolitical and environmental issues, such as resource depletion, climate change, and energy instability. The global energy industry is responding by shifting towards renewable energy sources, or ecologically beneficial, naturally replenishing alternatives.

12.1 Introduction

Renewable energy comes from natural sources such as sunshine, wind, water, tides, and geothermal heat, among others. Renewable energy sources are important because they replenish more quickly than they are consumed. An important source of energy that is more environmentally friendly and sustainable is renewable energy. Fossil fuels, such as coal, oil, and gas, are non-renewable resources that

G. Sabavath
CSJM University, Kanpur, Christ Church College, Kanpur, Prayagraj, Uttar Pradesh, India

M. Sharma
School of Engineering and Technology, Manav Rachna International Institute of Research and Studies, Faridabad, India

S. Rai
Department of Applied Science, United University, Prayagraj, Uttar Pradesh, India

Sonika
Department of Physics, Rajiv Gandhi University, Rono Hills, Doimukh, Papumpare, Arunachal Pradesh, India

A. Kumar (✉)
Department of Physics, CSJM University, Kanpur, Christ Church College, Kanpur, Uttar Pradesh, India

S. K. Verma et al. (eds.), *Sustainable and Natural Polymers*, Clean Energy Production Technologies, https://doi.org/10.1007/978-981-92-0092-4_12

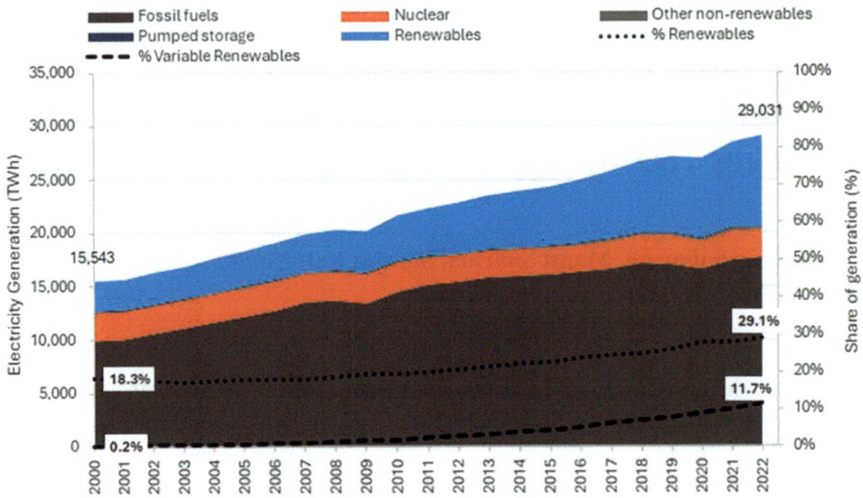

Fig. 12.1 Electricity generation by renewable and non-renewables from 2000 to 2022. (Source: IRENA highlights 2024) (IRENA and ILO 2024)

take millions of years to create, provide energy when they burn, and emit toxic gases like carbon dioxide. Because it is free of pollution, naturally occurring, and readily available at a reasonable price, renewable energy is the most crucial factor to take into account when considering population growth. Renewable energy is needed to minimise greenhouse gas emissions and balance production and consumption.

The International Renewable Energy Agency (IRENA) announced on July 11, 2024, that 29.1% of the world's electricity generation in 2022 came from renewable sources, while the total energy produced was 8440 TWh. The remaining 70.9%, or 20,591 TWh, is attributed to non-renewables, fossil fuels, pumped storage, and nuclear energy. 2022 will see the generation of 29,031 TWh of energy from all forms of energy sources (IEA 2024). The total amount of electricity generated has grown by 2.4% since 2011. The contributions from renewables and non-renewables were 6.1% and 1.3%, respectively. IRENA has also reported that electricity by renewables driven by wind and solar reached by 11.7% mix in 2020, with a growth rate of 18.2% from 2021. The advancement of renewable energy will lead to improved energy access, decreases the air pollution from fossil fuel, and improves the energy security (Figs. 12.1 and 12.2).

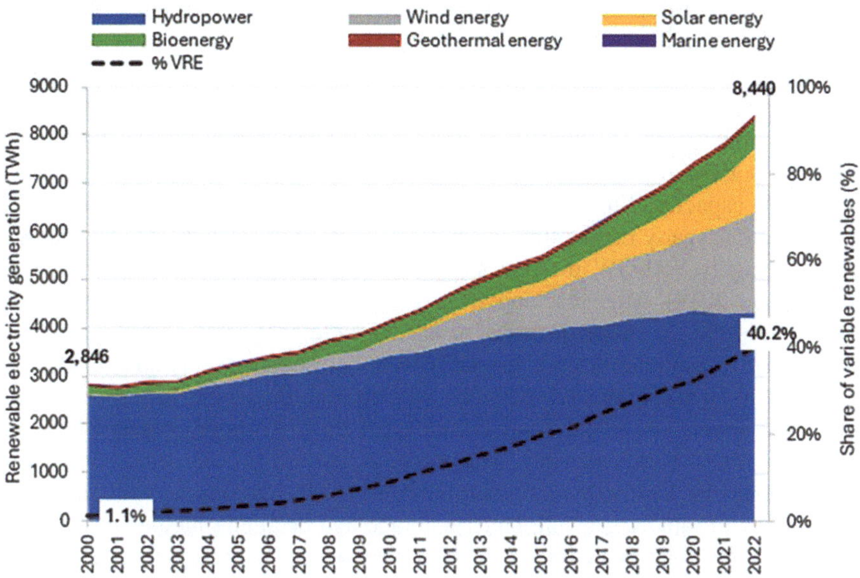

Fig. 12.2 Electricity generation using only renewables from 2000 to 2022. (Source: IRENA highlights 2024) (IRENA and ILO 2024)

12.2 Classification of Renewable Energy Resources

Natural replenishment processes, including sunshine, wind, river flow, geothermal heat, and biomass are examples of renewable energy sources. For optimal generation, storage, and distribution, contemporary professional applications of renewable energy incorporate cutting-edge technologies, computerised controls, and hybrid systems. Resources for renewable energy fall into the following categories:

12.2.1 Solar Energy

One of the primary drivers behind the switch to clean energy and a crucial tool for reducing environmental pollution is solar energy. Compared to conventional methods of electricity generation, solar installations require less water. Solar energy systems are cost-effective and provide freedom from electricity bills. Anywhere in a remote area, solar energy can be obtained. The sustainable source of energy is solar radiation from the sun. Photovoltaic cells and solar thermal systems are the two methods used to capture this energy. Photovoltaic (PV) cells, also referred to as solar cells, use the photoelectric effect to transform sunlight into electrical energy. The energy found in sunlight is in the form of photons, which correspond to various solar spectrum wavelengths. In general, semiconductor materials are used to make

PV cells. The PV solar panel's charge carriers from semiconductors absorb the energy when photons from the sun impact it. The charge carriers will recombine as a result of absorbed energy, releasing light as an output.

Among renewable resources, solar energy is growing at a rapid rate. According to some assessments, photovoltaic systems make up the majority of installations; as of 2022, solar will produce 3.6% of all electricity and about 31% of the world's renewable capacity. With projected growth rates of 22% between 2021 and 2022 and an earlier yearly rise of 40%, quantitative projections include installed capacities of roughly 300 GW in 2016 and 114,425.81 MW in 2018 (Upadhyay and Chowdhury 2014). After reaching around 1.6 TW of total global PV capacity in 2023 and seeing a strong increase in annual additions, solar photovoltaic (PV) energy has transitioned from a speciality to a mainstream energy source (Pourasl et al. 2023). It is anticipated that solar will become the single largest renewable electricity source and lead the way in the expansion of renewable capacity between 2024 and 2030. Simultaneously, onshore wind and utility-scale solar continue to rank among the least expensive new-build electricity options without subsidies. Together with rapidly growing battery storage and declining module costs, these trends position solar as a cornerstone of future low-carbon energy systems; but difficulties related to integration, storage, grid upgrades, and materials/recycling must be addressed.

Semiconductor characteristics, device architecture, and light-matter interaction processes all affect solar cell performance and efficiency. When sunlight (photons) strikes the solar cell surface, photons with energy $E = h\nu$ greater than or equal to the band gap energy (E_g) of the semiconductor are absorbed. The absorbed energy promotes an electron, leaving behind a hole. These charge carriers are generated mainly near or within the depletion region of the *p*–*n* junction. This excites electrons from the valence band to the conduction band, generating electron–hole pairs. The internal electric field at the *p*–*n* junction drives the electrons towards the *n*-side and holes towards the p-side. This separation prevents recombination and creates a potential difference. When the external circuit connects the *n*-side and *p*-side, electrons flow through the load, producing electric current. The resulting flow constitutes direct current electricity. The output current I and voltage V relationship is given by the diode equation:

$$I = I_{\mathrm{ph}} - I_0 \left(e^{\frac{qV}{kT}} - 1 \right) \tag{12.1}$$

where: I_{ph}: Photo-generated current, I_0: Reverse saturation current, q: Electron charge, k: Boltzmann constant, T: Absolute temperature.

The efficiency (η) of a solar cell is given by the ratio of electrical power output to incident solar power:

$$\eta = \frac{P_{\text{out}}}{P_{\text{in}}} = \frac{I_{\text{mp}} V_{\text{mp}}}{P_{\text{in}}} \tag{12.2}$$

where I_{mp} and V_{mp} are the current and voltage at the maximum power point (MPP).

12.2.1.1 Benefits of Solar Energy

(i) **Accessibility and Abundance:** Solar energy is a universal resource because it is present practically everywhere.
(ii) **Environmental Benefits:** By eliminating direct greenhouse gas emissions, solar energy generation lessens reliance on fossil fuels.
(iii) **Economic Growth:** Millions of jobs in production, installation, and maintenance have been created globally by the solar sector.
(iv) **Decentralisation of Energy:** By installing solar panels on rooftops, businesses and homes can generate their own electricity and lessen their reliance on the grid.

12.2.1.2 Challenges in Solar Energy Adoption

Despite its promise, solar energy has a number of drawbacks.

(i) **Intermittency:** Daytime and weather patterns affect the availability of solar energy, necessitating effective storage options.
(ii) **Land Use:** The substantial land requirements of large-scale solar farms might occasionally put biodiversity or agriculture in jeopardy.
(iii) **Infrastructure and Storage Costs:** Energy storage solutions are still somewhat costly, notwithstanding a decrease in PV module prices.
(iv) **Material Reuse:** End-of-life solar panels present waste management issues that call for environmentally friendly recycling techniques.
(v) **Comparing Solar Energy to Others Renewable Source**

Solar energy provides special benefits, although wind and hydropower are equally important. While hydropower is dependent on water supply and may disturb ecosystems, wind power is site-specific and seasonal. Solar energy, on the other hand, is more adaptable and scalable because it can be installed in both urban and rural areas, ranging from enormous solar farms to small rooftop installations.

12.2.1.3 Sustainability Aspects and Environmental Impact

Solar energy has a bright future owing to advancements in materials, storage, and grid integration. Higher efficiencies and reduced costs are promised by emerging technologies, including concentrated solar power, bifacial modules, and perovskite solar cells. Sustainable solutions to intermittency may be possible if solar systems are combined with next-generation batteries and green hydrogen production. Through sustained regulatory backing, research funding, and widespread acceptance, solar energy has the potential to overtake all other electricity sources worldwide by the middle of the century.

- **Greenhouse gas reduction**: The majority of research highlights how solar energy can help mitigate climate change by lowering greenhouse gas emissions.
- **Quantitative estimates**: Elavarasan (2019) is the only source of a quantitative estimate, pointing to a projected 98.22 million tonnes of carbon dioxide emissions reduction by 2020 as a result of greater usage of renewable energy sources (Elavarasan 2019).
- **Air pollution and public health**: A number of studies demonstrate how solar energy may replace fossil fuels, lowering air pollution and enhancing public health.
- **Lifecycle impacts**: Some studies, such as Ahmed et al. (2024), point out that some of these advantages may be outweighed by the environmental effects of producing solar technologies, especially the mining and processing of raw materials (Ahmed et al. 2024). In general, solar energy has a more favourable sustainability profile than fossil fuels; nevertheless, more research is needed to fully understand its lifespan consequences.

12.2.1.4 Scope of 3D-Printing Approaches for Solar Energy Applications

The field of solar energy research and manufacturing is changing as a result of additive manufacturing (AM) and digital printing processes. Perovskite and organic photovoltaics (PV) can be deposited with functional thin films using digital printing (inkjet, aerosol-jet, gravure/roll-to-roll), while structural components, optical textures, and integrated thermal-management hardware can be designed freely using true 3D AM (extrusion, stereolithography). Rapid prototyping and innovative device architectures are made possible by AM and digital printing, which—in contrast to subtractive manufacturing—offer mask-less, digitally controlled deposition of materials and complicated geometries. The development of printed active layers (perovskite and organic PV), printed interconnects and encapsulation, and 3D-printed thermal and optical components during the last 5 years has advanced ideas from specialised labs to pilot tests and business collaborations. A notable example of the promise of linked printing paths for scalable production was the 2024 report of fully roll-to-roll printed devices with competitive efficiency

(Weerasinghe et al. 2024). Printing technology related to solar energy is broadly classified into two category-

(i) **Digital Thin-Film Printing**
Inkjet printing is a drop-on-demand digital deposition technique that works well for functional interlayers and active layers like perovskites and organics. Benefits include minimum material waste, mask-free patterning, and compatibility with roll-to-roll (R2R) lines for scaling up. Ink rheology, drying dynamics, and flaws, such as coffee-ring effects are the main obstacles.
Aerosol-jet/direct-write: a non-contact technique that allows for conformal coatings and fine conductive traces on three-dimensional surfaces; especially helpful for printed encapsulation and interconnects. Aerosol-jet encapsulants have been shown to significantly increase the longevity of organic photovoltaic devices (Basu et al. 2023).
(ii) **True 3D Additive Manufacturing (Structural & Micro Architected)**
Housings, substrates, and structural components that incorporate PV devices into custom modules or building materials are made utilising extrusion (FDM) and material jetting.
Vat photopolymerization (SLA/DLP): microfluidic channels for heat control and optical components (microlens arrays, light-trapping texturing) with great accuracy.
Metal and ceramic AM (laser powder bed fusion, binder jetting, stereolithography + ceramics): create metallic frames and micro-architected heat exchangers with intricate internal channels and high thermal conductivity for concentrated solar thermal or PV cooling (Anwajler 2024).

Co-optimization of ink formulation, substrate compatibility, drying/curing techniques, and in-line quality control is necessary for successful printed solar systems. Important material factors to consider:

(a) Functional inks: printability (viscosity/surface tension) and final optoelectronic qualities require customisation of solvent systems, precursor chemistry (for perovskites), nanoparticle dispersions (conductive inks), and polymer binders.
(b) Encapsulation materials: UV-curable layers that can be conformally placed and printable barrier polymers are essential to the lifespan of the device. Direct-write encapsulants and aerosol-jet techniques have demonstrated promise.
(c) Thermal budgets for the substrate and process: a lot of printed active layers are sensitive to temperature and air exposure; low-temperature curing methods and formulations that are favourable to the environment are now being researched. Robust, fast drying/annealing techniques (such as infrared, photonic, and laser sintering) are necessary for integration into R2R (Basu et al. 2023).

Technical Challenges in 3D Printing Technology of Solar Energy

Although encouraging demonstrations, a number of challenges still exist:

(a) Long-term stability: To match the lifespans of crystalline silicon modules, perovskite and organic absorbers need improved encapsulation techniques and inherent material advancements. Although barrier research and aerosol-jet encapsulation are encouraging first steps, they require additional verification by expedited outside testing.
(b) Increasing yield and controlling defects: Low defect concentrations across square meters are required for high-throughput R2R systems; inline inspection and machine learning-driven process optimisation are being investigated. As part of the scale-up approach, the CSIRO demonstration used large-scale screening and optimisation techniques.
(c) Supply chain and material toxicity: Lead-based perovskites present environmental and regulatory issues; lead mitigation plans and recycling routes need to be created concurrently with production.
(d) Standardisation and certification: There is still work being done to set industry standards for mechanical robustness, printed modules, and expedited testing procedures.

12.2.2 Wind Energy Technologies

Wind energy is a clean and renewable energy source that is produced by employing wind turbines to convert the kinetic energy of the wind into mechanical power and ultimately electricity. This technology, which provides a clean and effective substitute for fossil fuels, is among the fastest-growing renewable energy sources. Wind energy boosts economic growth by creating jobs in production, installation, and maintenance while lowering greenhouse gas emissions and enhancing energy security. As a result of government regulations, technological improvements, and the growing demand for clean energy globally, wind energy is now one of the renewable energy sources with the quickest rate of growth. With more than 1097 gigawatts (GW) of installed wind generating capacity globally as of 2024, this is a major turning point in the shift to renewable energy. This number comprises both offshore and onshore wind farms, with the offshore sector accounting for an increasing portion of installations. Due to dropping technology costs, increased efficiency, and a dire need for reduced reliance on fossil fuels, wind power has expanded more quickly globally during the last 10 years. Over 123 GW of additional wind capacity were built globally in 2023 alone, representing a 12.6% annual growth rate. This consistent rise indicates the growing international dedication to using wind energy as a fundamental component of sustainable power production. Presently, more than 900 GW of the installed capacity is onshore wind energy, which leads the global wind capacity. Many areas choose onshore wind farms because they are simpler and less expensive to build than offshore projects. The offshore wind industry, on the other hand, has been expanding remarkably quickly; by 2023, its total installed capacity will have surpassed 75 GW. According to the Global Wind Energy Council (GWEC), wind power will keep growing quickly, with an extra 791 GW of capacity predicted to be deployed worldwide by 2028. This is a significant increase trend, amounting to an average of 158 GW of new wind power annually. Wind power has risen to prominence in the renewable energy industry thanks to a mix of regulatory backing, growing private sector investments, and technology breakthroughs. One of the main sources of electricity generation in the globe is expected to be wind energy as we move towards a cleaner and more sustainable future. As energy demands rise, wind power capacity will continue to grow, lowering greenhouse gas emissions and making a major contribution to the battle against climate change. Some benefits of wind energy are pointed out here (Pira 2020; Shoeib et al. 2021; Mills 2017):

(i) **Clean and Sustainable Source of Power**: Utilising wind energy can help society reduce carbon emissions and lessen the effects of climate change. Wind energy provides a long-term sustainable solution to the world's energy demands because it is renewable and limitless.
(ii) **Economic and Employment Opportunities**: One of the main attractions behind job development and economic expansion is the wind energy industry. The industry sustains millions of employments globally, from the production and installation of turbines to their maintenance and logistics. Since wind energy projects increase tax revenues in rural regions and provide cash for landowners, they help boost local economies.
(iii) **Energy Independence and Security**: Utilising native wind resources can help nations improve their energy security and lessen their dependency on imported fossil fuels. Because wind power broadens the country's energy mix, economies are protected from the geopolitical risks connected with oil and gas supply chains as well as the fluctuating global fuel costs.
(iv) **Low Operating Costs and High Scalability**: Wind turbines are relatively inexpensive to operate and maintain after they are constructed. As a result of technological developments, wind power is now more cost-effective overall since turbines are now more economical and efficient. Additionally, wind farms may be built on a variety of scales, ranging from modest offshore installations to massive community-based projects, which makes the technology extremely flexible to meet a range of energy requirements.
(v) **Minimal Land and Water Usage**: Due to the very limited land footprints of wind turbines, dual land use—such as grazing or agriculture—can coexist with wind energy generation. Furthermore, unlike traditional power plants, wind energy doesn't need water for cooling, which is a significant benefit in desert areas where water is scarce.
(vi) Technological Advancements and Innovation: Persistent advances in turbine design, materials, and digital monitoring systems have improved wind energy reliability as well as efficiency. Wind power generation is rising worldwide because of offshore wind farms, floating turbines, and hybrid renewable systems.

In order to achieve net-zero energy targets and promote environmental resilience, wind power integration into national grids will become more and more important as technology develops. The standard aerodynamic and electromechanical principles that govern modern wind turbines originated from fluid mechanics, rotor dynamics, and electrical engineering. The basic idea behind wind energy is that the wind's kinetic energy is transformed into rotating mechanical energy, which is then transformed into electrical energy by a generator. Aerodynamic lift and drag forces drive the rotor of a wind turbine to rotate when the wind blows over its blades. The shaft attached to an electric generator, which generates electricity, is driven by this circular motion. The available power of wind energy is given by:

$$P_{\text{wind}} = \frac{1}{2}\rho A V^3 \tag{12.3}$$

where: ρ: Air density (kg/m^3), A: Swept area of the turbine blades (m^2), V: Wind velocity (m/s).

This cubic relationship shows that small increases in wind speed greatly enhance power output. Aerofoil-shaped blades, which resemble aeroplane wings, are used in wind turbines.

Lift Force (F_L): Produced as a result of pressure variations on the blade surfaces, perpendicular to the direction of the wind.

Drag Force (F_D): Acts in opposition to rotation and parallel to the direction of the wind.

The lift-to-drag ratio (L/D) determines efficiency. The rotor extracts energy by reducing the wind speed behind it — a process limited by Betz's Law.

$$P_{\max} = \frac{16}{27} \times \frac{1}{2}\rho A V^3 \tag{12.4}$$

This means no turbine can capture more than 59.3% of the wind's kinetic energy. A low-speed shaft that is attached to the rotor sends torque to a gearbox. For the generator, the gearbox raises the rotational speed from about 30 to 60 rpm to about 1500 rpm. The most popular types of generators are permanent magnet synchronous generators (PMSGs) and induction generators (asynchronous). Through electromagnetic induction, the mechanical energy of rotation is transformed into electrical energy (Faraday's Law). By use of rectifiers and inverters to transform variable-frequency AC output into grid-compatible AC electricity. The power coefficient (Cp) represents a turbine's efficiency is given by:

$$Cp = \frac{P_{\text{Turbine}}}{P_{\text{Wind}}} \tag{12.5}$$

Modern turbines often reach values of power coefficient between 0.35 and 0.45, which is less than the theoretical Betz limit of 0.593. The practical output power P_{out} is given by:

$$P_{out} = \frac{1}{2}\rho A V^3 C_P \eta_m \eta_e \quad (12.6)$$

where: η_m: Mechanical efficiency (gearbox, bearings) and η_e: Electrical efficiency (generator, power electronics).

12.2.2.1 Current Aspects of Wind Energy

A key component of the global energy revolution, wind energy is growing quickly in both onshore and offshore markets after surpassing 1 terawatt (TW) of total installed capacity. In 2024, the industry recorded record yearly additions. Deployment is being driven by ongoing cost reductions, technological advancements (bigger turbines, digital operations), and regulatory incentives; but, improved market frameworks, grid investments, supply-chain resilience, and integration with storage and hydrogen systems will be necessary for quicker growth.

(i) **Global expansion and Deployment:** 2024 was a milestone year for the wind industry, with an estimated 117 GW of new capacity added worldwide, surpassing the ~1 TW cumulative capacity. Rapid offshore expansions combined with robust onshore deployment have propelled growth. Several European nations, Brazil, the United States, and China all made significant contributions.

(ii) **Competitiveness and Cost:** The levelized costs of wind have decreased due to economies of scale, long-term learning, and better turbine designs. Land-based wind costs are continuing to fall, while offshore wind costs are improving, according to NREL assessments and market reports. By 2035, additional cost reductions are anticipated through larger rotors, higher hub heights, and system optimisation.

(iii) **Technology Mix:** The majority of capacity is still accounted for by onshore wind, whereas offshore wind—which includes fixed-bottom and expanding floating technologies—is fast evolving and geographically expanding. Deeper oceans and new markets are becoming accessible for the deployment of floating offshore wind.

(iv) **Technological and Market Trends Shaping the Near Future:** Greater capacity factors and energy yield per turbine are the results of trends towards taller towers, larger rotor diameters, and modular manufacturing. Digitalisation lowers O&M expenses and increases availability (predictive maintenance, condition monitoring). Offshore wind is becoming more and more popular in East Asia, Europe, and the United States. It will open up new markets and deepwater resources (including areas of Asia, the U.S. West Coast, and Japan). An increasingly popular policy strategy to ensure long-term supply chains and jobs is the combination of supply-chain localisation and incentives, which countries are using to boost domestic manufacturing and reap the benefits of industry. At larger wind penetrations, combining wind with demand response, long-term storage, battery storage, and green hydrogen electrolysers improves value capture and aids in managing variability. Wind is increasingly included in integrated electricity/hydrogen systems in IEA and IRENA forecasts.

12.2.2.2 Leading Countries in Wind Energy

China is at the forefront of the onshore and offshore wind energy industries. China has over 430 GW of installed capacity overall, which makes up more than half of all offshore capacity worldwide. China cemented its supremacy by adding 76 GW of wind power in 2023 alone. One of the biggest wind farms in the world is Gansu Wind Farm in China. Almost 10% of all electricity produced in the United States comes from wind energy. About 150 GW is the total installed capacity. With large onshore installations, the United States remains a major provider of the world's wind energy capacity. The three states that produce the most wind energy are Texas, Iowa, and Oklahoma. Alta Wind Energy Center (California) is one of the significant wind projects in US. In terms of offshore wind energy and wind technology innovation, Germany leads the world. It has over 70 GW of installed capacity overall. The North Sea Offshore Wind Farms are located there. Latin America's top producer of wind energy is Brazil. About 30 GW is its overall installed capacity. With the addition of 4.9 GW of new wind capacity in 2023, rapid expansion has been accomplished. With 4.9 GW of new wind capacity added in 2023, Brazil has become a significant player and is now the third-largest market for new wind turbines worldwide. The Dogger Bank Wind Farm, the largest offshore wind project in the world, is one example of how the United Kingdom is leading the way in offshore wind energy. It has about 30 GW of total capacity. India's renewable energy objectives heavily rely on wind energy. About 45 GW is its total installed capacity. Located in Tamil Nadu, Muppandal Wind Farm is one of the biggest wind farms in Asia.

Japan had tremendous expansion in 2024, boosting its capacity from 5.2 to 5.8 GW and solidifying its position as Asia's third-biggest wind power market after years of stasis. Vietnam's wind market, which is close to 5 GW, is still the fourth largest in Asia. The percentage of wind power in more than 30 countries is currently higher than the global average, at over 10%. Eleven nations currently rely on wind power for more than 20% of their electricity, with seven of them having a share of 30% or higher. Denmark leads the pack with more than one in two kWh of wind-generated electricity, followed by Germany, the UK, Portugal, the Netherlands, Ireland, and Uruguay. These nations serve as evidence that, in the long run, the WWEA's (World Wind Energy Association) projected 40–50% wind power share of global electricity generation is achievable. In the future of the world's energy supply, wind power will be even more important given the trend towards electrification in the transportation and heating/cooling sectors. Energy from wind will also be used to heat and cool houses and power automobiles. According to the WWEA annual report 2024, the table below shows a comparative growth of installed capacity of wind energy till 2024 for leading countries (WWEA Annual Report 2024; Global Wind Energy Council (GWEC) 2024) (Table 12.1).

Table 12.1 Wind energy capacity of several countries (www.wwindea.org)

Country/region	2024	New capacity 2024	Growth rates 2024 (%)	2023	2022	2021
China	561′492	86′892	18.3	474′600	395′230	346′670
United States	154′609	4′154	2.8	150′455	144′053	135′177
Germany	72′683	3′208	4.6	69′475	66′242	63′924
India	48′163	3′427	7.7	44′736	41′600	39′800
Brazil	34′000	5′420	19.0	28′580	23′661	21′567
United Kingdom	32′360	2′324	7.7	30′036	28′683	25′668
Spain	32′007	1′210	3.9	30′797	30′158	28′143
France	24′383	909	3.9	23′474	20′915	19′084
Canada	18′435	1′449	8.5	16′986	15′212	14′206
Sweden	17′266	1′015	6.2	16′251	14′278	12′173
Australia	15′288	3′299	27.5	11′989	11′047	9′633
Italy	12′915	903	7.5	12′012	11′647	11′322
Turkey	12′864	1′167	10.0	11′697	11′405	11′100
Netherlands	11′720	705	6.4	11′015	8′215	7′846
Poland	10′233	850	9.1	9′383	7′950	6′570
Denmark	7′506	399	5.6	7′107	6′949	6′181
Japan	5′840	626	12.0	5′214	4′727	4′574
Norway	5′133	3	0.1	5′130	5′105	5′105
Ireland	4′934	221	4.7	4′713	4′527	4′332
South Africa	3′560	0	0.0	3′560	3′560	3′256
Rest of the World	33′464	1′317	4.1	32′147	29′402	26′445

12.2.2.3 Future Aspects of Wind Energy

Despite installed capacity has fallen short of projections and expectations for 2024, the world has set a new record in total installations with 1,173,581 MW. It is projected that some of the new installations that were initially planned for 2024 will actually happen in 2025. The future of the US markets, as well as the sustainability and dependability of global supply chains are both very questionable due to recent events in the US, even if growth is typically anticipated in the upcoming years due to new legislation. Consequently, while a tenfold rise by the middle of the century can still be seen as a feasible alternative, a tripling of wind power capacity before the end of this decade is not guaranteed for all countries. The expansion of the wind industry is also at risk from growing opposition in several nations. In certain nations, a contentious discussion concerning onshore and offshore wind farms is being fuelled by false information propagated by populist movements and interest organisations (International Energy Agency (IEA) 2024a).

By 2030 and beyond, wind energy is predicted to grow significantly due to factors like deeper coupling with storage and hydrogen, larger and smarter turbines, rapid offshore deployment (including floating), and further cost reductions. Resilient supply chains, socially conscious siting, improved policy frameworks, and grid upgrades are all needed to realise this promise. Recent international assessments highlight that although there has been significant progress, annual deployment needs to significantly increase in order to reach the challenging climate goals. By eliminating depth restrictions, floating foundations open up a plethora of new resources, such as exposed seaboards and deep coastal waterways. According to technical evaluations and commercial studies, floating concepts are fast-growing due to active research and development as well as pilot deployments. With scale, port improvements, and standardised designs, cost reductions are anticipated (Edwards et al. 2023a). In developed economies, one of the main drivers of near-term capacity and yield increase will be the repowering of ageing turbines with larger, more efficient versions. The advantages of policies that expedite repowering approvals will increase. To control unpredictability and ensure a steady supply at high wind shares, wind farms must be paired with both short- and long-duration storage (batteries, flow batteries, pumped hydro, hydrogen) and market reforms for flexibility. To meet decarbonisation goals, IEA guideline emphasises integrated planning for wind and solar energy. Innovations in long-duration storage, affordable green hydrogen electrolysis, enhanced grid management (such as wide-area control systems), and floating foundation lowering costs would all significantly increase wind's role (Edwards et al. 2023b). Inspite of all these, there are some environmental, technical and social conditions that must be considered (Hong et al. 2024):

(i) Marine impacts and biodiversity: Both onshore and offshore projects need to conduct thorough environmental monitoring and assessment (birds, bats, marine animals). Threats will be decreased and social license will be maintained through adaptive management and mitigation strategies.
(ii) Community benefits and regional jobs: Acceptance is significantly increased by local content, income sharing, and community funds.
(iii) Grid limitations: In order to prevent restriction and backlog delays, transmission and interconnection capacity must be in line with procurement and auction schedules.
(iv) Policy ambiguity—Abrupt modifications to auction regulations, subsidies, or permits may prevent investment; capital-intensive projects require solid, long-term frameworks.
(v) Specialisation of the supply chain: Diversified and localisation policies lessen the risk. Supplies, turbine parts, and port/installation capacity are occasionally concentrated in a few areas.
(vi) High Operating Costs: In comparison to onshore wind and solar energy, offshore wind farms cannot be considered economically viable due to their significant capital and operating costs.
(vii) Technical Failures: In large wind projects like Dogger Bank and Vineyard Wind, incidents like blade breakdowns have been documented, bringing

attention to dependability problems and engineering complexity. Frequent repairs are necessary for wind turbines due to gearbox issues, generator failures, and blade deterioration. Offshore wind farms have higher maintenance costs and downtime because they need specialised ships and skilled workers.

(viii) Damage to Wildlife: Lawsuits and delays in wind projects along the U.S. East Coast have resulted from worries that offshore wind farms might damage marine life, including the rare North Atlantic right whale.

(ix) Collisions with birds: Turbines can be dangerous to birds, although research indicates that the number of bird fatalities attributable to turbines is much smaller than that of other human-caused causes.

(x) Interference: Because wind farm projects may endanger the local landscapes and ways of life, they have been challenged in court in places like Galicia and Spain.

To ensure the sustainable development of wind energy, addressing these issues calls for community involvement, thorough environmental impact assessments, and technology developments. In order to meet the world's energy needs and reduce carbon emissions, wind power is essential. Wind energy is poised to emerge as one of the primary energy sources of the future, making a substantial contribution to a more sustainable and greener planet as long as investments and governmental support keep growing.

12.2.2.4 3D-Printing Technology in Wind Energy

In order to attain greater efficiency, reduced costs, and quicker deployment, the worldwide wind energy industry is constantly changing. 3D printing, sometimes referred to as additive manufacturing (AM), is one of the most revolutionary inventions advancing this development. 3D printing has the potential to completely transform turbine design, manufacture, maintenance, and supply-chain logistics by making it possible to produce intricate parts precisely and on demand. Recent developments in large-scale additive manufacturing have made it possible to produce enormous wind turbine components and blades. For example, scientists at GE Renewable Energy and the U.S. Department of Energy's Oak Ridge National Laboratory (ORNL) have shown how to use composite materials to 3D-print blade moulds longer than 20 m (GE Renewable Energy and Oak Ridge National Laboratory (ORNL) 2023). The time needed to make full-scale blades is drastically cut down by these moulds, going from weeks to a few days. Custom aerodynamic designs are another benefit of 3D printing, which increases energy capture effectiveness and decreases material waste (Siemens Gamesa Renewable Energy 2022).

Large plants and intricate international logistics are required for the production of conventional turbines. On the other hand, replacement parts can be manufactured at wind farm locations using on-site 3D printing. This lowers the carbon emissions, downtime, and shipping expenses related to long-distance transportation. Systems for moveable additive manufacturing are being tested for on-site hub replacements, blade repairs, and other maintenance requirements. Topology optimisation is made possible by additive manufacturing, which creates complex shapes that are not achievable with conventional casting or forging (National Renewable Energy Laboratory (NREL) 2024). By minimising the total weight of nacelles, gearboxes, and support structures, lightweight lattice frames can increase turbine stability and save installation expenses. By employing exactly the precise amount of material needed, 3D printing reduces material waste. The usage of composites and recycled polymers improves the sustainability of the environment even more. The transition to digital manufacturing models could eventually result in a considerable reduction in wind energy infrastructure's carbon footprint as well as production expenses. Considering its potential, a number of obstacles still exist (Liu and Montalbo 2023; Vestas Wind Systems 2024):

1. Material certification: Verifying that, for wind applications, 3D-printed materials satisfy strict fatigue and environmental resistance requirements.
2. Scaling constraints: It is still difficult and expensive to print very big turbine blades or towers.
3. Standardisation and quality assurance: At the moment, 3D-printed turbine parts lack universal certification.
4. Expense of high-performance materials: Metal powders and sophisticated polymers used in printing can be costly.

The design, construction, and maintenance of wind turbines have undergone a radical change because of 3D printing. The objectives of next-generation renewable energy systems are closely aligned with additive manufacturing's capacity to combine flexibility, efficiency, and sustainability. Its long-term potential to revolutionise the wind energy sector is evident, even though more study, standardisation, and funding are needed.

12.2.3 Hydropower

Using the kinetic and potential energy of flowing water to create electricity is known as hydropower, or hydroelectric power. Among many traditional and popular renewable energy sources, it offers clean, dependable and adaptable electricity generation. Hydropower plants use turbines to shift the gravitational potential energy of water in reservoirs or moving rivers into mechanical energy, which is subsequently converted into electricity by generators. The Greeks employed waterpower more than 2000 years ago to power wheels for grinding grain; nowadays, it is one of the most economical ways to produce electricity and is frequently the method of choice when it is available. Hydropower has been used in India for over 120 years. The emergence of hydropower in India began in 1897 with the commissioning of the first 130 kW hydro project in the Darjeeling hills. The subsequent project to be built in the Mysore district of Karnataka in 1902 was the 4500 kW Sivasamudram project, which was intended to bring electricity to the Kolar gold mines. After that, several small hydro projects were established in the nation's hilly regions. 508 MW of hydroelectric installations, mostly small and medium-sized, made up the nation's installed capacity of 1362 MW prior to its independence in 1947. For instance, hydropower provides 99% of Norway's electricity. The 22.5 GW Three Gorges Dam in China is the largest hydropower facility in the world. It generates 80–100 terawatt-hours annually, which is sufficient to power 70–80 million households. Utilising water to power turbines is the fundamental idea behind hydropower. There are two main types of hydropower plants: those with dams and reservoirs and those without. Big reservoir hydropower dams can store water for brief or extended periods of time to satisfy the highest levels of demand. Additionally, the facilities can be separated into smaller dams for various uses, for example, pumped-storage reversible plants for both pumping and electricity generating, seasonal storage, or during the day or night use. Without dams and reservoirs, hydropower is produced on a smaller scale, usually from a plant built to function in a river without obstructing its flow. Because of this, many people believe that small-scale hydro is a more ecologically beneficial choice. Hydropower has been a major contributor to the world's electrification beginning in the end of the nineteenth century and continues to be a vital component of sustainable energy systems. By 2025, hydropower will account for almost 60% of all renewable power output and 16% of the world's electrical generation (International Energy Agency (IEA) 2024b). The core concept behind hydropower is the transformation of water's potential and kinetic energy into mechanical and eventually electrical energy.

(a) **Potential Energy**: Because of gravity, water kept in a dam or reservoir at a greater altitude has potential energy.
(b) **Kinetic Energy**: The water's potential energy is transformed into kinetic energy once it circulates through penstocks at an extremely rapid speed.
(c) **Mechanical Energy**: The turbine rotates as a result of the water's impact on the blades.
(d) **Electrical Energy**: By electromagnetic induction, a generator connected to the turbine shaft transforms this mechanical spinning into electrical energy.

The following formula can be used to determine how much electricity a hydropower system produces:

$$P = \eta \rho g Q H \tag{12.7}$$

where:

- P = Power output (Watts)
- η = Efficiency of the system
- ρ = Density of water (kg/m^3)
- g = Acceleration due to gravity (9.81 m/s^2)
- Q = Flow rate (m^3/s)
- H = Effective head or height difference (m)

This theory emphasises that the water height and flow rate are important factors that determine how much power is produced.

12.2.3.1 Current Status of Hydropower

An inexpensive way to generate electricity is by using hydropower. Hydropower uses moving water, an abundant and readily available natural resource, to generate electricity. States like Idaho, Washington, and Oregon that rely mostly on hydropower can reduce their citizens' energy costs. When maintenance, operations, and fuel expenses are taken into account, hydropower also offers comparatively cheap costs over the course of a project when compared to other electricity sources. Considerable upfront costs are inevitable for any large project. However, the average yearly cost of facilities is reduced by hydropower's extended lifespan (usually 65–85 years) and cheap operating and maintenance expenditures. This is because hydropower operations' equipment frequently runs for longer stretches of time instead of requiring restorations or repairs, resulting in cost savings over time.

The development of dams, tunnels, and other essential infrastructure, as well as the cost of electromechanical equipment (machines that produce electricity), make up the majority of the installation costs for big hydropower plants. Because hydropower is a region-specific technology, it is feasible to reduce these expenses at the planning phase by carefully choosing its location and design (International Hydropower Association (IHA) 2024). Excluding hydropower in the energy mix, no nation has even approached reaching 100% renewable energy. Through 15 GW of increased capacity in 2024, the world's hydropower capacity surpassed 1283 GW (not including pumped storage). The global calculated average levelized cost of electricity (LCOE) for hydropower decreased by 2% between 2023 and 2024, however as a well-established renewable technology, hydropower has remained a commercially appealing choice that offers not only inexpensive power production but additionally important grid services like storage and large-scale versatility. The global calculated average LCOE for 2024 was USD 0.057 kWh (International Hydropower Association (IHA) 2024). The top five hydropower-producing countries by installed capacity are:

China: China is the biggest maker of hydroelectric electricity in the world. It produced over 1303 terawatt-hours (TWh) of hydropower in 2022, or roughly 31% of the world's total. Important infrastructure projects, such as the well-known Three Gorges Dam on the Yangtze River, sustain this enormous output.

Brazil: A total of an estimated 427 TWh of hydropower produced annually in 2022, Brazil is the world's second-largest hydropower generator. Brazil's energy mix relies heavily on hydropower, which supplies a significant amount of the country's electricity. One of the most important hydropower projects in the nation is the Itaipu Dam, which is situated on the border between Brazil and Paraguay.

Canada: Around 62% of Canada's total electricity generation comes from hydropower. The nation generated over 398 TWh of hydroelectric electricity in 2022.

However, water levels in important reservoirs have been impacted by recent, protracted droughts, which have decreased hydropower production and made it necessary to import electricity from the US. One of Canada's largest hydropower projects is the Churchill Falls.

United States: In 2022, the United States produced over 249 TWh of hydroelectric electricity, ranking among the world's top producers. Hydropower is still an important renewable energy source even though it makes up a lesser portion of the U.S. energy mix than in certain other nations. Grand Coulee Dam is a significant hydropower project in the United States.

India: With a present capacity of over 50 GW, India is one of the main producers of hydropower, accounting for around 10% of the nation's total electricity generation. The Bhakra Nangal Dam (1.3 GW) in Himachal Pradesh and the Tehri Dam (2.4 GW) in Uttarakhand are two significant hydroelectric development projects. To promote stability, the Indian government has been aggressively increasing hydropower capacity, concentrating on pumped-storage hydropower (PSH). However, issues including local community displacement and seasonal water scarcity have sparked worries about large-scale hydropower expansion. India plans to increase its hydropower capacity in the future, with an installed capacity of 70,000 MW by 2030. About 13,000 MW of hydropower projects are presently under development as part of this ambitious strategy, while another 8000 MW are in the pipeline and should start building in the next 6–8 months (Table 12.2).

Modern hydropower technology has evolved far beyond conventional large dams. Today, run-of-river systems, pumped-storage hydropower, and small- and micro-hydropower plants have expanded the sector's reach.

Table 12.2 Some major hydropower projects and their capacity is enlisted here (World Bank 2023)

Name	Capacity (MW)	Location	Completed	Reservoir surface (Km^2)
Three Gorges Dam	22,500	China	2009	1000
Itaipú Dam	14,000	Brazil/ Paraguay	2003	1350
Xiluodu	13,860	China	2014	
Belo Monte	11,233	Brazil	2019	441
Guri Dam (Simón Bolívar)	10,235	Venezuela	1986	4250
Grand Coulee Dam	6809	USA	1991	324
Sayano-Shushenskaya	6400	Russia	1989	621
Krasnoyarsk Dam	6000	Russia	1972	2000
Robert-Bourassa GS	5616	Canada	1981	2835
Churchill Falls	5429	Canada	1974	6988

12.2.3.2 Future and Challenges of Hydropower

The combined needs of scientific advancement and environmental responsibility will shape hydropower's future. Hydropower will continue to play a crucial role in the upcoming decades thanks to the global energy shift towards low-carbon systems. The infrastructure for hydropower is being updated by developing technologies such as-

1. **Scientific Progress**

Digitalisation and AI-driven forecasting techniques are used in smart hydropower plants to track and preserve operation instantaneously. By integration of floating solar-hydro hybrid systems with current reservoirs for hydropower with drifting solar panels, land and water use are maximised. Innovative designs of turbine improve performance and lessen the environmental impact include variable-speed and fish-friendly turbines.

2. **Adaptation to the Environment and Sustainability**

Sustainable activities are the focus of future hydropower plants. Project authorisation and execution now depend heavily on integrated community planning, preservation of biodiversity, and environmental flow management. In order to construct hybrid renewable systems, hydropower will increasingly be combined with other renewable energy sources. Because of its quick reaction time, it is a perfect complement to sporadic energy sources like solar and wind, improving grid stability and dependability.

3. **Progress in Developing Countries**

There is still an enormous unrealised possibility of hydropower in many developing nations. Just over half of the world's economically viable hydropower resources, according to the International Hydropower Association (IHA 2024), have been developed (International Hydropower Association (IHA) 2024). In order to fulfil growing energy demands in a sustainable manner, nations in Asia, Africa, and Latin America are giving new hydropower developments top priority.

Despite its advantages, hydropower faces several challenges (Tomé Robles et al. 2024), such as—Large preliminary capital costs for construction of infrastructure and dam building, which demand large sums of money. Sedimentation, greenhouse gas emissions from reservoirs, and disturbance of river ecosystems are persistent problems. Changes in precipitation patterns and glacial melt may have an impact on the output and dependability of hydropower. So that in order to preserve efficiency and safety requirements, many current power plants must be modernised. Aquatic ecosystems can be negatively impacted by dams and reservoirs because they can hinder fish migration patterns, modify river flow patterns, and disrupt natural water temperatures. For example, building dams can prevent certain species, like salmon, from reaching their spawning sites, which can result in population decreases (Bănăduc et al. 2024). Although hydropower is typically seen as a low-emission energy source, the breakdown of organic substances underwater in reservoirs can release greenhouse gases, including carbon dioxide and methane. The precise levels of these emissions are unknown and rely on a number of site-specific variables.

Massive regions are frequently flooded in order to create reservoirs for hydropower, which has the potential to sink entire villages, farms, and forests, causing local populations to be uprooted and habitats to be lost. Dams can change water temperatures, nutrient cycles, and sediment loads, which affect human societies as well as downstream ecosystems that depend on these water resources sources. Furthermore, variations in water levels brought on by hydroelectric activities can have an impact on local economies and leisure activities. Geopolitical conflicts, ecological compliance, finance and political uncertainties, trade regulations are some legal obstacles faced by water energy. Large-scale hydropower projects have the potential to cause geopolitical conflicts, particularly when rivers that are across borders are involved. For example, China's intention to construct a huge hydroelectric dam in Tibet on the Yarlung Tsangpo River has aroused worries in India and Bangladesh over potential postponements to water flow necessary for both drinking water and agriculture. To lessen their environmental effects, hydropower projects must comply with strict environmental regulations. Finance for hydropower developments can be influenced by political choices. As an illustration of how political unrest can obstruct the creation of projects, the European Investment Bank postponed financing for the Ruzizi III hydropower plant in Central Africa because of territorial disputes. Hydropower's economics can be impacted by global trade policies (Othman et al. 2025). In order to ensure that hydropower's advantages as a renewable energy source are harmonious with ecological preservation and socioeconomic factors, it is imperative that these environmental and governmental challenges be addressed.

Hydropower continues to be one of the most dependable and popular renewable energy sources despite these obstacles, making a substantial contribution to the production of electricity worldwide. It is essential to the shift to sustainable energy because it can offer grid flexibility and reliable, large-scale power generation.

12.2.3.3 3D-Printing in Hydropower

Most hydropower equipment (especially huge runners and wicket gates) is complex to construct, sometimes made overseas, and can take months or years to be repaired when broken. Additive manufacturing promises to cut lead times, enable complicated customised designs, and facilitate on-demand local production, possibly boosting plant availability and reducing lifecycle expenses. Applications of 3D-printing in hydropower are listed below:

Creation of Effective Micro-Hydro Turbines Micro-hydro systems, which usually range from a few kilowatts to several megawatts, produce power from tiny water flows (rivers, canals, or pipes). These systems are perfect for decentralised power generation, isolated settlements, and rural electrification. The casting and machining processes used in traditional turbine manufacturing are costly and time-consuming. Faster iterations and efficiency gains are made possible by the rapid and accurate production of turbine components, made possible by 3D printing. Because the water flow and pressure at hydropower sites vary, custom turbine designs are needed. For optimal efficiency, site-specific turbine blades and casings can be made via 3D printing. Turbine blades can be optimised to decrease energy losses and increase efficiency using intricate aerofoil and hydrodynamic designs that are 3D printed. In order to improve turbine durability while lowering weight, 3D printing enables the integration of various materials, including metal alloys and polymer composites. For applications on a small traditional turbine production is costly; however, 3D-printed turbines lower labour and expenses for materials, making micro-hydro cheaper (Yong and Wong 2025). Highly effective hydro turbine models have been successfully 3D printed by ETH Zurich, cutting waste and production time in half. In order to increase micro-hydro potential in off-grid areas, the University of Laval in Canada is creating 3D-printed Kaplan and Francis turbines that are optimised for low flow circumstances. The massive-scale 3D printing of concrete makes the on-site construction of spillways, penstocks, and dam walls, which eliminates the need for expensive prefabricated part shipping. Highly durable polymer composites and strengthened concrete can be used with 3D printing to create stronger, lighter materials. Wear and tear result from hydropower components' continual exposure to water and environmental conditions. Turbine housings and structural supports have a longer lifespan thanks to 3D-printed composites with corrosion-resistant coatings (Advanced Casting Cell: A Leap in 3D Printing Technology 2023). A 30% reduction in production time has been demonstrated by GE Hydro Solutions' testing of 3D-printed hydropower turbine casings. For large-scale dam infrastructure upgrades, Sandia National Laboratories is investigating 3D printed hydraulic components. By creating modular spillway and dam structures, ETH Zurich's concrete 3D printing initiative is cutting down on construction time and material waste. As 3D printing technology develops, it will be essential for updating hydropower infrastructure and increasing the availability, dependability, and sustainability of renewable energy.

12.3 Conclusion

With growing global capacity and enhanced efficiency, technologies including solar, wind, hydropower, bioenergy, and geothermal energy are always growing. These developments are essential for attaining energy security, slowing down environmental deterioration, and fostering economic expansion. The development, production, and implementation of energy systems will undergo a radical change when 3D printing technology is integrated into renewable energy solutions. Customised components, cost savings, material efficiency, and quick prototyping are all made possible by additive manufacturing, which makes it possible to build lightweight, high-performing structures for a variety of renewable energy applications. In the future, a carbon-neutral and sustainable world could be achieved through the combination of sophisticated manufacturing technology and renewable energy. The efficiency and resilience of renewable energy systems will be further improved by the integration of smart grids, IoT-based monitoring, and energy storage advancements.

3D printing and renewable energy will be crucial in creating a more sustainable and environmentally friendly future with sustained investment in research, material development, and large-scale deployment.

References

Advanced Casting Cell: A Leap in 3D Printing Technology (2023). 3Dprinting.com, 12 Oct 2023 (GE Research & voxeljet collaboration on large sand binder-jet casting)

Ahmed R, Khubnani R, Reddy GV, Singla A, Maan P, Dwivedi S (2024) A review paper on current state of the worldwide solar energy generation. E3S Web Conf

Anwajler B (2024) Potential of 3D printing for heat exchanger heat transfer optimization—sustainability perspective. Inventions 9(3):60

Bănăduc D, Curtean-Bănăduc A, Barinova S, Lozano VL, Afanasyev S, Leite T, Branco P et al (2024) Multi-interacting natural and anthropogenic stressors on freshwater ecosystems: their current status and future prospects for 21st century. Water 16(11):1483

Basu R, Siah KS, Distler A, Häußler F, Franke J, Brabec CJ, Egelhaaf H-J (2023) Aerosol-jet-printed encapsulation of organic photovoltaics. Adv Eng Mater 25(17):2300322

Edwards EC, Holcombe A, Brown S, Ransley E, Hann M, Greaves D (2023a) Evolution of floating offshore wind platforms: a review of at-sea devices. Renew Sustain Energy Rev 183:113416

Edwards EC, Nishino T, Willden RHJ (2023b) Evolution of floating offshore wind platforms: a review. Renew Sustain Energy Rev 182:113326. https://doi.org/10.1016/j.rser.2023.113326

Elavarasan R (2019) The motivation for renewable energy and its comparison with other energy sources: a review. Eur J Sustain Dev Res

GE Renewable Energy and Oak Ridge National Laboratory (ORNL) (2023) Large-scale 3D printing for wind turbine blade molds. U.S. Department of Energy, Office of Energy Efficiency and Renewable Energy. https://www.energy.gov/eere/articles/ornl-ge-3d-printed-blade-mold

Global Wind Energy Council (GWEC) (2024) Global Wind Report 2024. GWEC, Brussels. https://gwec.net/global-wind-report-2024

Hong S, Kim D, Lee Y (2024) Floating offshore wind farm installation: challenges and economics. Energy Policy 187:113476. https://doi.org/10.1016/j.enpol.2024.113476

IEA (2024) Renewables 2024. IEA, Paris. https://www.iea.org/reports/renewables-2024. Licence: CC BY 4.0

International Energy Agency (IEA) (2024a) Renewables 2024: analysis and forecast to 2029. IEA, Paris. https://www.iea.org/reports/renewables-2024

International Energy Agency (IEA) (2024b) Hydropower Special Market Report 2024: outlook for global hydropower development. IEA, Paris

International Hydropower Association (IHA) (2024) Hydropower Status Report 2024. IHA, London

IRENA and ILO (2024) Renewable energy and jobs: annual review 2024. International Renewable Energy Agency/International Labour Organization, Abu Dhabi/Geneva

Liu Y, Montalbo TJ (2023) Additive manufacturing for renewable energy systems: a review. Renew Sustain Energy Rev 184:113425. https://doi.org/10.1016/j.rser.2023.113425

Mills S (2017) Wind energy and rural community sustainability. In: Handbook of sustainability and social science research. Springer, Cham, pp 215–225

National Renewable Energy Laboratory (NREL) (2024) Advanced manufacturing and wind energy integration. U.S. Department of Energy, Golden, CO. https://www.nrel.gov/wind

Othman MEF, Sidek LM, Basri H, El-Shafie A, Ahmed AN (2025) Climate challenges for sustainable hydropower development and operational resilience: a review. Renew Sustain Energy Rev 209:115108

Pira S (2020) The importance of renewable energies with emphasize on wind power. Int J Eng Res Technol (Ijert) 9(6)

Pourasl HH, Barenji RV, Khojastehnezhad VM (2023) Solar energy status in the world: a comprehensive review. Energy Rep
Shoeib EAH, Infield EH, Renski HC (2021) Measuring the impacts of wind energy projects on US rural counties' community services and cost of living. Energy Policy 153:112279
Siemens Gamesa Renewable Energy (2022) Additive manufacturing applications in wind turbine production. Siemens Gamesa Press Release, Hamburg. https://www.siemensgamesa.com/en-int
Tomé Robles DJ, Maurer F, Nøland JK, Panwar M, Mishra S, Øyvang T (2024) Paradigmatic outlook of hydropower toward 2050 and beyond—a comprehensive survey. Available at SSRN 5004720
Upadhyay A, Chowdhury A (2014) Solar energy fundamentals and challenges in Indian restructured power sector. Int J Sci Res Publ 4(10):1–13
Vestas Wind Systems (2024) Additive manufacturing in offshore wind operations. Vestas Technical Report, Aarhus. https://www.vestas.com/en/media
Weerasinghe HC, Macadam N, Kim J-E, Sutherland LJ, Angmo D, Ng LWT, Scully AD et al (2024) The first demonstration of entirely roll-to-roll fabricated perovskite solar cell modules under ambient room conditions. Nat Commun 15(1):1656
World Bank (2023) The future of hydropower: a clean energy pathway. World Bank, Washington, DC
WWEA Annual Report (2024) World Wind Energy Association (Global Wind Statistics)—www.wwindea.org
Yong CY, Wong JZZ (2025) A novel 3D-printed turbine to harvest renewable energy from water pipeline system for street lighting. J Renew Energy Environ 12(4):53–63

FSC
www.fsc.org
MIX
Papier aus verantwortungsvollen Quellen
Paper from responsible sources
FSC® C105338

If you have any concerns about our products,
you can contact us on
ProductSafety@springernature.com

In case Publisher is established outside the EU,
the EU authorized representative is:
Springer Nature Customer Service Center GmbH
Europaplatz 3, 69115 Heidelberg, Germany

Printed by Libri Plureos GmbH
in Hamburg, Germany